GC   Hersey, John Brackett, 1913-
60
.H4   Deep-sea photography

DATE DUE			

THE JOHNS HOPKINS OCEANOGRAPHIC STUDIES

Number 3

The Johns Hopkins Oceanographic Studies, published by The Johns Hopkins Press, is a series designed for the publication of monographs and long technical papers in the field of oceanography. Manuscripts may be submitted to any member of the Editorial Board, which has the responsibility for appraisal and selection of papers for publication.

DEEP-SEA PHOTOGRAPHY

THE JOHNS HOPKINS OCEANOGRAPHIC STUDIES

Deep-Sea Photography

edited by

John Brackett Hersey

Woods Hole Oceanographic Institution
Woods Hole, Massachusetts

THE JOHNS HOPKINS PRESS
BALTIMORE AND LONDON

Preface

This is a book about the use of photography as a tool for the scientific study of the deep sea. It presents the development of the art of underwater photography at great depth and illustrates its use in several branches of oceanography by reviews of past work, and by original contributions.

Photography as a means of observing and recording data beneath the sea has been an especially lively interest in recent years among scientists and amateurs devoted to free diving. The thousands of superb photographs, including many thousands of feet of motion picture films of marine life and the sea floor in shallow water, scarcely require another general description. But photography in the deep sea, either near the sea floor or in mid water, has developed more slowly because there has been far less opportunity for it, the expense is greater, and it is fundamentally more difficult.

In the present day of fast-moving change in methods and interests, it is difficult to know when a review of the past may be helpful toward shaping the future and hence make another book worth while. Such a time seems to have arrived for deep-sea photography, and I feel that it would be helpful to many groups just starting scientific programs to have a thorough review of past technology and sea-going experience as a guide.

Beebe was a pioneer in mid-water photography during his work with the bathysphere, as was E. Newton Harvey in his attempt to photograph pelagic animal life in the open ocean by random selection methods. While this work resulted in no identifiable pictures of animals the same method, many years later, has quite unintentionally produced some remarkable photographs.

The first successful deep-sea photographs were those made by Maurice Ewing and his co-workers during cruises on *Atlantis* in 1940–48. Their subject was the sea floor and their method of selection was to trigger the camera mechanically when its mounting struck bottom. This remained the only method of triggering cameras to photograph the sea floor until the work of Edgerton and Cousteau with the sonar pinger. Both methods are used today, along with several others such as tripping weights and mounting cameras on sleds which ride on the bottom. These and other special methods of controlling cameras are reviewed below.

The group at the Woods Hole Oceanographic Institution under Maurice Ewing turned their peacetime experience in photography to good account by making many successful identifications of wrecks, including sunk submarines, during World War II.

During the twenty years since that war, multiple-exposure cameras have been designed, stereographic combinations of cameras have been used, special underwater lenses have been designed to enhance the quality of the photographs; and side lighting, increased energy in strobe flash lights, and many hours of testing in artificial pools and at sea have made it possible to take good photographs at distances from a few inches to nearly fifty feet from the subject in clear ocean water. For the past seven or eight years it has been possible to make several hundred exposures per camera lowering. Recently this ability has been used from ships proceeding under control at very low speeds to photograph strips of the bottom over a mile long, from which continuous photomosaics have been made. Good progress has been made toward triggering cameras, both by optical and acoustical means, to photograph animals in open water out of contact with the bottom.

All of these advances have been achieved in an atmosphere of both co-operation and friendly competition. Many scientists have participated and a number have made contributions to this volume. The advancement of technique in underwater photography is necessarily married to its contributions to the solution of scientific problems. In the chapters that follow, some history is reviewed in greater detail than in this preface, and technical design and special applications are described, followed by contributions illustrating the scientific uses to which photography can be put in the study of the processes and structures of the ocean floor, observations of water currents, and in the study of animal life.

Acknowledgments

I am grateful to all the authors for their friendly co-operation in completing this book. I am particularly grateful to Maurice Ewing, E. M. Thorndike, K. O. Emery, and Harold Edgerton for their support, encouragement, and helpful criticism of various parts of the book, and for their considerable personal contributions in writing and illustrative material. David M. Owen has been most helpful, not only by his several contributions but also through his quiet enthusiasm and support of the work. R. H. Backus has helped greatly by reviewing several manuscripts and by repeated helpful discussions.

Frances Dakin, who acted as assistant editor, is clearly the one person most responsible for the completion of the whole work. She has carried out preliminary editing of all manuscripts and has attended to the myriad details involved in compiling such a book competently and efficiently. I am most grateful for her truly large contribution.

In addition to the considerable effort of many who cannot be individually acknowledged, I especially wish to express appreciation for the assistance of the following Woods Hole staff in the final preparation of figures and text: C. Innis for guidance in both photography and drafting; A. T. Johnston, K. M. Welby, and Kurt Fuglister for photographic reproductions; Helen Hays for final preparation of the photographs; and Florence Mellor and Alice Broadbent for what must have seemed endless typing and retyping of the edited manuscripts.

                                                                    J. B. H.

Contents

1. Early development of ocean-bottom photography at Woods Hole Oceanographic Institution and Lamont Geological Observatory[*†]

Maurice Ewing and J. Lamar Worzel *Lamont Geological Observatory, Palisades, New York**

Allyn C. Vine *Woods Hole Oceanographic Institution, Woods Hole, Massachusetts†*

Abstract

This chapter traces the development of deep-sea photography from its beginning, in 1939, up to the end of 1948 when the camera had finally proved itself an important tool for the study of the structure and life of the sea floor. Despite initial obstacles, namely lack of enthusiasm from other oceanographers, and difficulty in obtaining financial support, the authors in 1939 and 1940 constructed the first free-floating, remotely triggered underwater cameras. High quality photographs of the sea floor were obtained at depths up to 150 fm (270 m), and possibilities for the use of photography in studies of geology, light scattering, and ocean currents were opening when World War II intervened. The war stimulated the progress of the photographic work, but in changed directions. Special techniques were developed for locating mines and identifying wrecks. These included the use of underwater television, location and manipulation by sonar, and provision for taking a large number of exposures on a single lowering. Color and stereo photography were used successfully but were found to have no special value in this work.

These wartime services proved the usefulness of underwater cameras, particularly in their ability to photograph

selected targets. When the war was over efforts were concentrated on the study of the ocean bottom and cameras were developed for photographing submarine canyons, continental slopes, and basin floors. Successful surveys were carried out on the Mid-Atlantic Ridge, on the continental slope off New England, and in the Mediterranean Sea, and excellent photographs were obtained to a depth of 3,026 fm (5,534 m).

This pioneering work provided the basis for the extensive developments, improvements, and range of applications of deep-sea photography that are described in the following chapters of this book.

1–1. Introduction

The development of ocean-bottom photography up to 1946 was reviewed by the authors in an earlier paper (Ewing, Vine, & Worzel, 1946; see also bibliography at the end of this chapter). In brief, until 1939 results had been obtained only in depths accessible to waders and divers. The authors began to work on photography of the deep-sea floor in 1939, as a minor accessory to the seismic-refraction program planned as the principal work of an expedition of the U.S. Coast Guard cutter *Campbell* to the mid Pacific. Our photographic work was sponsored by the National Geographic Society. One of us (Ewing) was then on leave of absence from Lehigh University under a Guggenheim Fellowship and the other two were students at Lehigh. However, we found little or no enthusiasm for the photographic project among the oceanographers consulted. In general they expressed the view that the water near the deep-sea floor was probably an opaque soupy suspension. The *Campbell's* Pacific expedition was postponed abruptly in September, 1939, when Hitler's troops marched into Poland.

* Lamont Geological Observatory, Columbia University, Contribution Number 1002.

† WHOI Contribution Number 1749.

The studies reported in this paper were supported by the Office of Naval Research, the Bureau of Ships, the National Geographic Society, the National Science Foundation, the Woods Hole Oceanographic Institution, and the Geological Society of America. Work on the paper itself was supported by the Office of Naval Research under contract Nonr 266(48) with Columbia University. Many persons at Woods Hole Oceanographic Institution and in the United States Navy played vital roles in the development and operation of the cameras and in the manning of the ships from which the early ocean-bottom photographs were taken. Their help was indispensable and is deeply appreciated.

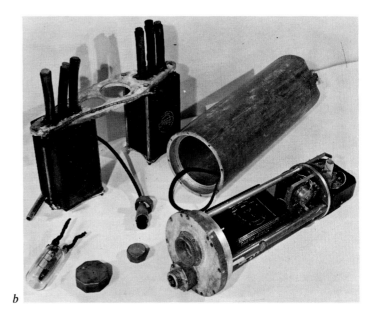

Figure 1-1. (*a*) Tests of the ocean-bottom camera were made prior to the departure of *Atlantis* cruise 94 from Jacksonville, February, 1940; A. C. Vine, crewman, M. Ewing. (*b*) Components of the 16-mm, deep-sea camera made at Lehigh University by the authors in 1939. (*c*) Preparing to release the camera on *Atlantis* cruise 94; Vine, Ewing, and crewman. (*d*) A. C. Vine holding the camera.

1–2. Camera No. 1

For our first camera we used a 16-mm camera that could take either single shots or motion pictures. It was placed behind a glass window in an aluminum-alloy, cylindrical, pressure vessel. The window could withstand 10,000 psi external pressure. A 12-v, 1000-watt, incandescent lamp, provided by the Nela Park Laboratories of the General Electric Company, had an envelope that could safely be exposed to the water pressure at oceanic depths. The apparatus was powered by a 12-v battery (also pressure-compensated) and was to be taken down, free-floating, by ballast and returned to the surface by a float made of a

gasoline-filled neoprene fabric tube. The ballast was released by a salt-solution releasing device. Figs. 1–1 and 1–2 show the apparatus.

A system for using seismographs, bombs, and cameras undisturbed on the ocean floor, and retrieving them by flotation had been developed by Ewing during the previous winter in Bermuda, but was relatively untested. Columbus Iselin, as Director of Woods Hole Oceanographic Institution, granted us the opportunity for further development on voyages of *Atlantis;* but the terms of our grant made it mandatory that the principal emphasis be on seismic refraction measurements.

The first test of the camera from the surface to bottom

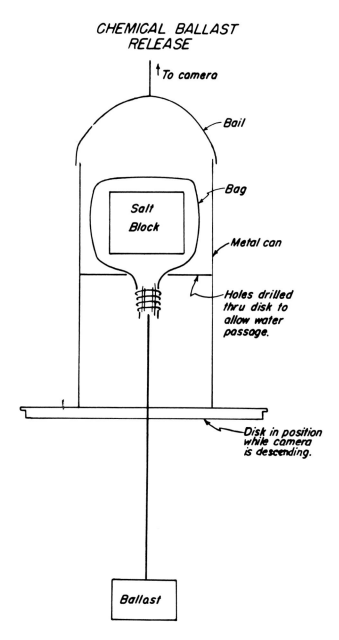

CHEMICAL BALLAST RELEASE

↑ To camera

Bail

Bag

Salt Block

Metal can

Holes drilled thru disk to allow water passage.

Disk in position while camera is descending.

Ballast

Figure 1–2. The salt-block ballast release mechanism used for the first ocean-bottom camera and for subsequent cameras.

in deep water was on *Atlantis* cruise 94, station C–12, made from 0916 to 1731, March 1, 1940, at 29°51′N, 72°20′W in about 2,900 fm (5,300 m). It produced daylight measurements only, since the storage battery had been poisoned the night before by addition of formalin instead of distilled water. A small, strong-motion seismograph in it showed 2¼-hours submergence and about 2¾ hours on the surface, but no indication of bottom contact or ballast release. On the second test (Station C–13; March 2, 1940), the camera was released with

50-lb ballast. The seismograph showed that it reached bottom at 1012, and ran until 1142 without indication of ballast release. The lamp was about 3 ft (1 m) and the camera 13 ft (4 m) from the ballast. The aperture was f/16, and was programmed to cycle at 0904, 0905, 0914, 0915 .... On each cycle the incandescent lamp was turned on for about 17 sec, during which three exposures of 1.5, 0.3, and 15 sec were made. Several deductions may be made from the film for this station (shown in fig 1–3).

(a) The camera started down after the 0904 cycle and reached bottom shortly after 1005. The mud cloud disappeared entirely between contact and 1014. The camera left bottom between 1044 and 1045 (evidence of a mud cloud in the last exposure of 1044) and surfaced just before 1325.

(b) Time-lapse series of bottom photographs were obtained on each cycle. The boundary between strong and weak exposure in each frame is the edge of the area screened by the reflector. One half of the field is recorded on the short exposures. the other half on the long ones, but the quality is poor—as is evident in fig 1–4 (*a* and *b*) a print of frame 58. At the time we did not know whether the poor quality of the photographs resulted from poor quality of the subject or of the technique. Having seen many good bottom photographs in the intervening years, we now attribute it to slight motion of the camera during exposures, due either to current or to slight motion of the release mechanism.

(c) Back-scattering from the water was recorded in all frames of longest exposure during descent and ascent. The increase in back-scatter near bottom, discussed by Ewing & Thorndike (1965, see also chap. 10) on the basis of recent data, is evident in frames 43 and 46, taken in 1940. This film was not developed until brought to the National Geographic Society after the voyage, so the almost insuperable handicap of experimentation without prompt knowledge of results had to be borne.

Subsequent lowerings on this cruise were primarily to study penetration of daylight and the means to overcome a serious defect of the salt releases (fig. 1–2) namely, time of release was uncertain because the rate of solution was much greater during descent than during the stay on bottom.

Measurement C–14 (1:00 P.M to 4:22 P.M., March 4, 1940) was a light-penetration experiment in which the camera was attached to the hydrographic wire, pointing upward. No light source was used. An improvement to the programming gave the following exposures for each 10 min of cycle: 1.5, 0.3, 15, 43, 1.5, 0.3, 15 and 523; or ratios of 1, 5, 50, 143, 1743—a range of about 2,000 to 1. Fig. 1–3 shows the film made during this lowering, with Eastman Super XX film at f/1.9. Of the indications of

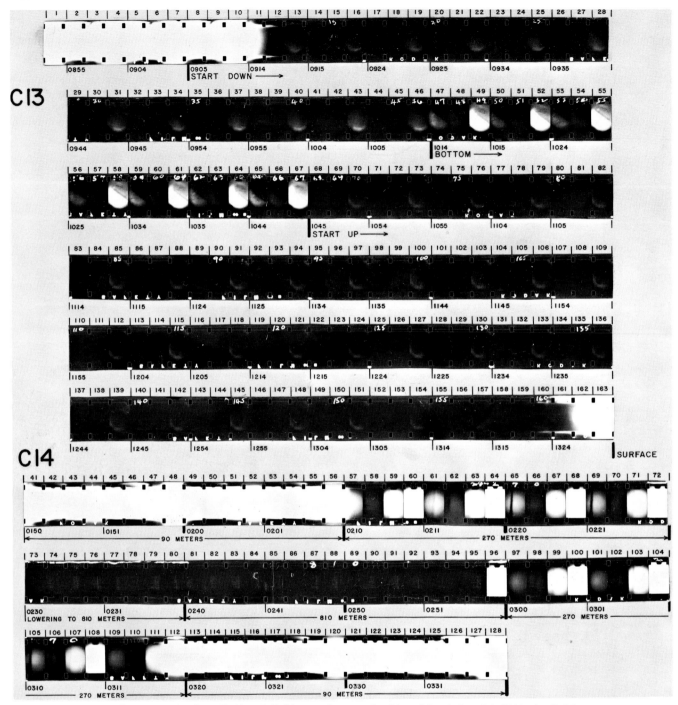

Figure 1–3. Stations C–13 and C–14 on *Atlantis* cruise 94 on March 2 and 4, 1940: At C–13 seven bottom photographs were obtained at an exposure of 15 sec, and seven at 1.5 sec. The seven at 0.3 sec showed nothing. At C–14 the camera was used on a bright day without artificial light. The camera pointed upward and was lowered to the indicated depths on a wire. The purpose was to measure the penetration of daylight. (The time marks on this figure are wrong, and 1200 should be added to each. The time of the segment shown is from 1350 to 1531 hours.)

the intensity (I) made at 0, 10, 30, 90, 270 and 810 m of wire out, only that at 270 m was on scale, but it is possible to infer that $\frac{1}{50} I_{90} > I_{270} > 388 I_{810}$; hence the extinction coefficient is greater than 0.02 per m. A repetition of this experiment with depths varying by multiples of 50 m on either side of 270 m would have given a very

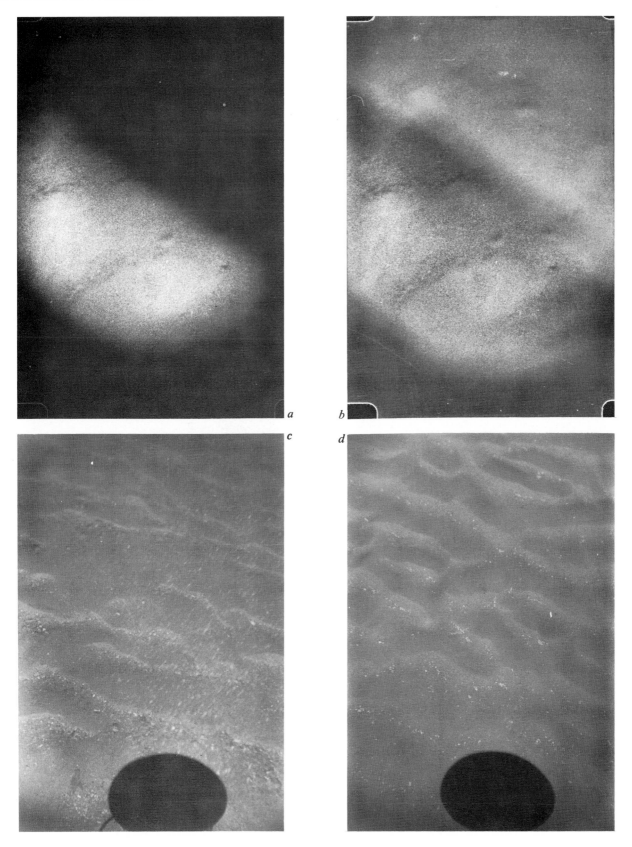

Figure 1–4. (*a*) and (*b*) Station C–13, frame 58, 15-sec exposure. The print on the left is a simple enlargement in which the heavily exposed part only was printed. On the right, the print was hand-dodged to equalize the exposure. The photographs at this station were the first bottom pictures taken in the deep sea. (*c*) and (*d*) Stations C–28 and C–29, showing a sandy bottom with current ripples oriented across the camera field. *Atlantis* cruise 98, June 6, 1940. C–28: 41°44′N 66°38′W, 32 fm (58 m). C–29: 41°55′N 66°55′W, 32 fm (58 m).

useful determination, but here again we paid a high price for the delay in information from not developing the film on board ship, and also from the fact that most of our attention was on the accomplishment of the far more complicated seismic measurements. Tests C–15 and C–16, quick tests of rate of solution of the salt block in the ballast release, were the last possible on *Atlantis* cruise 94. These tests were made on March 5, 1940.

The next opportunity for tests was on *Atlantis* cruise 97 (May 7–16, 1940). Two seismic-bomb shooting devices had each made two successful trips to bottom before the camera was launched in 1,050 fm (1,920 m). The lamp had been rearranged to provide side lighting. This excellent camera was not recovered, and when we asked for funds to build another, the sponsoring agency declined to provide them. But it was so apparent to us that ocean-bottom photography was worth while that we resolved to continue as well as we could.

1–3. Camera No. 2

Camera No. 2 was constructed quickly from parts that could be found around the laboratory or purchased for a total of less than fifty dollars. It was to be used on the next cruise, *Atlantis* cruise 98, May 28 to June 8, 1940, a biological survey of Georges Bank. For our next operation there was the restriction that the camera operation must not consume any station time. Since the biological survey work required the use of the only available winch, we made the camera to operate under ballast and flotation and used a hand-tended cod line to bring it to the ship after it surfaced. This camera produced excellent results on *Atlantis* cruises 98 and 100, with experiments on its use in deeper water on *Atlantis* cruises 101 and 102. It was lost during *Atlantis* cruise 102 in about 1,500 fm (2,700 m) of water, which was near its designed collapse depth. Camera No. 2 was used on four cruises and was modified during each time in port. Hence the designations "Camera 2A" etc., refer to the same camera with different accessories or rigging.

The camera used iron ballast — usually 20 or 30 lb — and a glass-tube float that doubled as an instrument case. It contained an Argus C–1 with an improvised clockwork drive and photoflash synchronizer like that shown in fig. 1–5. The camera operated on a cycle of three shots at intervals of about 20 sec, started by a time clock. The synchronizer was used to shoot a photoflash bulb exposed to sea-water pressure. The ballast was dropped off by solution of a block of salt (fig. 1–2), allowing the camera to float to the surface and be pulled to the ship by the cod line. The camera was mounted on a pole, about 12 ft (3.5 m) above the ballast, pointing just far enough off

vertical to see only the edge of the lamp holders. The lamp holders were put well below it on the same pole to provide side lighting of the ocean floor and to minimize the effect of light scattered by the water. This camera is shown in fig. 1–5 (*a* and *b*).

Photographs of the ocean bottom were taken at seventeen stations on *Atlantis* cruise 98 on Georges Bank, in depths ranging from 21 fm (38 m) to 83 fm (152 m). Usually two photos of the same field were obtained at each station. The white sand of Georges Bank, so familiar to generations of American navigators and fishermen (Dana, R. H., 1840) was seen to have current ripples that were systematically oriented across the camera field (figs. 1–4*c* and *d*). This observation in a region of rotary currents (U.S.C. & G.S. Chart 1107) proved that the camera assembly was oriented by the existing current, and that the sand-ripple pattern was continually being reoriented.

Before *Atlantis* cruise 100, Camera No. 2B was modified to start its cycle upon contact with the bottom and to drop ballast instead of flashing a light source as the third exposure of each series was made. A compass to show camera orientation, a pennant to show current direction (stations C–41 and later), a pendulum to show current strength, and a tube to take a sample of the sea floor were added (figs. 1–5*c*, and 1–6, *a*, *c* and *d*). On station C–42 and later, a large spot of aluminum paint was put on one side of each photoflash bulb in an effort to equalize illumination over the field. A clockwork safety release was used to aid in recovery of the camera, in case the programming mechanism failed to operate correctly.

On *Atlantis* cruise 100, photographs (normally two per station) were made at twenty-seven stations, ranging in depth from 21 fm (38 m) to 113 fm (207 m). Many of these photographs are of very high quality. There is abundant evidence that current is in the direction concordant with the sand ripples photographed (fig. 1–6*a*). Except in a few cases where the camera was evidently dragged between shots, the field of view and the current direction indicated by the pennant are in agreement on the two photographs at each station.

One exception is at C–48, where a unique change of current direction of almost 90° in the 20 sec between shots is recorded. Another remarkable aspect of this pair of photographs (fig. 1–6*c* and *d*) is that a cord or line, not part of our equipment, crosses the field of view near the compass and, on the first of the two photographs, a part of the head of a large fish, lying on bottom, appears at the lower left near the current vane indicator. Evidently the abrupt departure of this fish after the first photograph caused the apparent change in current direction.

Not only were the photographs obtained on *Atlantis*

a

b

c

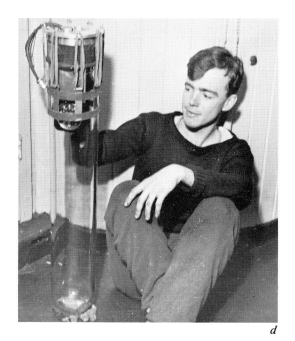

d

Figure 1–5. (*a* and *b*) Shows Camera 2A being assembled by Roger Merrill, Jr. and Maurice Ewing, and (*b*)—over the side. (*c*) Shows Vine, Wilson, and Ewing, with Camera 2B— pendulums, compass, and bottom sampler having been added. (*d*) The glass tube float with the works installed, exhibited by R. J. McCurdy.

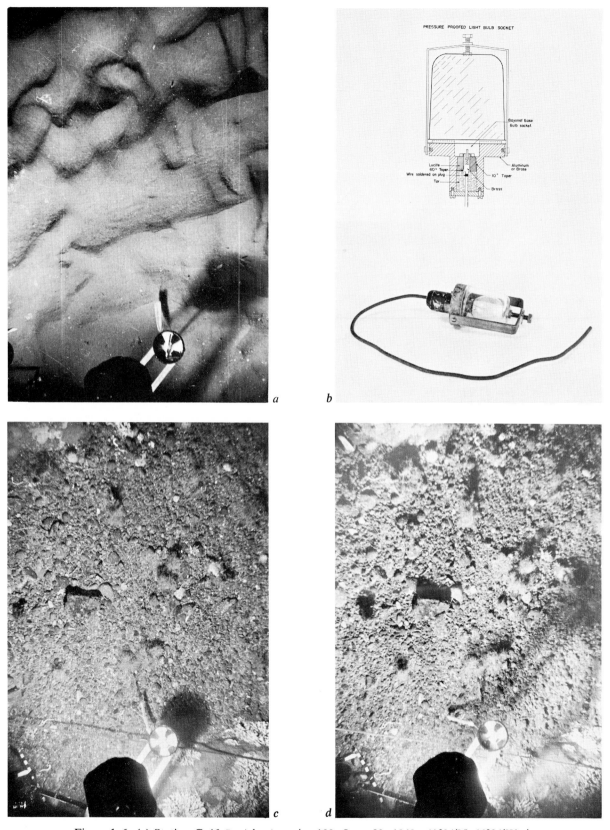

Figure 1–6. (*a*) Station C–46–2. *Atlantis* cruise 100, June 23, 1940; 41°34′N 66°24′W, in 51 fm (93 m). (*b*) Pressure proof lamp socket for ⚡5 flashbulbs, made from an automobile carburetor sediment bulb lapped onto an aluminum base. (*c*) and (*d*) *Atlantis* cruise 100, station C–48, 1 and 2. June 24, 1940; 42°09′N 66°30′W, in 54 fm (99 m).

cruise 100 of excellent quality but the camera also had evolved into a convenient and effective data-gathering instrument. The indicator of current strength required improvement, for its design provided no means for correcting for deviation from vertical of the camera pole, and several photos showed that tilt was present to the extent of putting the current meter into the sediment.

Before the next use of underwater cameras (*Atlantis* cruise 101) a frame was added on which the compass was supported, and from which were suspended by single wires as pendula three spheres of equal diameter, but widely different density. One sphere was a ping-pong ball, freely flooded. The second was similar except that it contained twelve steel air-rifle pellets. The third was solid iron. In preparation for working to greater depths, a lamp housing for No. 5 flashbulbs was built of an automobile carburetor sediment bulb, with a lapped joint to an aluminum alloy base plate that had an insulated electrical connector in it (fig. 1–6*b*).

Eight stations, C63–C70, were occupied on *Atlantis* cruise 101, which was primarily devoted to current measurements by E. E. Watson at anchor stations south of Montauk Point, in depths of 70 fm (128 m) to 455 fm (832 m). On the last six of these stations, the camera was lowered on the hydrographic wire, with a piece of ½-in manila line between it and the hydrographic weight. The photographing cycle was started by a clock. Hence there were some demands for rigid maintenance of a schedule of lowering. The bottom photographs obtained at station C–65 were the deepest known at that time (fig. 1–7). At all stations made on this cruise, numerous large light-scattering particles in the water were more abundant than have been found on any other occasion (station C–66, fig. 1–7).

On *Atlantis* cruise 102 (July 22–27, 1940) camera stations C71–C73 were made successfully in depths up to 155 fm (284 m). Toward the end of this cruise Camera No. 2 was lowered to about 1,500 fm (2,700 m) with the cod-line technique. Contact with bottom could not be felt. The cod line was broken, the camera lost, and the ocean-bottom photography had to be suspended. We had resolved to try to supplement the stations already made to constitute a photographic survey of Georges Bank and adjacent areas.

Atlantis cruise 106 (September 17–28, 1940) was made as the final attempt to make seismic-refraction measurements by the ballast-float process — the work that had really been the main object of the entire campaign (Ewing & Vine, 1938). Immediately following this cruise, Vine, Worzel, and Ewing moved to Woods Hole to begin work for the National Defense Research Committee and were principally engaged in bathythermographic and underwater sound work until the end of World War II.

1–4. Cameras 3, 4, and 5 — built in summer, 1941

In the summer of 1941 a fellowship was given to David Hagelbarger for continuation of the survey of the Georges Bank area by ocean-bottom photography. Three cameras were built — each designed for a special purpose and all embodying the experience gained in the work described above. The first of these cameras had a safe operating limit of 100 fm (180 m). It was known as the "Hagelbarger camera" or "Camera No. 3" (fig. 1–8*a*). It used roll film and produced a 3⅜-in (8.6 cm) square negative. It was mounted without buoyancy, 7 ft (2.1 m) from the foot of a pole that had a trigger and ballast weight at the bottom. A photoflash bulb mounted 2 ft (0.6 m) above bottom had to be replaced on deck after each shot. The film could be advanced by turning a small crank without opening the watertight case. The pole was provided with a "sail" to reduce rotation. The shutter was a trap-door device, mechanically operated by the trigger, that closed the watertight window from the outside. This shutter, when in wide-open position, actuated a switch that shot the flashbulb. This camera could be operated from a bathythermograph winch or a hydrographic winch. With Camera No. 3, Hagelbarger made various developmental tests in small boats, on day trips near to Woods Hole. As soon as good results were obtained he joined *Anton Dohrn* cruise 6, August 14–18, 1941, to Portland, Maine, and obtained good bottom photographs at fourteen stations in Cape Cod Bay and Massachusetts Bay, and parts of the Gulf of Maine. On August 22, enroute to New London on *Asterias*, he made seven more stations, on a line southwest of Martha's Vineyard, extending to 30-fm (55-m) depth. On *Anton Dohrn* cruise 8, August 25–September 1, M. Ewing and D. Hagelbarger occupied ten stations on a line south of Block Island, extending to a water depth of 40 fm (73 m). After Hagelbarger's departure at the end of the summer, there was no one whose time was not fully committed to acoustical problems. Specialized underwater photography was done under a small commitment to study the detection of mines in harbors and in their approaches — regions of very dirty water. The mine studies were not very fruitful for obtaining good bottom photographs because of rather arbitrary operating conditions. There could be no contact — even of a trigger — with bottom, and the restrictions against magnetic disturbances which might set off a mine were so severe that we were led to use rubber bands instead of steel springs to transport the film in cameras. One of the few interesting devices produced by this program — otherwise devoted to various light sources, filters, and films — was a hydrofoil camera which will be discussed later.

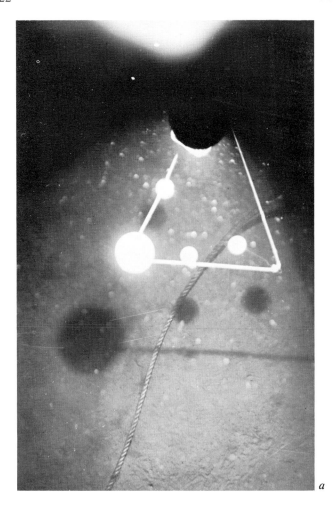
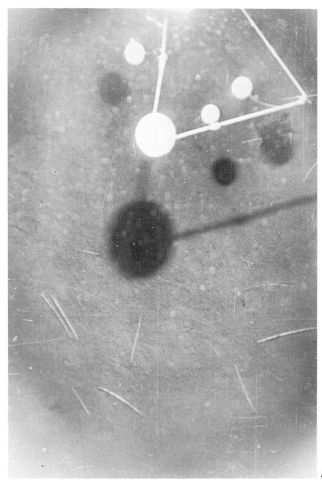

a

b

Figure 1–7. (*a*) Station C–65, *Atlantis* cruise 101, July 16, 1940. 39°35′N 70°57′W, in 455 fm (832 m). With the exception of the poor quality photographs at station C–13, the photographs at this station were the deepest ocean-bottom photos obtained at that time. (*b*) Station C–66 shows quill-worm tubes in abundance on the sea floor and light-scattering objects suspended in the water on *Atlantis* cruise 101. It was taken one hour and twenty minutes after C–65, in 325 fm (594 m).

The ocean-bottom photographic stations occupied in the Georges Bank area prior to final interruption by World War II are shown in fig. 1–9. Many excellent photographs were obtained and the great possibilities of bottom photography in the Georges Bank Canyon, on the continental slope, and in the deep-ocean basins had been thought over carefully. The authors had produced cameras No. 4 and No. 5 during the summer of 1941 for this purpose, but no opportunity to use them could be found.

Camera No. 4 (fig. 1–8*b*) was a single-shot camera, using round-cut film. It was designed for use in water up to 3,000 fm (5,500 m) deep. After only a few tests, we decided that the means of loading and unloading were too cumbersome.

Camera No. 5 was a time-lapse camera, using a Robot camera at the top of a tripod, with a means for taking a photograph each time an electric signal was received, and twelve flashbulbs (to be used one at a time) around the base of the tripod (fig. 1–8*c*). This camera contained a clock that could program twelve exposures at intervals of $\frac{1}{2}$, 1 or 2 hours, and was intended for surveying — rotary currents or biologic events requiring time-lapse photography — on Georges Bank and on other shelf areas. Demand for defense work prevented its use beyond tests in Woods Hole harbor, and it was soon adapted for mine photography. It was convenient for this adaptation because it used 35-mm film, for which a wide range of emulsions and auxiliary equipment was available. It had a built-in synchronizer, and a spring motor for repeated exposures. Representatives of the Bureau of Ordnance insisted that this be only an interim device on account

a

c

b

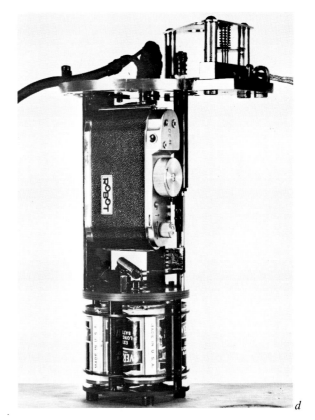

d

Figure 1–8. (*a*) Camera 3, designed for use in water depths less than 100 fm (180 m). (*b*) Camera 4, for use in water depths up to 3,000 fm (5,500 m). (*c*) Camera 5, a time-lapse camera. (*d*) Camera 6, which used a Robot camera.

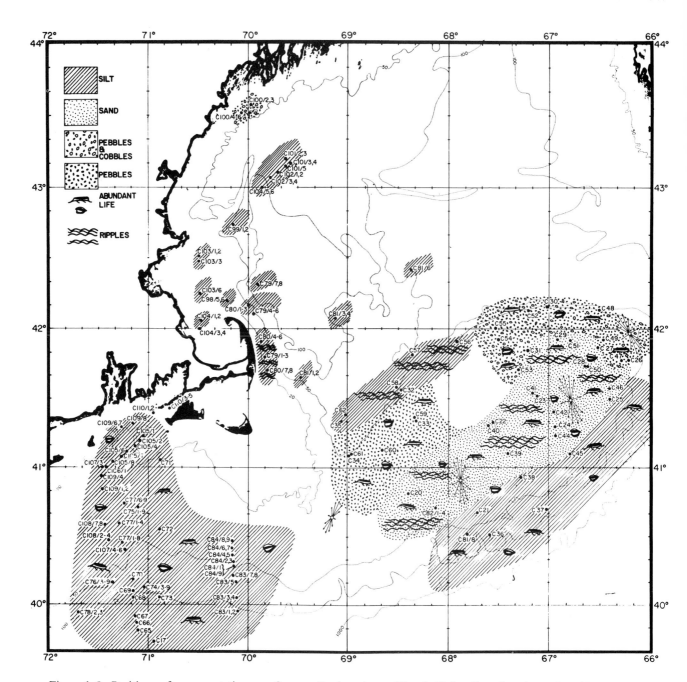

Figure 1–9. Positions of camera stations on Georges Bank and near Woods Hole, all made prior to World War II.

of its German origin and the possibility that it later might become unavailable.

The principal results achieved in the photographic campaign by the end of 1941, when war intervened, were:

1. It was demonstrated that the water clarity was such that photographs could be taken at mid-depths and of the bottom in all ocean depths with comparatively simple and lightweight equipment, from any kind of vessel.

2. A survey was carried out in the Gulf of Maine, on

the continental shelf and slope south of Cape Cod and Georges Bank, and over Georges Bank, largely as a byproduct of other work.

3. It was demonstrated that the southern part of Georges Bank between about 20 fm (35 m) and 40 fm (70 m) was a clean sand region exhibiting much life and many sand ripples; that the northeast corner of Georges Bank was an area of pebbles and cobbles, changing to a pebbly region on the northwest side; that the Gulf of Maine was a dark, silty region with little bottom life; and

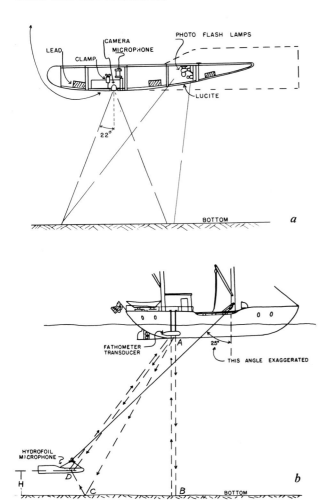

Figure 1–10. (a) The construction of the hydrofoil and (b) its operation.

that the shelf and slope south of Cape Cod and Georges Bank was a dark, silty region with abundant life and traces of bottom dwellers to depths up to 450 fm (820 m) (fig 1–9).

4. Photographic measurements of light scattering by particles in the water could be made.

5. Photographic measurements of light penetration as a function of depth were feasible.

6. Measurements of current direction and velocity near bottom could be made by photographic means.

7. Samples of the ocean floor which was photographed could be obtained for comparison with the photographs.

1–5. Mine photography, the hydrofoil camera and the diver's camera

In the early autumn of 1941, heavy demands were made on our group for photographic means of locating, localizing and identifying mines in the dirty water of harbors

and harbor approaches. Drs. E. M. Thorndike and D. E. Kirkpatrick joined the group and built transparency meters and periscopes for viewing submerged objects from a surface ship. The group experimented with an underwater television system that used an orthicon tube for viewing harbor bottoms. Other members of our group — J. L. Worzel, J. I. Ewing and G. B. Tirey — tested various films, light sources, and filters in a large number of harbors from Maryland to Maine. Camera No. 3, the shallow-water model, was used for comparison, with our earlier one being used as a monitor, and the 12-shot Robot II repeating camera (No. 5) was modified (and renamed Robot No. 6) (fig. 1–8d) to gain the flexibility and convenience of 35-mm systems. There were objections to this camera from the Bureau of Ordnance representatives on two grounds: (1) the steel mainspring might trigger magnetic mines and (2) the Robot was of German manufacture and might become unavailable. Therefore alternatives, one made from a Kodak-35 (Nos. 9 and 10) and powered by a rubber band, and the other, a Sept-35-mm (No. 7) were provided. Perhaps the most interesting part of this whole development was the requirement that the camera be added as an auxiliary to a hydrofoil bearing am electromagnetic detection system that could be towed near bottom. The idea was that any target found by the detector could be identified at once by photographs. This idea resulted from experience in answering an emergency call (July 23–26, 1942) to identify suspected mines in clear water off Virginia Beach. The newly developed hydrofoil detector was to be towed from a 50-ft wooden launch (minimum magnetic effect) and we were to identify the target by photography, without touching it. It was to be the first attempt to use the hydrofoil in open water.

We obtained a hydrofoil for experiments at Woods Hole, and by the end of October, 1942, had proved out our system and written a report on it. Some months later, we were asked to deliver it to the R/V *Howard* at Port Everglades, Florida, and to instruct personnel of that vessel in its use. No further reports of it were heard. The construction and method of operation of the hydrofoil camera are shown in figs. 1–10a and b. The camera proper was a Robot, electrically activated as a 12-shot repeater, in a case like that shown in fig. 1–8d. The vessel from which it operated had a recording echo sounder that could record echoes *ABA* and *ADA* of fig. 1–10b as shown in fig. 1–11a. From these echo times, the distance of the camera off bottom could be determined (since the angle between the two beams was known) and adjustments made to the towing cable to keep the camera at the proper focal distance from the ocean bottom. A clinometer was used to monitor the angle of the towing cable. A second means of maintaining proper distance of camera from bottom was provided by a hydrophone, in the hydro-

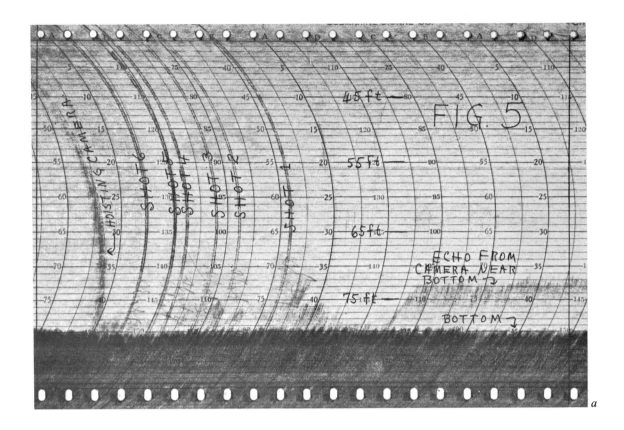

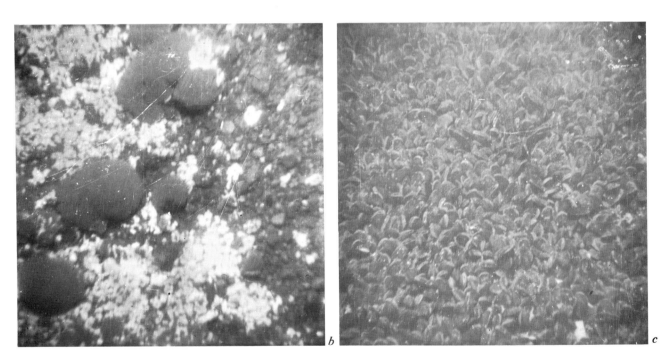

Figure 1–11. (*a*) Fathometer record, showing the echo from the camera near bottom. (*b* and *c*) Two photographs taken by the hydrofoil camera.

foil, connected to an oscillograph on the ship. The time difference between paths ACD and AD was measured directly and could be used to calculate the distance with high resolution. We found that this combination of apparatus — a fathometer on the ship and a microphone near bottom — provided a very direct method for measuring the acoustic reflection coefficient of the bottom.

In the few days during which we operated the hydrofoil camera, excellent photographs of bottom in 30 ft (9 m) to 70 ft (21 m) of water were obtained at speeds of 3 to 6 knots. Examples of these are shown in fig. 1–11b and c. Some useful results were also obtained using a 16-mm multiple-shot camera and a Kodatron electronic flash — an arrangement that could provide a very large number of photographs during a single lowering.

1–6. Wreck photography, including color, stereo, and adjoining frames

Another interesting application of ocean-bottom photography was made when we assisted in a salvage operation off Morehead City, N.C. Early in May, 1942, we were asked, on an emergency basis, to assist in the problem of identifying a wreck, believed to be an enemy submarine. As soon as possible we modified a camera according to a pattern compatible with the working conditions which we imagined might prevail. The cameras were No. 10 and No. 11, modified by Vine for work in harbors. He used a Kodak-35 camera with a Kodatron illumination — the lamp only being submerged. This camera, which was lowered on a bathythermograph wire, was near the top of a pole. Its lamp and reflector were near the bottom. A trigger was hung below the camera at the focal distance.

Our work was done from the *Anton Dohrn*, on board of which we arrived in Morehead City on May 26, 1942, to find that several antisubmarine vessels and a salvage tug had been over this wreck for some weeks, trying unsuccessfully to get divers on it to make positive identification and to obtain certain information about equipment. Their slow progress was at least partly attributable to some operating requirements that the tug take daily refuge in Morehead City during hours of darkness. We had a conference with the group commander on the evening of May 28, to extract information that would allow us to devise tactics and make plans. He said that several large buoys marked the area but that the position of the wreck relative to them was unknown. Predawn sailing of *Anton Dohrn* was necessary in order to make rendezvous at the working site with the faster vessel, scheduled to leave in good daylight. The *Dohrn*'s only special equipment was a recording echo sounder which produced

records like that of fig. 1–11a and a bathythermograph winch — hopelessly inadequate. While awaiting our turn, which came in mid-afternoon, Ewing visited aboard the armed trawler *HMS Icarus*, whose ASDIC sonar equipment was said to be very efficient for locating the wreck, and he observed the efforts to aid the tug in getting its divers on the wreck. During this visit the operational plan for the wreck photography was born.

It was arranged that the *Icarus*, a twin-screw vessel, would obtain sonar contact on the wreck from the azimuth and distance of her choice, then point her sonar permanently dead ahead and hold continuous sonar contact at constant range by use of her twin screws. *Anton Dohrn* then ran in on the bow of *Icarus*, from a distance well beyond that of the wreck, using the masts of *Icarus* as ranges. During this approach a small marker buoy was held at the rail, rigged for instant release, and the echo sounder was closely monitored for indication of sudden shoaling. Owing largely to the precision with which *Icarus* was held pointing at the target, *Anton Dohrn* found an echo-sounder indication of the wreck on the first attempt, and planted her marker buoy near to it. A few runs past this buoy defined (relative to the buoy) the area in which echo soundings from the wreck could be obtained. In the third and final phase of the operation *Anton Dohrn* was held in fixed position relative to the buoy by use of her propeller and rudder, which could be done easily by taking proper account of wind, sea, and tidal current. Actually, for a given suitable setting of rudder and throttle the vessel could usually be made to move very slowly with respect to the marker until it had passed across the wreck. A suitable small adjustment would then cause it to move slowly back across the wreck. The camera was then lowered repeatedly until a number of hits on the wreck had been made. In such a series of exposures — since the bottom is always found to have no irregularities that are large compared to the wreck — the operator can quickly determine what length of wire paid out corresponds to contact with the sea floor, and confine his further hits to the wreck by simply stopping a few feet short of bottom and moving the ship a little, in case he appears to be off the wreck. At 1740, about one hour after our operation began, the order was given to head for Morehead City. By then we had had twenty camera hits on the wreck. Our arrival in port was long after dark, the Commandant apparently being unaware that our maximum speed was about 8 knots.

During the race to reach Morehead City before dark, the film was developed, dried, and prints made. To the conference that followed in the wardroom of the salvage tug, we were able to submit prints of twenty-two frames, of which nine showed views of the wreck and definitely established that it was the German submarine U–352,

Figure 1–12. Wreck photos. (*a*) The German submarine U–352; (*b*) unidentified; (*c*) *Atlas*; (*d*) *Proteus*.

with conning tower intact and a hull or superstructure of riveted construction. Several views of limber holes, deck lockers, and hand railings were shown. In a conference on his vessel late May 29, the Commander of the tug was able to identify, attached to the conning tower railing, a grapnel and line that he had lost some days before.

A sample of the photographs obtained is shown in fig. 1–12*a*. Additional photographs were taken occasionally, as the schedule of salvagers would permit, on May 30, and June 2, 3. Conditions were far from ideal; the Kodatron lamp gave so little light that its effect added little to that of the daylight and the lamp circuit was extremely unreliable. Many of the usable negatives have been exposed by daylight alone. One set of films that recorded interesting views of the submarine was badly reticulated while drying. The photographic quality was poor because the distance from camera to bottom was chosen greater than that which permits full clarity, in order to get as great area coverage as possible. Only one

photograph (of minimal quality) showed identifiable damage, but the area of photograph 92W–11 was a mass of wreckage. The ever-recurring question of whether the damage caused the loss of the ship or whether it resulted from subsequent attacks is further complicated here by the fact that what appears to be a hip-length rubber boot protrudes — foot first — from the wreckage.

The Cape Lookout–Cape Hatteras area was an exceptionally suitable one for working out the technique of wreck photography, for the water there is far clearer than any other part of the continental shelf known to us. Presumably an eddy brings the Gulf Stream water near shore here.

While awaiting permission to leave the U–352 operation, we requested the services of a sonar-equipped escort to permit examination of other wrecks in the area. Three wrecks were located, photographed, and two of them identified, between June 7 and June 10, 1942, using the same technique and equipment as described for the U–352. On the first (unidentified) wreck, we saw only a propeller shaft heavily encrusted with marine growth (fig. 1–12b, table 1–1) and some heavy hull planking; on *Atlas*, superstructure, deck machinery (fig. 1–12c, table 1–1) and one heavily damaged area; and on *Proteus*, many shots of welded hull plates, moderately coated with marine growth (fig. 1–12d, table 1–1).

Only the clarity of the water had allowed us to work with the Kodatron light source, therefore, a short cruise, *Anton Dohrn* cruise 25, July 1–3, 1942, into Block Island Sound, was made to compare Camera 11 with Camera 3 and to convert No. 11 to flashbulb operation.

As word of this capability to identify wrecks got around, requests for the service came in. The next call was to identify a target that had been located by magnetic detection, attacked and sunk by a blimp off Manasquan, N.J. This led to the novelty rendezvous of *Anton Dohrn* with the blimp on July 12, 1942, at 40°01.8′N, 74°54.8′W, in 12 m (22 fm). The blimp pilot spotted the target for us, but the photographs showed that the wreck was old and

encrusted, and had suffered extreme damage (fig. 1–13a). Investigation proved that it was the *Mohawk* that burned and sank a few years earlier after a collision. Additional damage had resulted from attempts to reduce it with explosives to where it would not be a menace to navigation.

The next such call was an urgent request from the headquarters of the Third Naval District, to go south of Nantucket Shoals and assist a Coast Guard cutter and a converted yacht in the prolonged depth-charge attack upon a submerged target that though "heavily damaged and making an oil slick, was still moving." With a pick-up crew (*Anton Dohrn* carried five hands) and three scientists, in which the chief scientist ultimately doubled as cook and the other two scientists as dishwasher and messman, *Anton Dohrn* proceeded to the position given. It turned out to be a familiar position, that at which the Nantucket Shoals lightship had been run down on May 15, 1934, by *Olympic* (a ship that habitually passed too close aboard

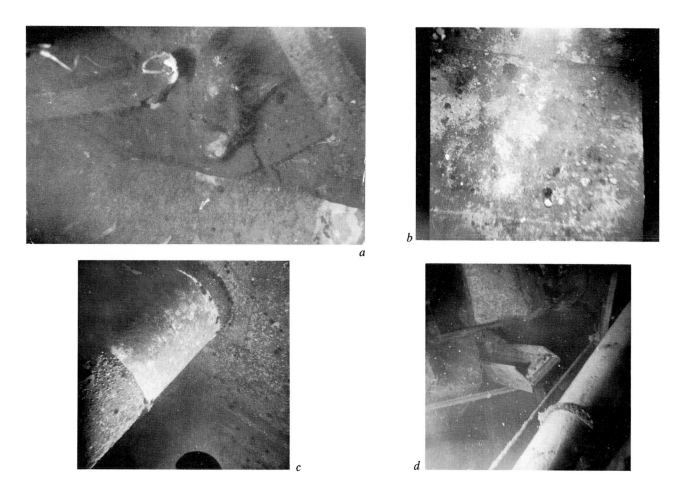

Figure 1–13. Wreck photos. (a) *Mohawk*, (b) *Nantucket Lightship*, (c) *Manuela*, (d) *Bidevind*.

in bad weather) with the loss of seven lives, mostly New Englanders. Radio traffic between the vessels was, of course, forbidden at this stage of the war. By the time *Anton Dohrn* reached the field of battle (a most inappropriate location for that little vessel in November), the depth-charge salvos were rather infrequent, but the two attackers were following their prey closely, guns ready for a quick *coup de grace* when the target could no longer postpone surfacing. The Commandant eventually found time to speak to us. He had little notion of our capabilities or why we had been sent. His main concern was that we should not interfere with action at the decisive moment. His target was changing course, trying to creep away at 1 or 2 knots. Despite the fact that his sonar equipment had been reduced almost to uselessness by the depth charges, he knew his target was changing course because it was leaking oil that rose to the surface and formed an oil slick. Without that sonar it might be very difficult to deal with the target if it surfaced after dark.

In a conference aboard the yacht, our method of identification by photography was explained and Ewing asked for an hour or two to locate the target by echo sounder running up the oil slick, since the cutter now had no sonar. The suggestion that the speeds and courses of target indicated by the oil slick could all be accounted for by the rotary tidal currents (see chart, fig. 1–9) was received without enthusiasm, possibly even without comprehension.

Anton Dohrn was eventually allowed to take photographs. The marker buoy was set on an echo sounder indication found easily by running up the oil slick. It was immediately apparent that the source of the oil slick did not move relative to the buoy and that the target rose to 36 ft (11 m) above bottom, rather too much beam for a submarine. The photographs showed several views of smooth hull plating, covered with marine growth that included a number of small starfish — obviously an old wreck that was leaking oil afresh as the result of pounding by about 100 depth charges (fig. 1–13*b*).

The attack ended there. In breaking off, the cutter got up to full speed, spent all her remaining depth charges in a salvo at the spot marked by our buoy and then asked us to lead her to Gay Head Light, reporting that the last salvo had knocked out all of her compasses.

The waiting time for *Anton Dohrn* during this operation was considerable. It was used to photograph the bottom in the area and to try some improvements of technique, but not much was accomplished. The Commander did not want interference by unauthorized ship movements, the scientists had to double in the galley, and the weather was quite rough. The Chief Scientist dreamed of Cervantes on the way back to Woods Hole, and *Anton Dohrn* received a Unit Citation.

The finale of the first phase of the wreck photography campaign was a trip to Ram Island, near Marblehead, Massachusetts, where airplane wreckage was photographed April 8–10, 1943. In the meantime we had been asked to mount a campaign to locate, photograph, and identify wrecks in the Fifth Naval District, working on board military vessels that had their own sonar equipment. The first such test was on SC664. Then the project shifted to the 180-ft, single-screw U.S. Coast Guard buoy tender *Gentian*. An official project officer was assigned, and the vessel engaged in systematic sonar search and photographic identification from May to September, 1943. The only modification in technique was in setting the marker buoy. When a sonar target was detected, an approach was made as though to attack with depth charges, but the buoy was dropped instead of charges. Standard procedure was to take twenty photographs of the wreck before developing, to leave the marker buoy in place and make a dead reckoning run to some fixed position and back while developing the films, and to lift the marker only after seeing the photographs.

Some difficulty arose from the lack of skill and experience of the watch officers. Wartime training naturally was confined to the bare fundamentals of routine operation. Some skill was required to hold a single-screw ship stationary relative to a buoy in various conditions of wind and current. Occasionally the camera would become caught in the wreckage and considerable skill was required to retrieve it. The wrecks listed in Table 1–1 were located and identified in this campaign, and a number of interesting photographs taken, several of which are shown in figs. 1–13*c*, 1–13*d* and 1–14.

The cameras used in this campaign were Robots, essentially like that shown in fig. 1–8*d*. These cameras proved very convenient, reliable, and adaptable.

Several tests of wreck shots in color were made, but it was decided that, except in very special situations, the subjects were not very colorful, and the advantages of color did not compensate for the difficulties of processing and the need for uniform intensity of lighting that were entailed.

Several tests of stereoscopic photography were made by mounting two of these cameras to view a common field at a proper separation on the same pole and linking the triggers together. These tests were thoroughly successful, but the stereo work was judged to be of less value than the doubled area coverage that could be obtained by pointing the two cameras so that they covered adjacent fields.

In the summer of 1944 a campaign of wreck photography in the Fifth Naval District was begun. The project officer, Harleston R. Wood, was trained in the operation and took some excellent photographs, locating and

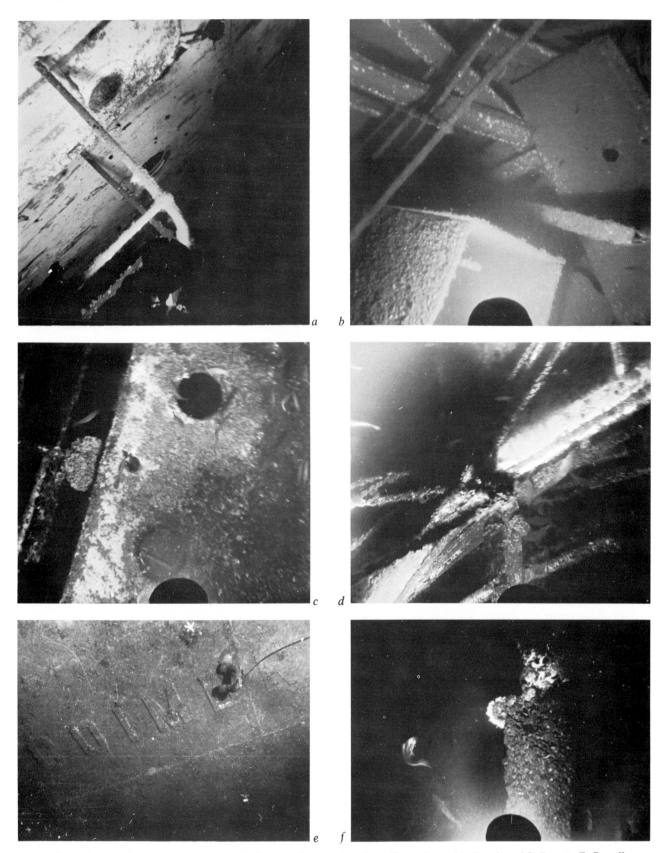

Figure 1–14. Wrecks. (a) *Proteus*, (b) *City of Atlanta*, (c) and (d) *Tamaulipas*, (e) *Coimbra*, (f) *Francis E. Powell*.

TABLE 1–1. Wrecks surveyed.

Wreck	Date surveyed	Lat. north		Long. west		Water depth		Height above bottom	
		deg.	min.	deg.	min.	feet	meters	feet	meters
U–352	5/28/42–6/3/42	34	13	76	34	103	31.4	27	8.2
Unidentified	6/7/42								
Atlas	6/9/42	34	34	76	14	116	35.4	32–55	9.8–16.8
Proteus	6/10/42	34	49	75	54	82	25.0	42–45	12.8–13.7
Mohawk	7/13/42	40	02	73	55	75	22.9	Disintegrated (10)	3.1
S.S. Strathbone	11/9/42	40	26	69	29	222	67.7	36	11.0
Off Ram Island	4/8/43–4/10/43	42	28	70	51	42	12.8		
Equipoise	4/17/43	36	14	74	51	225	68.6	55	16.8
Arundo	5/20/43	40	11	73	41	126	38.4	63	19.2
Tolten	5/29/43	39	55	73	47	90	27.4	60	18.3
R.P. Resor	5/29/43	39	48	73	26	120	36.6	90	27.4
Bidevind	6/12/43	39	48	72	49	192	58.6	49	14.9
Coimbra	6/13/43	40	22	72	20	180	54.9	45	13.7
John Morgan	6/24/43	37	00	75	25	97	29.6	67	20.4
Lillian Luckenback	6/25/43	36	58	75	25	99	30.2	33	10.1
E. M. Clarke	7/20/43	34	49	75	33	222	67.7	78	23.8
Proteus	7/21/43	34	49	75	54	90	27.4	36	11.0
U–352	7/31/43	34	13	76	34	107	32.6	15	4.6
Naeco	8/2/43	34	03	76	35	130	39.6	39	11.9
Tamaulipas	8/4/43	34	31	76	02	142	43.3	56	17.1
Manuela	8/7/43	34	39	75	48	145	44.2	57	17.4
Nordal	8/11/43	34	41	75	35	402	122.6	72	22.0
Papoose	8/15/43	34	09	76	40	110	33.6	48	14.6
Francis E. Powell	6/8/44	36	50	75	23	89	27.1	39	11.9
Marore	6/12/44–7/6/44	35	33	75	15	94	28.7	34	10.4
City of Atlanta	7/7/44–7/14/44	35	24	75	21	78	23.8	18	5.5
Venore	7/10/44	35	05	75	24	94	28.7	28	8.5
U–85	8/5/44	35	53	75	17	84	25.6	9	2.7
Lancing	8/7/44	35	38	74	52	186	56.7	42	12.8
Harpathian	9/3/44	36	26	74	58	132	40.3	15	4.6
Malchace	9/5/44	34	45	75	48	122	37.2	52	15.9
Ario	9/6/44	34	19	76	27	89	27.1	43	13.1

identifying nine additional wrecks. This work was concentrated in inshore regions where there was a possibility that the wrecks might be menaces to navigation.

On all of these cameras, photoflash bulbs were used as light sources, and a "sail" like that shown in fig. 1–15b was used to prevent rotation of the camera and consequent blurring of the pictures. The success achieved in striking targets as small as wrecked ships naturally suggested that rich rewards could come from geologic study of submarine canyons or seamounts by photography, and it was resolved that such work would have high priority in the post-war years. A camera with greater depth capability than Camera No. 6 was required. It was decided that the most convenient form would be a modification to put a film-transport mechanism like that of No. 3 into a case like that of Camera No. 4, and the production of such a camera was carried along as a "spare-time" project.

Various problems in underwater sound began to demand all of our time, with the result that, except for a

diver's model of our basic Robot camera case (for use by frogmen in underwater demolition, etc., see fig. 1–15a), little more work on underwater optics was done until the end of World War II.

1–7. Ocean-bottom photography: its use in geological and biological studies

In August, 1944, a camera was loaned to F. P. Shepard who successfully photographed the submarine canyon off the California coast (Shephard & Emery, 1946).

In the spring of 1945 David Baldwin did some ocean-bottom photography on U.S.S. Saluda in the Gulf of Mexico, using a Robot type of camera, after gaining some experience in the Woods Hole area.

In the summer of 1946 most of the effort was on completion of the seismic sections across the continental shelf, using the crash boat SC-446, and Balanus, a converted 42-net-ton fishing vessel 72 ft long. During this

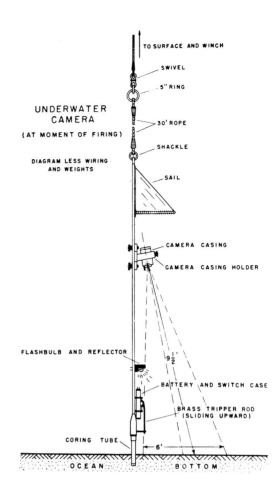

UNDERWATER
CAMERA
(AT MOMENT OF FIRING)

DIAGRAM LESS WIRING
AND WEIGHTS

TO SURFACE AND WINCH

SWIVEL

5" RING

30' ROPE

SHACKLE

SAIL

CAMERA CASING

CAMERA CASING HOLDER

FLASHBULB AND REFLECTOR

9½'

BATTERY AND SWITCH CASE

BRASS TRIPPER ROD
(SLIDING UPWARD)

CORING TUBE

6'

OCEAN BOTTOM

a b

c d

Figure 1–15. (*a*) Diver's model camera. (*b* and *c*) Diagram of the deep-sea Argoflex
and its use on *Atlantis* cruise 150. (*d*) Housing for No. 21 flashbulbs, using an O-ring.

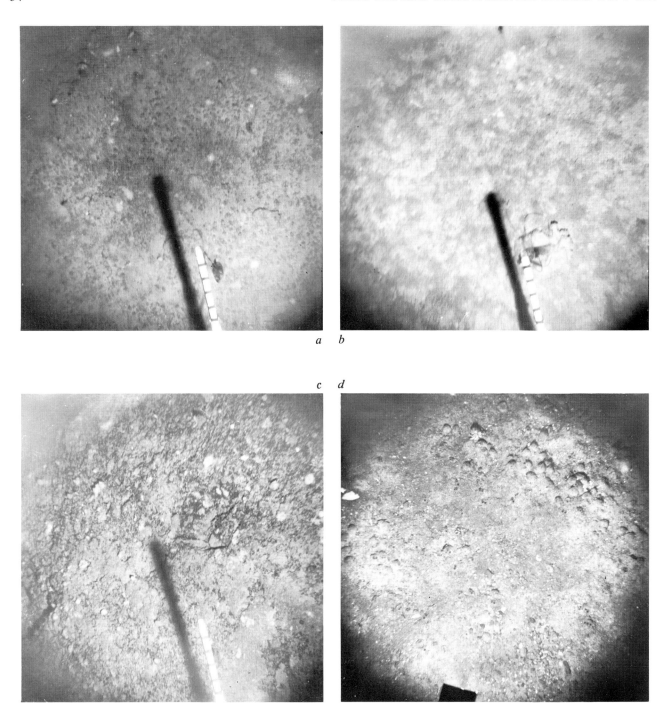

a   b

c   d

Figure 1–16. Photographs taken on *Atlantis* cruise 150. (*a* and *b*) are on Atlantic seamount at 34°02′N 30°15′W, in 400 fm (730 m); and (*c*) is just off the same seamount in 1,100 fm (2,000 m). (*d*) A photograph taken with the Argoflex camera used on *Balanus*, summer, 1947.

campaign J. L. Worzel and G. B. Tirey made several photographic traverses across the continental shelf using a Robot camera of the type used for most of the wreck photographs (fig. 1–8*d*). These lines formed a valuable addition to our basic project of surveying the shallow water area from Cape Hatteras to Cape Sable.

By the spring of 1947 two cameras had been constructed that were suitable for photography in deep water, and the work on submarine canyons, continental slopes, and basin floors that had been our goal for many years could begin. This camera was made from an Argoflex and put in a cylindrical pressure case tested for

3,000-fm (5,500 m) submergence like that of Camera No. 4. Its parts and the means of its use are illustrated in fig. 1–15b and c. After about one year it was decided that the No. 5 flashbulbs, for which suitable waterproof housing had been made from carburetor sediment bulbs, gave too little light, and the housing for No. 21 flashbulbs was used instead (fig. 1–15d). This was our first use of O-rings for deep-sea pressure vessels.

There were ship facilities available for attacking two parts of the problem simultaneously during the summer of 1947, so two groups were formed. One group went on *Atlantis* cruise 150, the first expedition to the Mid-Atlantic Ridge sponsored jointly by the National Geographic Society, the Woods Hole Oceanographic Institu-

tion, and Columbia University. The camera work was only a minor portion of the program, which included seismic refraction measurements, sediment coring, rock dredging, bottom trawling, and several detailed topographic investigations (Tolstoy & Ewing, 1949).

In the work of the previous year results had been plagued by failure of the electric switch, designed to trigger the camera on bottom contact. In the spring of 1944 a new type of bottom switch had been designed by Worzel to operate from a magnet brought into the proper position near a pressure case in which a microswitch was triggered by the magnetic attraction. Unfortunately, construction of this device had not been completed in time for *Atlantis* cruise 150 because of all the other equipment

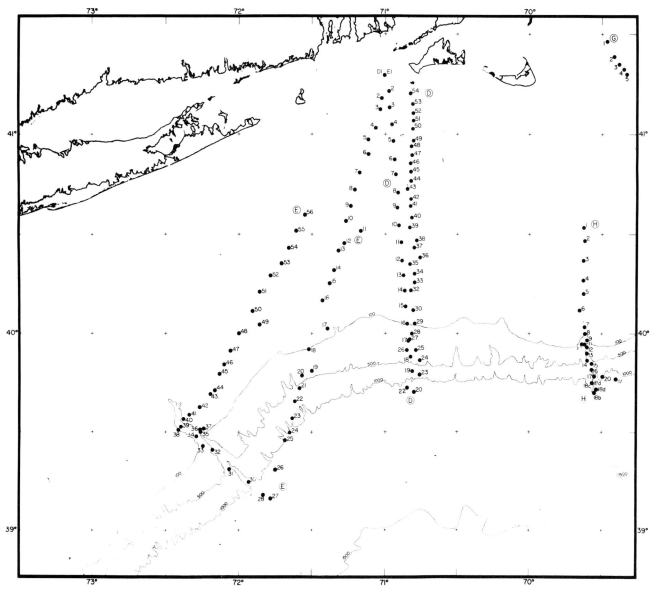

Figure 1–17. Stations occupied by *Balanus* in the summer of 1947. The F-series in the Gulf of Maine, and part of the G-series are not shown.

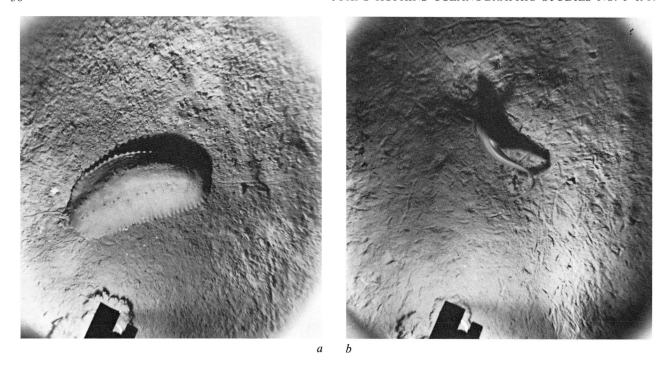

a   b

c   d

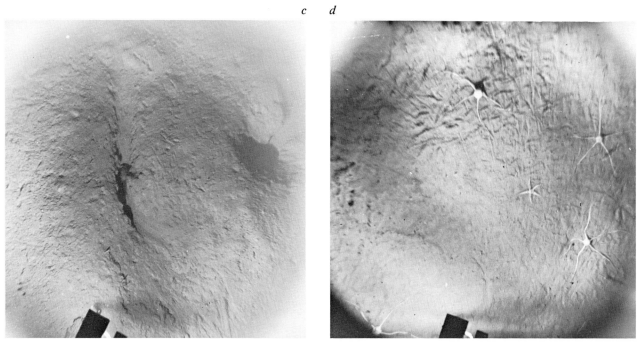

Figure 1–18. Photographs from *Balanus*, 1947. (*a*) E–25, 39°28′N 71°41′W, 1,420 fm (2,600 m). (*b*) E–22, 39°39′N 71°37′W, 1,160 fm (2,120 m). (*c*) E–34, 39°29′N 72°17.5′W, 1,240 m. (*d*) H–27, 39°56.5′N 68°58′W, 1,071 fm (1,958 m).

construction required, so the former type of oil-filled bottom switch was attached to the camera. This cruise was plagued with failures of this switching apparatus so that only a few bottom photographs were taken. The principal results were a series of three photographs taken down the side of the Atlantis Seamount (fig. 1–16*a*, *b* and *c*).

The second group fitted the new magnetic switch to a

similar camera and undertook a series of five short cruises on R/V *Balanus* with John Ewing, Bruce Heezen (then an undergraduate at the University of Iowa), and John Northrop (a graduate student at Columbia) — John Ewing instructing the others in the use of the camera. During these cruises more than 200 photographs were taken. Positions of the photographs, the *D, E, F, G,* and *H*-series, are shown in fig. 1–17.

The *D*-survey series consisted of a line of fifty-four stations, extending south from the entrance of Vineyard Sound to about the 1,000-fm (1,800-m) curve; and back. It might be noted that in Northrop (1951) a published photograph identified as *D*-2 was said to be at 41°12′N and 70°59′W, whereas it was actually taken earlier in Vineyard Sound.

The *E*-survey (fifty-six stations) extended south by west from the entrance of Vineyard Sound to about the 1,500-fm (2,750-m) curve, and then moved up into the deep part of the Hudson Canyon, and returned from there to Vineyard Sound.

F (forty-one stations) covered a triangular course starting about 10 miles (18.6 Km) north of Race Point, at the tip of Cape Cod, and going east, then north, then back to the starting point. On this cruise they studied the Murray Basin and Stellwagen Bank in the Gulf of Maine (not shown in fig. 1–17).

G (forty-three stations) started about 25 miles (46.5 km) east of Monomoy Point, the southeastern tip of Cape Cod, and followed a southeasterly course to about 1,400 fm (2,560 m) in the region of Welker Canyon.

H (thirty stations) started near the Nantucket Shoal Lightship; went south through the Veatch Canyon to about 1,000 fm (1,800 m) and then over to Hydrographer Canyon.

This *Balanus* survey made a major contribution to our campaign to survey the area because it extended the survey with many photographs into deeper water and because the work was done along lines, providing data more useful than that taken at scattered points. Unfortunately, the value of the work was seriously impaired by the necessity to navigate by dead reckoning alone, by the use of an inadequate sounder, by the necessity of using a winch of extremely limited capabilities, and by poor data recording. Much of the time the depth reported for the photographs was determined by the amount of wire paid out. There was, however, the advantage that attention could be fully devoted to photography.

The use of a device for sampling sediment, attached to the camera pole (which had started with Camera No. 2C) was resumed here, under the philosophy that both photographs and bottom samples were greatly enhanced in value if they represented exactly the same area. In fact, it was this point that caused us to use single-shot cameras

and flashbulbs instead of multiple-shot cameras and electronic flashes for several years after.

The results of this campaign greatly extended, both in areal coverage and in depth, the survey of the continental shelf and slope south of Cape Cod and Georges Bank, as well as south of Long Island and in the Gulf of Maine. The earlier conclusions were fully corroborated. Figs. 1–16*d* and 1–18 (*a, b, c,* and *d*) show some of the more interesting photographs. Perhaps the most noteworthy finding was the discovery of an Eocene sediment outcrop. Three photographs of this outcrop were made and described by Northrop & Heezen (1951).

In the winter and spring of 1947–48, one of these Argoflex cameras was lent to the Woods Hole Oceanographic Institution for use on *Atlantis* cruise 151. This cruise, led by M. J. Pollak, crossed the Atlantic, cruised the Mediterranean, passed through the Strait of Sicily, into the Aegean Sea. The camera with its bottom sampling tube (operated by D. M. Owen) made one of the most important contributions to the results of this cruise with seventy-five photographs and accompanying samples. Four of the photographs are shown in fig. 1–19.

This cruise extended the photographic coverage of the ocean floor and provided the deepest picture taken up to that time — a photograph of manganese nodules in 3,026 fm (5,534 m) (fig. 1–19*d*).

The second Mid-Atlantic Ridge expedition extended from June to September, 1948, on *Atlantis* cruise 152. This cruise also was supported jointly by the National Geographic Society, the Woods Hole Oceanographic Institution, and Columbia University. With the magnetic switch now in use, the camera functioned satisfactorily and many useful pictures and sediment samples were obtained to supplement the other work.

Fig. 1–20 shows some of the pictures from the Mid-Atlantic Ridge which show effects of current scouring. The picture of the sand ripples at 410-fm (750-m) depth was the first of many which have been taken since, demonstrating that current ripples may be found in water depths much greater than the 100-fm (180-m) maximum previously thought to be the greatest depth for their formation.

1–8. Summary

By the end of 1948, the underwater camera had been shown to be a very effective tool for the study of organic geology and biology at all depths of the ocean and as a tool for measurement of currents, light scattering by particulate matter, and light penetration into the sea. The intervening years have greatly expanded the coverage using these tools. The camera had been used to obtain stereoscopic and color photographs, as well as the more conventional single black-and-white pictures.

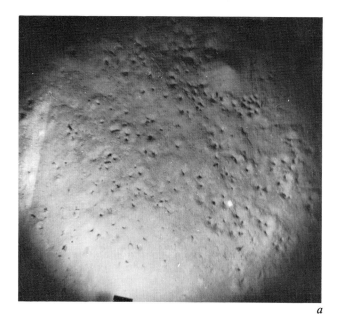

a   b

c

d

Figure 1–19. Photographs from *Atlantis* cruise 151. (*a*) 151–3, 39°46′N 70°50′W, 1,000 fm (1,800 m); (*b*) 151–19, 39°06′N 25°53′E, 113 fm (207 m); (*c*) 151–14, 37°50′N 26°25′E, 603 fm (1,103 m); (*d*) 151–73, 30°37′N 59°07′W, 3,026 fm (5,534 m) the deepest photograph at that time.

The principal advances in photographic equipment in the intervening period have been the ability to photograph repeatedly at one lowering by the improvement of the electric flash devices (of which the Kodatron was a forerunner) and the use of the camera in conjunction with other apparatus to supplement their information.

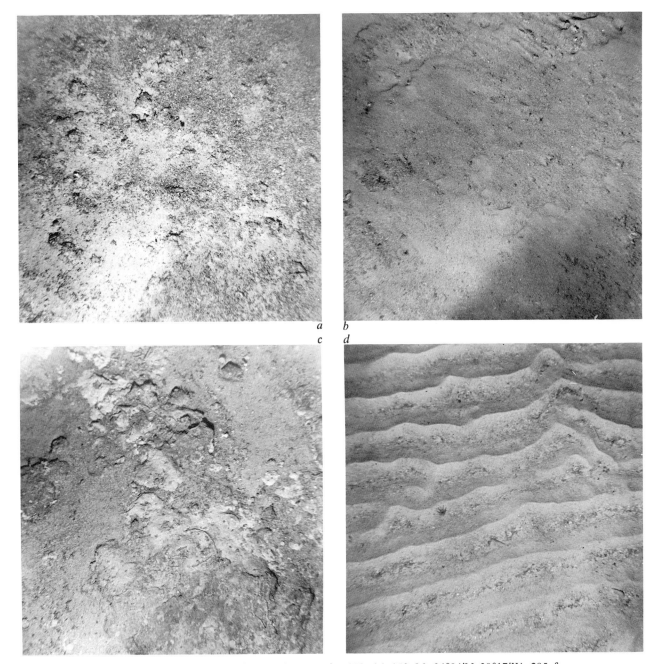

Figure 1–20. Photographs from *Atlantis* cruise 152. (*a*) 152–96, 34°04′N 30°17′W, 205 fm (375 m); (*b*) 152–98, 34°05′N 30°11′W, 225 fm (412 m); (*c*) 152–101, 34°07′N 30°13′W, 190 fm (348 m); (*d*) 152–110, 34°14′N 30°12′W, 410 fm (750 m).

References

Dana, Richard H., 1964: *Two years before the mast.* Dodd, Mead and Co., New York (reprint of 1840 edition), 338 p.

Ewing, Maurice, & Allyn Vine, 1938: Deep-sea measurements without wires or cables. *Trans. Amer. Geophys. Union*, Nineteenth Ann. Meeting 1938, 248–51.

——, ——, & J. L. Worzel, 1946: Photography of the ocean bottom. *J. Opt. Soc. Am.*, **36**, 307–21.

——, & Edward M. Thorndike, 1965: Suspended matter in deep ocean water. *Science*, **147**, no. 3663, 1291–94.

Northrop, John, 1951: Ocean-bottom photographs of the neritic and bathyal environment south of Cape Cod, Massachusetts. *Bull. Geol. Soc. Am.*, **62**, 1381–84.

——, & Bruce C. Heezen, 1951: An outcrop of Eocene sediment on the continental slope. *J. Geol.*, **59**, 396–99.

Shepard, F. P., & K. O. Emery, 1946: Submarine photography off the California coast. *J. Geol.*, **54**, 306–21.

Tolstoy, Ivan & Maurice Ewing, 1949: North Atlantic hydrography and the Mid-Atlantic Ridge. *Bull. Geol. Soc. Am.*, **60**, 1527–40.

Bibliography: a summary of the early development of underwater photography[*]

Anonymous, 1920: Exploring with a camera; searching for German traps in the Lens coal mines. *Sci Am. Monthly*, **2**, 172–73.

————, 1920: A submarine camera. *The Engineer*, 129, 658.

Atkins, W. R. G., G. L. Clarke, H. Pettersson, H. H. Poole, C. L. Utterback, & A. Angstrom, 1938: Measurement of submarine daylight. *J. du Conseil*. Conseil Permanent International pour l'Exploration de la Mer, 13, 37–57.

Barton, Otis, 1930: The bathysphere. *Bull, N.Y. Zool. Soc.*, **33**, no. 6, 233–34.

————, 1935: Five hundred fathoms deep. *Natural History, J. Am. Mus. Nat. Hist.*, **35**, 144–45.

Bartsch, Paul, 1927: Movie camera for use on the sea bottom. *Sci. Am.*, **136**, 127–28.

————, 1930: A quarter mile down in the open sea. *Bull. N.Y. Zool. Soc.*, **35**, no. 6, 200–31.

Beebe, William, 1927: The Haitian expedition. *Bull. N.Y. Zool. Soc.*, **30**, no. 5, 123–29.

————, 1932: A half-mile dive in the bathysphere. *Bull. N.Y. Zool. Soc.*, **35**, no. 5, 143–72.

————, 1932: A wonderer under sea. *Nat. Geog. Mag.*, **62**, no. 6, 740–58.

Boutan, L., 1893: Mémoire sur la photographie sous-marine. *Arch. Zool. Exper. Gen.*, **21**, 281–324.

————, 1898: L'instantané dans la photographie sous-marine. *Arch. Zool. Exper. Gen.*, **26**, 299–330.

————, 1900: *La photographie sous-marine et les progrès de la photographie*. Schleicher Freres, Paris, 332 p.

Clarke, George L., & Harry R. James, 1939: Laboratory analysis of the selective absorption of light by sea water. *J. Opt. Soc. Am.*, **29**, 43–55.

————, 1941: Observations on transparency in the southwestern section of the North Atlantic Ocean. *J. Mar. Res.*, **4**, no. 3, 221–30.

Du Pont, A. Felix, 1940: *Under sea with helmet and camera*. Dodd Mead and Co., New York, 87 p.

Hartman, H., 1917: in: An electric camera for deep-sea photography. *Sci. Am.*, **116**, 483.

Harvey, E. Newton, 1939: Deep-sea photography. *Science*, **90**, 187.

Johnson, E. R. F., 1939: Undersea cinematography. *J. Soc. Motion Picture Engrs.*, **32**, 3–17.

Longley, W. H., 1916: The significance of the colors of tropical reef fishes. *Carnegie Inst. Washington Yearbook*, **15**, 209–12.

————, & Charles Martin, 1927: The first autochromes from the ocean bottom. *Nat. Geog. Mag.*, **51**, no. 1, 56–60.

Miner, Roy Waldo, 1924: Hunting corals in the Bahamas. *Natural History, J. Am. Mus. Nat. Hist.*, **24**, 594–600.

Murray, John, & Johan Hjort, 1912: *The depths of the ocean*. Macmillan and Co., London, 821 p.

Pettersson, Hans, & Svante Landberg, 1934: Submarine daylight. *Göteborgs Kungl. Vetenskaps-och Vitterhets-Samhälles Handlingar*. Femte följden, Series B, Band 3, no. 7. (Meddelanden från Göteborgs Högskolas Oceanografiska Institution) 1–13.

————, 1936: The transparency of sea water. *Rapports et Procès-Verbaux. Conseil Permanent International pour l'Exploration de la Mer*, **101**, part 2, no. 6, 1–7.

————, & Horace H. Poole, 1937: Measurements of submarine daylight. *Göteborgs Kungl. Vetenskaps-och Vitterhets-Samhälles Handlingar*. Femte följden, Series B, Band 5, no. 5. (Meddelanden från Göteborgs Högskolas Oceanografiska Institution) 1–34.

Poole, H. H., & W. R. G. Atkins, 1929: Photo-electric measurements of submarine illumination throughout the year. *J. Mar. Biol. Assoc. U.K.*, 16, 297–324.

Reighard, Jacob, 1907: Photography of aquatic animals in their natural environment. *Bull. U.S. Bureau of Fisheries*, **27**, 43–68.

Svendrup, H. U., Martin W. Johnson, & Richard H. Fleming, 1942: *The oceans*. Prentice-Hall, Inc., New York, 1,087 p.

Tee-Van, John, 1928: Methods in submarine photography. Appendix in William Beebe: *Beneath tropic seas*. G. P. Putnam's Sons, New York. 211–15.

————, 1928: A submarine motion picture camera. *Bull, N.Y. Zool. Soc.*, **31**, no. 2, 41–45.

Utterback, C. L., 1936: Spectral bands of submarine solar radiation in the North Pacific and adjacent inshore waters. *Rapports et Procès-Verbaux. Conseil Permanent International pour l'Exploration de la Mer.*, **101**, part 2, no. 4, 1–15.

Ward, F., 1911: *Marvels of fish life as revealed by the camera*. Cassell and Co. Ltd., London.

————, 1919: *Animal life under water*. Cassell and Co. Ltd., London.

————, 1928: in: *Aquatic life in its own setting*. *Sci. Am.*, **107**, 552–53.

Williamson, J. E., 1913: in: *Photographing under water*. *Sci. Am.*, **109**, 6.

————, 1936: *Twenty years under sea*. Hale, Cushman, and Flint, Boston.

Young, J. C., 1925: Walking the ocean floors. *Worlds Work*, **50**, 155–60.

[*] This bibliography is based on that of Ewing, Vine, & Worzel, 1946: Photography of the ocean bottom, *J. Opt. Soc. Am.*, **36**, 307–21.

2. Physics of underwater photography*

Edward M. Thorndike   *Lamont Geological Observatory, Palisades, New York; and Queens College, Flushing, New York*

Abstract

In underwater photography, light is transmitted through water and through a window into air where the camera is located. Thus, knowledge of the refractive index, dispersion, absorption, and scattering of the water is needed for the design of an underwater camera system and for the evaluation of its capabilities. Refractive index and dispersion must be known, for they determine the bending of light rays of different colors on passage from water into air. Knowledge of the absorption and scattering of the water is essential for the evaluation of the losses and redirection of the light beam as it traverses the part of the path that is in water.

The brief discussion of these factors presented in this chapter is intended to provide a background for the discussions of camera designs and uses which follow.

2–1. Refraction and dispersion

As a light ray passes from one medium to another, it is refracted in accordance with the equation

$$n_1 \sin \phi_1 = n_2 \sin \phi_2 \tag{1}$$

in which n_1 and n_2 are the refractive indices of the two media and ϕ_1 and ϕ_2 are the angles between the ray and the normal to the surface in the corresponding cases. If light passes from water into glass or plastic and then from the glass or plastic into air, this equation is used successively at the two surfaces. The situation is simplified if the glass or plastic has surfaces that are plane and parallel to each other. Then the properties of the window do not affect the final result which becomes

$$\sin \phi_a = n \sin \phi_w \tag{2}$$

in which n is the refractive index of water relative to air, approximately 4/3. For an optical system viewing an object in water through a plane window, refraction gives four effects of interest:

* This work was supported by the Office of Naval Research, under contract Nonr 266(48).

1. Rays diverging from a point at a distance s_w from the window appear, after they have passed through the window, to come from a distance s_a, such that

$$s_a = s_w/n \tag{3}$$

Thus, the optical system must be focused for an air distance approximately ¾ of the actual distance in water.

2. A ray coming to the window at any angle of incidence other than zero will leave it at a larger angle. Thus, an optical system such as a camera, which has a given angular field of view in air, will have a smaller angular field in water. In addition, an object which would produce an image of a certain size if it were in air produces a larger image if it is in water. The lens appears to have a longer focal length, approximately 4/3 that in air, and a correspondingly modified focal ratio or f value, making the lens slightly slower.

3. Rays from points forming a rectangular grid in water will not seem to come from a rectangular grid after they have passed into air. There will be distortion.

4. Light of different wavelengths has different values of n; thus, there is dispersion in water, and the three effects above will all vary with the color of the light. The most important effect of dispersion is the introduction of lateral chromatic aberration in the outer part of the field. White light passing obliquely from water to air is dispersed into a spectrum along a line extending radially out from the axis of the system.

In order to calculate the magnitudes of these effects, one must know the numerical values of the refractive index of water under the conditions that exist where the equipment is located. The refractive index of distilled water throughout the visible part of the spectrum and for temperatures from 0° C to 60° C is well known from the work of Tilton & Taylor (1938). The effect of pressure on distilled water at 25° C has been studied by Rosen (1947) for four wavelengths. The effect of dissolved materials has been investigated by many workers. Sea water is, of course, the case of interest here. The refractive index has been measured under various conditions by several investigators, among them Utterback (1934), Bein (1935), and Miyake (1939). Tables 2–1, 2–2, and 2–3, taken from these sources, show the variation of refractive index with temperature, chlorinity, salinity, and pressure.

The range in values of the refractive index for light of wavelength 0.5893 μ for conditions that may be encountered in underwater photography is from 1.3325 for fresh water at 25° C and 1 atm pressure to 1.3545 for sea water at 0° C and a pressure of 1,000 atm corresponding to a depth of 10 km. For fresh water at 20° C and 1 atm pressure, the value varies from 1.3312 at wavelength 0.6563 μ to 1.3371 at 0.4861 μ. From these data and equations (2) and (3) it is evident that an optical system designed for observing an object in water through a plane window must allow for the differences in focus and angular field, for the distortion mentioned above, and must be corrected for lateral chromatic aberration unless the region of the spectrum used is sharply restricted. It is also evident that such a system designed for the water of one location will probably also operate satisfactorily at any other location.

TABLE 2–1. Refractive index of distilled water relative to air at atmospheric pressure, for wavelengths (λ) of 0.6563, 0.5893, and 0.4861μ (data from Tilton & Taylor, 1938).

Tempera-ture (°C)	Refractive index		
	$\lambda = 0.6563$	$\lambda = 0.5893$	$\lambda = 0.4861$
0	1.332 09	1.333 95	1.338 11
1	09	95	11
2	09	94	10
3	07	93	09
4	06	91	07
5	03	88	04
10	1.331 84	69	1.337 84
15	55	39	53
20	15	1.332 99	12
25	1.330 67	50	1.336 63

TABLE 2–3. Refractive index of distilled water at 25° C, for wavelengths (λ) of 0.579, 0.546, 0.436 and 0.406 μ (data from Rosen, 1947).

Pressure (atmo-spheres)	Refractive index			
	$\lambda = 0.579$	$\lambda = 0.546$	$\lambda = 0.436$	$\lambda = 0.406$
1	1.333 0	1.334 0	1.339 8	1.342 2
500	1.340 1	1.341 3	1.347 2	1.349 6
1,000	1.346 2	1.347 4	1.353 1	1.355 8
1,500	1.351 6	1.352 8	1.358 8	1.361 3

2–2. Absorption and scattering

Light is attenuated as it passes through water in accordance with the relation

$$I/I_o = e^{-kx} \qquad (4)$$

where I_o is the initial intensity of the light; I the intensity after scattering and absorption in the water; and k, the attenuation coefficient, is the sum of the absorption coefficient α and the scattering coefficient β. Both α and β depend upon the wavelength of the light and upon the material dissolved and suspended in the water. Many investigators have studied the transmission of light through water. Our present knowledge of this important field has been presented by Duntley (1963) and Jerlov (1963), and an annotated bibliography of older work has been given by Du Pré & Dawson (1961). Some studies of light-scattering in deep-sea water are discussed in chap. 10.

Observers agree that the absorption and scattering in clear ocean water are essentially the same as in clear distilled water, that some dissolved matter increases the absorption, and that suspended matter increases the scattering. However, there is rather wide variation in the

TABLE 2–2. Refractive index of sea-water for wavelength 0.5876 μ at atmospheric pressure for temperatures from 0° C to 25° C (data from Bein, 1935; and Tilton & Taylor, 1938).

ρ	Chlorinity Cl (%)	Salinity S (%)	Refractive index					
			0° C	5° C	10° C	15° C	20° C	25° C
0	0	0	1.334 00	1.333 94	1.333 74	1.333 44	1.333 04	1.332 56
16	11.6	20.9	8 18	8 02	7 73	7 36	6 91	6 38
20	14.5	26.2	9 22	9 03	8 72	8 34	7 87	7 34
24	17.4	31.4	1.340 26	1.340 04	9 72	9 32	8 84	8 30
28	20.3	36.7	1 29	1 05	1.340 71	1.340 30	9 81	9 26
32	23.2	41.9	2 32	2 06	1 70	1 28	1.340 77	1.340 21

$$\rho = 1,000 \times \left(\frac{\text{Density of sea water at } 17.5° \text{C}}{\text{Density of distilled water at } 17.5° \text{C}} - 1 \right)$$

Approximate chlorinities and salinities calculated from:

$$Cl = \rho/1.38 \text{ and } S = \rho \times 1.309$$

quantitative results obtained, particularly in the case of scattering. The values given in tables 2–4 and 2–5, obtained from data of Hulburt (1945), give an idea of the magnitude of the coefficients, their variation with wavelength, and the effects that they have on the light transmitted through various distances in water.

Both absorption and scattering present difficulties when optical observations are made over appreciable distances in water. Scattering is the more troublesome as it not only removes useful light from the beam but also adds background illumination. Compensation for the loss of light by absorption can sometimes be made by the use of stronger light sources. The background illumination caused by scattering can be reduced by keeping unnecessary light out of the water between the object being viewed and the viewing instrument. Objects a few meters from a camera can be clearly imaged in ocean water but the clarity is sharply reduced for distances even as small as 10 m. In coastal water, the distance at which usable photographs can be obtained, or visual observations made, may be as small as a meter or two. Conditions in harbors, rivers, and lakes are often such that optical observations are next to impossible.

TABLE 2–4. Attenuation of light by absorption in distilled water for distances from 1 m to 30 m (absorption coefficients from Hulburt, 1945).

Wavelength in microns	Color	Absorption coefficient α, in m^{-1}	I'/I_o				
			1 m	5 m	10 m	20 m	30 m
0.40	(violet)	4.4×10^{-2}	0.96	0.80	0.64	0.42	0.25
.45	(blue)	1.7	.98	.92	.84	.71	.60
.50	(green)	2.3	.98	.89	.80	.63	.50
.55	(yellow green)	3.7	.96	.83	.69	.48	.33
.60	(orange red)	19.0	.83	.39	.15	.02	.00
.65		30.3	.74	.22	.05	.00	.00

I_o is the initial intensity of the light.

I' is the intensity after the light has been attenuated by absorption in water.

TABLE 2–5. Attenuation of light by scattering in distilled water for distances from 1 m to 30 m (scattering coefficient from Hulburt, 1945).

Wavelength in microns	Color	Scattering coefficient β, in m^{-1}	I''/I_o				
			1 m	5 m	10 m	20 m	30 m
0.40	(violet)	3.57×10^{-2}	0.96	0.83	0.70	0.49	0.34
.45	(blue)	2.25	.98	.89	.80	.64	.51
.50	(green)	1.47	.99	.93	.86	.75	.64
.55	(yellow green)	1.00	.99	.95	.91	.82	.74
.60	(orange red)	0.708	.99	.96	.92	.87	.81
.65		.507	.99	.97	.95	.90	.86

I_o is the initial intensity of the light.

I'' is the intensity after the light has been attenuated by scattering in water.

References

Bein, Willy, 1935: Physikalische und Chemische Konstanten des Meerswassers. Veroffentl. *Inst. Meeresk. Univ. Berlin, Nene Folge*, A, **28**, 36–186.

Duntley, Seibert Q., 1963: Light in the sea. *J. Opt. Soc. Am.*, **53**, 214–33.

Du Pré, Elsie F., & Leo H. Dawson, 1961: Transmission of light in water: an annotated bibliography. *U.S. Naval Research Laboratory Bibliography No. 20.*

Hopkins, Robert E., & Harold E. Edgerton, 1961: Lenses for underwater photography. *Deep-Sea Res.*, **8**, 312–17.

Hulbert, E. O., 1945: Optics of distilled and natural water. *J. Opt. Soc. Am.*, **35**, 698–705.

Jerlov, N. G., 1963: Optical oceanography. *Oceanography and Marine Biology, an Annual Review*, **1**, 89–114. George Allen and Unwin, Ltd.

Miyake, Y., 1939: Chemical studies of the western Pacific Ocean, IV, The refractive index of sea water. *Bull. Chem. Soc. Japan*, **14**, 239–42.

Rosen, Joseph S., 1947: The refractive indices of alcohol, water and their mixtures at high pressures. *J. Opt. Soc. Am.*, **37**, 932–38.

Thorndike, E. M., 1950: A wide-angle, underwater camera lens. *J. Opt. Soc. Am.*, **40**, 823–24.

Tilton, Leroy W., & John K. Taylor, 1938: Refractive index and dispersion of distilled water for visible radiation, at temperatures from 0° to 60°. *J. Res. Natl. Bur. Standards.* **20**, 419–77.

Utterback, C. L., Thomas G. Thompson, & Bertram D. Thomas, 1934: Refractivity, chlorinity, temperature relationships of ocean waters. *J. Cons. Int. Explor. Mer.*, **9**, 35–38.

3. The instruments of deep-sea photography[*]

Harold E. Edgerton *Massachusetts Institute of Technology, Cambridge, Massachusetts*

Abstract

Cameras provide a superior means of inspecting the sea floor and other objects of the under-sea world. At great depths special attention must be given to the design of pressure-proof containers, windows, and lighting systems. Containers are usually spherical or cylindrical in shape and made of metals such as steel or aluminum. Lenses ordinarily used in air give a distorted image under water, because a portion of the optical path is through sea water and the thick window of the pressure case. Therefore specially corrected lenses are preferred. Light, at great depths, is provided either by continuous floodlight or by flash units. A typical camera system consists of a light unit combined with a camera (or two cameras for stereoscopy), and is usually arranged to operate at regular, pre-set intervals. Several different camera assemblies have been designed for the specific interests of various deep-sea research projects.

3-1. Introduction

The camera is an essential tool for the scientist working under the sea, as well as on land and in space. All underwater research, in such fields as biology, geology, archaeology, and oceanography employs photography in many ways today, and will continue to do so in the future.

The photographic process excels all others for the collection and storage of information, and in the conveyance of that information. It is not a matter of chance therefore that great effort has been put into the design, development, and exploitation of camera systems.

The first person to develop and use underwater photographic devices was Boutan (1893, 1898) in France. He solved many of the problems encountered in working underwater and used his systems for his research on the wildlife in the sea. He was followed by many experimenters; mainly those who worked to record, in still and motion pictures, shallow underwater scenes observed by divers. Those who wish to trace the full background of underwater photography are referred to the bibliography at the end of chap. 1.

Research into deep-sea photography, using unmanned, automatic cameras, was initiated and greatly stimulated by the work of Ewing, Vine, & Worzel (chap. 1). Practical solutions for the problems caused by pressure on the optical windows, electrical seals, triggering, and other components, were devised, and are still in active use today. Since these early developments, many others have contributed to the design, modification, and use of underwater photography. Subsequent contributions to the development of underwater photography are described or referred to throughout the present volume.

The camera is as useful to underwater observers as it is in air. An ichthyologist, for instance, will use photographs to show his fishes in their natural conditions of life, just as an ornithologist shows his birds, and both face problems due to the nature and awareness of their subjects as well as to the different media in which the animals live. Bird photographers use unmanned automatic cameras for their work, and photocells, microphones, electrical contacts, and other devices have been used to trigger the cameras when the subject is in the desired position. Likewise, the user of an unmanned underwater camera can arrange for his exposures to be made at some desired condition. Sea-bottom photography is a common example of this. Here camera, light, and trigger mechanisms are lowered on a line from the surface to the bottom of the sea. The exposures are made by some automatic method, such as a bottom-contact switch, or at regulated intervals of time. Alternatively, as discussed in chap. 2, it is possible for the camera operator, using a television system as view finder, to sit at ease on the deck of his ship and push a button when he sights a subject in his camera range.

A camera is a poor substitute for a human observer because of man's ability to look and to study. Since there are many technical difficulties in placing a man at a particular spot in the ocean depths, the camera or television system can be used as an intermediary information

* The work on deep-sea photography at MIT has received support for many years from the Research Committee of the National Geographic Society of Washington, D.C.

link. When a human observer goes down into the sea in a bathyscaph or other diving machine he usually takes a camera with him to record what he sees. In this way he can study the resulting photographs, often seeing details that escaped his attention while at the site. Furthermore, he can refresh his memory when he looks at the pictures. A most important use of the photographs is their impact on others who did not go on the dive. There is no word combination that carries the visualization of a simple photograph of an unusual scene. Photography has no substitute. The information content of a photograph is tremendous and the information is immediately conveyed to a person without any language barrier.

3-2. Technical problems of underwater photography

Successful underwater photography requires apparatus that fulfills the technical specifications listed and discussed below:

1. The casing must be strong enough to withstand the hydraulic pressure of sea water. The pressure increases by one atmosphere for each 10 m of depth, and, at the greatest known depth, a pressure of about 1200 kg/cm² (1,160 atm) must not cause damage.

2. Provision must be made in the casing for a transparent, distortionless optical window for photography.

3. Control levers, or wires, are needed to carry triggering information, or electrical power, into the casing. These parts must also be able to withstand the great pressures experienced with depth, and their insertion must not weaken the casing. The smallest leak cannot be tolerated.

4. Light is required for deep investigations since daylight is almost completely absorbed at depths greater than 200 m in the clear water of the open sea. Light sources, such as electric lamps with batteries, must be able to operate efficiently at high pressure and low temperatures.

5. The optical aberrations discussed in chap. 2 must be compensated in order to obtain the maximum information from the photographs.

6. Triggering of the exposure is important.

7. Positioning of the camera at a known spot, or at a given height above the sea bottom, is very important, and often the most difficult problem in deep-sea photography. The various solutions for this problem are discussed in chaps. 1 and 4.

The details regarding the above items will now be discussed more fully.

3-3. The watertight case

Many who have attempted underwater photography have started with cameras designed for use in air which were then installed in a watertight case. Such cases may have irregular shapes in order to fit the camera. The pressure-resisting ability of square or oblong boxes is very unsatisfactory in comparison to symmetrical cylindrical or spherical designs.

It has been suggested that cameras, batteries, and lights be flooded with oil or some other liquid, so that the casings would not be required to withstand the tremendous pressures of the deep ocean. It appears, however, that no successful cameras have been built on this principle, although oil-flooded batteries, motors, and relays are very common aboard the *F.N.R.S. III*, *Trieste* and *Archimede* bathyscaphs, and the *Soucoupe* of Capt. Jacques-Yves Cousteau. The problems of film immersion in oil are messy, and the design of oil immersion lenses will be difficult. Certainly the development of such a camera offers interesting challenges.

What should be the shape of the casing to resist the enormous pressures exerted by deep submersion in the sea? The strongest and simplest is the sphere, which is the easiest to calculate, because of radial symmetry. The usual design is a pair of mating hemispheres that resemble the Magdeburg sphere of ancient times. Picard's bathyscaph used a sphere for the man-bearing volume of the craft, and nearly all deep-diving devices have since followed this design. Windows and electrical cables must be put through penetrations in the hull of the bathyscaph sphere. Picard used thick, cone-shaped, Plexiglas windows. The distortion caused by the pressure causes the plastic to extrude slightly, thus disturbing the optical quality. A correction for this distortion consisted of a flat glass surface with a thin liquid-filled volume between the glass and Plexiglas.

Glass is not used for windows on the three operational bathyscaphs of today, since it is unpredictable in strength. Glass is very strong under compression but exceedingly weak under tension, especially if the surface has any scratches or defects. However, since glass is less distorted than other optical material for optical windows, it is used on most cameras. An optical window designed for use in deep-sea cameras is shown in fig. 3-1.

Another simple form for pressure-resistant housings is the cylinder. This has the advantages of being a more convenient shape in which to arrange equipment, and of having flat ends which are suitable for insertion of windows and access ports.

Aluminum and steel are the favorite metals for the construction of deep-sea cases for cameras and lamps. An aluminum case for use at a certain depth will be somewhat

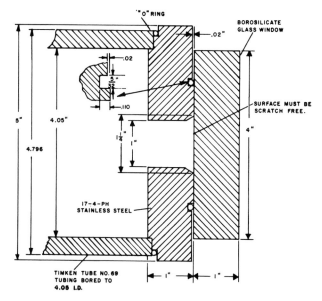

Figure 3–1. An optical window for use in an underwater camera at pressures up to 1,360 atm (1300 kg/sq cm). (Edgerton & Hoadley, 1960).

larger than, and of about the same weight as, the corresponding steel case. Of course, there are many different kinds of aluminum and steel materials, and these have different abilities to resist pressure and corrosion. One material favored by the author is the stainless steel type 17–4 PH since it can be centrifugally cast, machined, and finally tempered for hardness. Casting the steel in a rotating mold tends to produce a uniform metal composition. The tempering process occurs at a temperature that does not produce an oxide scale or a change of dimensions. Edgerton & Hoadley (1955, 1960) have considered the design of cases for cameras and lights in some detail. Figs. 3–2, 3–3 and 3–4 summarize some of their calculations.

It is very important that cases, windows, electrical seals, and other parts be pre-tested at the high pressures that will be found in the sea. A convenient test chamber was first made by Ewing out of a 16-in (40 cm) naval shell, which has an 8-in (20 cm) diameter hole in the center. The device to be pressure tested is placed inside the shell in water, the end cap is screwed in and pressure is exerted with a hydraulic pump. Details of a system based on the above design, and using the modern air-hydraulic pump, are described by Edgerton & Hoadley (1961).

3–4. Optical requirements and lights

The factors affecting lens design were discussed in chap. 2. Except where wide-angle coverage is a prime requirement, the Hopkins f/11 and f/4.5 lenses are most commonly used today in underwater cameras.

Almost all underwater photography, except for some

work at very shallow depth, requires artificial light. Natural sunlight is absorbed with depth and is completely useless for photography in the deep sea. Even Boutan, in his pioneering work, used underwater lamps. Several types of light sources are in use today to meet different requirements. Factors to consider in the choice of light source are:

1) whether a continuous light or flash is required;
2) convenience;
3) efficiency;
4) temperature, or color, of source;
5) flash duration, if a flash source is used.

Continuous lights are required for motion pictures. Overvolted tungsten lamps, with direct-contact water cooling to the bulbs, have been used with great success at shallow depths. The new iodine-quartz tungsten lamp is especially useful under water because of its small size. In deep water a glass housing is necessary since the iodine-tungsten cycle requires the quartz wall of the lamp to become hot.

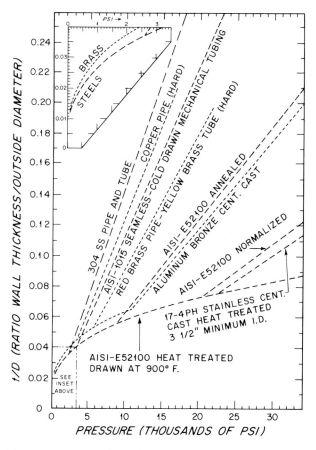

Figure 3–2. Comparison of the collapsing pressures of various materials in the form of long cylinders, such as would be used for housing underwater cameras and lights. The ratio of wall thickness to outside diameter is shown as a function of the pressure at which the cylinder collapses. (Edgerton & Hoadley, 1960.)

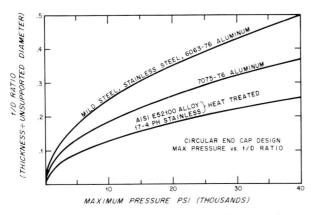

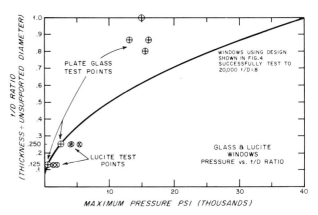

Figure 3–3. Guide for circular-end-cap design, showing the ratio of wall thickness to outside diameter as a function of the collapsing pressure. (Edgerton & Hoadley, 1960.)

Figure 3–4. Guide for design of glass and lucite windows, showing the ratio of thickness to diameter as a function of the collapsing pressure. (Edgerton & Hoadley, 1960.)

Flash sources are used for still photography, since they eliminate the need for a shutter except in shallow water where daylight penetrates. There are two types: the expendable flashbulb which must be replaced after each flash, and the electronic flash. A repeating electronic flash is most often used, since many photographs are needed, and it has the following advantages:

1) many flashes can be efficiently obtained from a small battery;
2) the instant of flash can be electrically controlled with great accuracy;
3) the color is of daylight quality, so that daylight types of color film can be used;
4) the flash duration is short, effectively arresting the motion of the camera or the subject.

The repeating flashlights (or strobes) provided by Edgerton, Germeshausen, and Grier have been supplied in 100- and 200-watt-sec capacities, and a representative flash rate would be one flash in 15 sec. A strobe light of this type is shown in fig. 3–5, and fig. 3–6 shows a series of graphs that serve as a guide for calculating the appropriate exposure time for various camera-to-subject distances. It is worth noting that two such 200-watt-sec repeating flashlights have been used to obtain exposures of the sea floor from a height of over 60 ft (18 m).

Rapidly repeating strobe lights, having 24 flashes per sec and 20-watt-sec per flash were once used by Cousteau and the author on the *Calypso*. The short exposure time (20 μ sec) was valuable, since it arrested the motion of fast moving subjects, but the disadvantage was the compli-

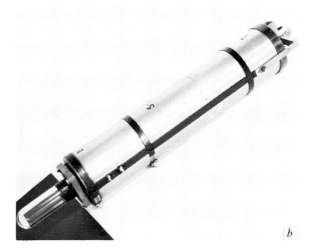

Figure 3–5. Strobe light used in an Edgerton, Germeshausen, and Grier camera assembly: (*a*) open view, showing internal circuitry; (*b*) closed view, showing pressure-resistant housing tube and flashbulb.

cation and weight of the electrical gear required to drive the lamps.

3–5. Underwater camera design

The author has developed an underwater camera and repeating flashlight combination which has undergone considerable evolution and change over about a decade. Several cameras of this design are now in use, and some of them are manufactured as models 200 and 204 by Edgerton, Germeshausen, and Grier. All models of this series employ 35-mm film, will store up to 100 ft of film, and are designed to mount either the Hopkins f/11 or f/4.5 lens. A typical camera of the series is illustrated in fig 3–7, which shows the camera itself, a functional diagram, and a replica of the data board. The last is a most valuable addition to the camera and automatically records on the film the depth, time, and approximate geographic position of each exposure. Because of the layout of the camera the picture of the data board lies nine frames in advance of the corresponding bottom photograph and due account must be made of this fact when determining the time of a photograph.

3–6. Automatic trigger

The camera and light-source combination may be wired either as an automatic, continuously running unit which starts cycling when a time-delay switch closes, or to operate upon application of an external signal, such as from a push button inside a bathyscaph. The first mode of operation, automatic triggering, is most commonly used, and the circuit is shown schematically in fig 3–8. The adjustable time-delay switch (S_1), which is in series with the battery, delays the start of the operating cycle. When S_1 closes, the camera and light commence to cycle automatically, they continue to do so for about 100 min, at which time the film supply will be exhausted.

The timing motor M_1 in the light-source casing turns at 5 rpm, driving a cam which momentarily actuates timing switch S_2 every 12 sec. While the timing switch is in position NO, current is fed to the various charging circuits in the camera.

Each time switch S_2 drops into position NC, three separate circuits operate in the camera:

1. The shutter capacitors (C_5) discharge through the shutter solenoid (L_2) causing the shutter to open momentarily. In opening, the shutter closes the synchronizing switch (S_3) momentarily.

2. The data-lamp capacitors (C_6) discharge through the incandescent bulb (B_2) exposing the data-chamber frame.

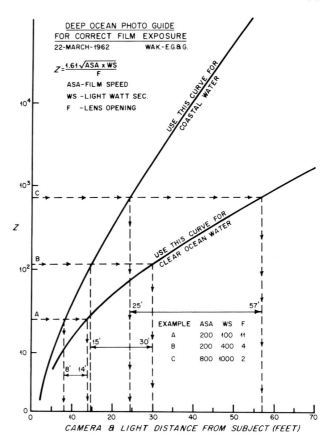

Figure 3–6. Guide for calculating the correct exposure for various distances between an underwater camera and the subject.

3. The motor-starting capacitor (C_7) discharges. The camera capacitors have diode-resistor combinations in series with them. This allows the capacitors to charge up slowly (through the resistors) and discharge quickly (through the diodes). If the diodes were shorted out, the data lamp and shutter solenoid would be actuated once when the capacitors were charged and again when they were discharged.

High dc voltage is obtained from the battery for the flashtube by a dc-to-dc converter, and alternating current is generated by an oscillator circuit containing transistors Q_1 and Q_2. The dc output is stepped up to 420 v by the toroidal power transformer T_1. Secondary current is rectified to 840 v by four diodes in a voltage doubler circuit. To avoid exceeding the inverse voltage rating of the diodes, two diodes are used in series in each leg. Each of the flashtube discharge capacitors (C_2, C_3) becomes charged to about 420 v.

A charge of about 190 v, which builds up on the trigger capacitor (C_4), is discharged to ground when the synchronizing switch (S_3) closes. This discharge generates a high-voltage pulse in the secondary of the trigger transformer (T_2), triggering the flashtube and causing the

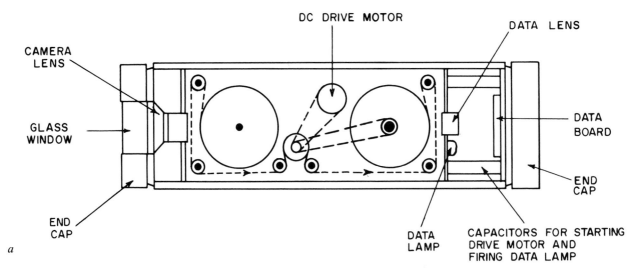

a

Figure 3–7. A typical underwater camera, as produced by Edgerton, Germeshausen, and Grier, Inc.: (*a*) schematic drawing showing component parts of camera, and film-winding mechanism; (*b*) disassembled camera, showing steel housing, interior components of camera, and data board; (*c*) data board. This is photographed with each frame to show date, time, depth, number of camera lowering, and other information necessary to identify the photograph. Due to the film-winding arrangement the data board is nine frames in advance of the corresponding photograph.

capacitors (C_2, and C_3) to discharge through the flash-tube. Flash duration is about 200 μ sec.

When the timing motor cam returns S_2 to position NO, current is again fed to the shutter capacitors (C_5), the data-lamp capacitors (C_6), and the motor-starting capacitor (C_7). As capacitor C_7 starts to charge up again, transistor Q_3 conducts, and current flows into the base of Q_4. Transistor Q_4 starts conducting, and operating current is applied to film-drive motor (M_2). This current is transitory, lasting about 1 sec until the voltage on C_7 rises to 6 v. During this short period the film-drive motor drives cammed switch S_4 into position NO, the position in which the film drive motor is connected directly to the 6-v circuit line. The film-drive motor advances the film one frame. When the cam moves S_4 back into posi-

tion NC, power is removed from the motor, the film stops advancing, and the operating cycle is complete. Resistor R_{11} limits over-travel when current is removed from the motor by its dynamic braking action on the motor.

3–7. Cameras for use in deep-sea research

At least three commercial manufacturers have produced cameras for use in the deep sea: Alpine Geophysical Associates, 55 Oak Street, Norwood, New Jersey; Edgerton, Germeshausen, and Grier, 160 Brookline Ave., Boston, Massachusetts; GM Manufacturing and Instrument Corporation, 2417 Third Avenue, New York.

In addition several designs of underwater cameras

have evolved at laboratories where there is an aggressive program in photography in the deep sea. In following chapters one British and several American organizations describe their cameras as applied to their particular research programs. From the U.S.S.R., H. L. Zenkevitch has sent the following information in response to a request for details of underwater cameras and their uses in the Institute of Oceanology of the Academy of Sciences:

The first camera was manufactured in 1953 on the basis of a Robot apparatus. For illumination of the bottom with this device, incandescent lamps are used. Power for the lamps and control of the shutter is accomplished at the side of the ships by a cable. With this camera it was possible to photograph the bottom at depths of 150 m.

In 1955 the second submarine camera, with which it was possible to work at depths of 3,500 m, was made at the Institute, completely autonomous and automatic. In this camera expendable flashbulbs are used; in this way it is possible to obtain only one photograph of the bottom at one lowering of the device. In this camera opening of the objective shutter and operation of the flashlamp were triggered by turning the cases of the camera and flashbulb at some angle, at the moment of the contact with the bottom.

Both these cameras were used successfully in the work of the expeditionary ship *Vityaz* in the Far East by Russian sailors, and in northwestern parts of the Pacific Ocean, up to the year 1957. A large number of very interesting photographs of the bottom were obtained.

Before the beginning of the International Geophysical Year the Oceanology Institute was designing and manufacturing a new and more perfect camera intended for ocean depths. In this camera illumination was provided by an electronic flash. In one lowering this camera permitted taking up to twenty photos of the bottom. The flash of the lamp and the exposure of the film is accomplished in this device by the turning of a mercury switch at the moment of contact with the bottom. The working of the film motor is controlled by a simple electronic circuit.

Since 1957 this camera has been used for all research on the *Vityaz*. In 1958 a large series of these cameras was manufactured, and given to other Soviet scientific research ships working in the Pacific, Atlantic, and Indian Oceans and in the Mediterranean Sea for the IGY and IGS program.

In 1958, at the Oceanology Institute, a stereoscopic camera was created with two wide-angle objective lenses, 3.5-cm diameter, in one case. The distance between objectives is 10 cm. The working of the film motor in this camera is controlled by an electronic circuit with transistors. Up to the present time, the stereoscopic camera has been used only in the expedition of the *Vityaz*. A series of these cameras is now being manufactured which will be provided to many of the Soviet research ships as soon as possible.

Stereophotographs, obtained in the expeditions of the *Vityaz*, are processed, at present, for the study of microrelief forms on the bottom of the ocean.

In all Soviet deep-sea cameras Plexiglas is used to protect the lamps instead of glass and, in my opinion, makes the device more reliable and durable. Not once have we had any trouble with illumination out of Plexiglas. In general, in designing cameras for deep-sea photography I set myself a fundamental task; simplicity of construction, mechanical durability, and convenience in operating. It seems to me that if a camera meets these three conditions, then it is not really so difficult to achieve its aim of taking pictures of the bottom. Our camera is so strong that with it one may, at will, hammer nails without injuring it. I am sure that for a daily sea device the quality of durability is very valuable.

In the Soviet Union there are several systems of corrective lenses, most successful of which is probably the specially-

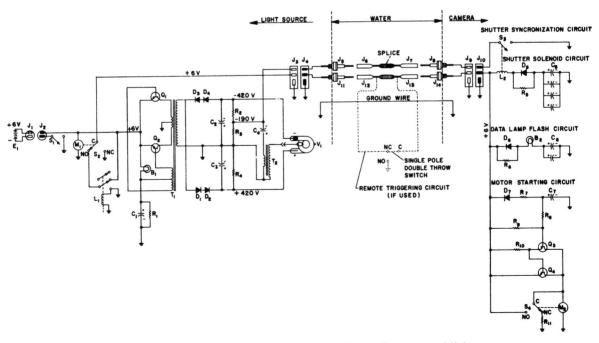

Figure 3–8. Circuit diagram for automatic trigger of camera and lights.

corrected-for-water objective of Prof. Rucihova in Leningrad. These corrected objectives are used mainly in shallow water cameras, controlled devices, and for special topographical surveying. In deep-water cameras of the Institute corrective lenses are not used, in order not to complicate construction and increase the cost.

Deep-sea cameras are used in the oceanographic programs of various other nations, but, so far as the author is aware, these cameras are based on British, French, or American designs.

References

Boutan, L., 1893: Mémoire sur la photographie sous-marine. *Arch. Zool. Exper. Gen.*, **21**, 281–324.

——, 1898: L'instantané dans la photographie sous-marine. *Arch. Zool. Exper. Gen.*, **26**, 299–330.

Ewing, Maurice, Allyn Vine, & J. L. Worzel, 1946: Photography of the ocean bottom. *J. Opt. Soc. Am.*, **36**, 307–21.

Edgerton, Harold E., & Lloyd D. Hoadley, 1955: Cameras and lights for underwater use. *J. Soc. Motion Picture and Television Engrs.*, **64**, 345–50.

——, & ——, 1960: *Protection for underwater instruments.* A.I.E.E. District Conference Paper no. DP 60–662.

——, & ——, 1961: Pressure-testing facility. *Underwater Engineering*, **2**, no. 2, 29–30.

Hopkins, Robert E., & Harold E. Edgerton, 1961: Lenses for underwater photography. *Deep-Sea Res.*, **8**, 312–16.

Laughton, A. S., 1957: A new deep-sea underwater camera. *Deep-Sea Res.*, **4**, 120–25.

Piccard, Auguste, 1954: *Au fond des mers en bathyscaphe.* Arthaud, Paris, 291 p.

——, 1956: *Earth, sky and sea.* O.U.P., New York, 192 p.

Thorndike, E. M., 1950: A wide-angle, underwater camera lens. *J. Opt. Soc. Am.*, **40**, 823–24.

——, 1955: Color-correcting lens for underwater photography. *J. Opt. Soc. Am.*, **45**, 584–85.

4. The manipulation of deep-sea cameras*

J. B. Hersey   *Woods Hole Oceanographic Institution, Woods Hole, Massachusetts**

Abstract

Deep-sea cameras are remotely manipulated either by suspending them on a long cable or by mounting them on a free submersible. Suspended cameras have been fitted with mechanical, optical, or acoustical sensing devices for their control. This chapter is mainly concerned with acoustical devices; mechanical and optical devices are discussed in other chapters of this book. Acoustic echo-location systems have proven successful in locating and photographing free-swimming fishes at shallow depth in the sea, but the development of these systems, not now actively pursued, occurred at a time when long electrical suspension cables were not available. The technique continues to be promising but it may not be competitive with the use of free, manned submersibles.

Sea-floor photography has used acoustical controls, variants of echo-sounding techniques, and one device which employed a mechanical bottom sensor to switch off an acoustic signal telling the operator thereby that the camera was the required distance from the bottom. The most used control is the sound-pulse generator attached to the camera. This radiates one short sound ping per second during a camera lowering, and the difference in time of arrival of this ping and its bottom echo is a measure of the height of the camera above bottom. Display of the acoustic data on a graphic recorder is required both to identify weak signals in noise and to facilitate interpretation of the echoes. Using such controls thousands of photographs of the sea floor have been made, from where the bottom was smooth and flat to the other extreme on the sides of rugged sea mounts and oceanic islands. Continuous coverage of a strip along the bottom over a kilometer long has been developed as a technique and has been used not only for geological

reconnaissance but also for deliberate search as in the well-known search for the wreck of the submarine *Thresher.*

4-1. Introduction

Deep-sea cameras must be remotely manipulated for the selection of the subject, for the focusing of the image, for lowering and raising the camera, and for protecting it from damage or loss. In the past the manipulation has been accomplished mostly by suspension on a metal cable or on a rope; at present, photography is increasingly being carried out from submersibles. Manipulation by cable or rope requires remote sensors, and these have received considerable attention. Other chapters deal with mechanical bottom-contact devices for focusing the sea floor (chaps. 1 and 8), with optical sensors for selecting and focusing free-swimming animals (chap. 20), and with cameras on bathyscaphs (chap. 25). This chapter deals mainly, but not exclusively, with the use of acoustical sensors and with some of the special problems of remote manipulation of suspended cameras.

4-2. Echo location in the photography of free-swimming animals

Historical notes

Photography was attempted unsuccessfully by Beebe during his observations from the Bathysphere during the 1930's, and during the same era Harvey (1939) likewise was unsuccessful in a series of attempts to photograph free-swimming animals with a camera suspended by cable. Beebe's failure was largely due to lack of proper camera and lighting, even though he could plainly see his subjects; Harvey had no means of selecting subjects, and he was not lucky enough to find a single sizable animal in several thousand randomly directed exposures.

Much later on, in the late 1940's, Owen, Hersey, and others attempted to identify the scatterers forming deep scattering layers. They lowered a repeating camera to

* WHOI Contribution Number 1750.

The material in this chapter is based on several years of work supported under various contracts of the Office of Naval Research, particularly Nonr-1367(00) NR 261–102, Nonr-2866(00) NR 287–004, and Nonr-4029(00) NR 260–101. I am grateful to A. T. Johnson for his early contributions to the preparation of this chapter, to Jacques Yves Cousteau for providing figs. 4–12 to 4–14, and to K. O. Emery for his critical review of the manuscript.

layers known from echo sounding, and followed supposedly careful procedures in order to obtain exposures above, within, and· below the layer or layers. In no instance were animals recognizable on the film, and only one clear image was recorded. It proved to be unidentifiable. Several times, however, the film showed blurred blobs of scattered light which might have been animals badly out of focus. Even these were not found consistently in a scattering layer, and our attempts only served to show the need for techniques enabling more specific identification of the subject.

In the meantime many observers were recording individual animals or isolated groups by means of echo sounders, either those fixed to the hull of a ship or those lowered on a cable (Kanwisher & Volkmann, 1955).

For these recordings, instruments similar to the Precision Graphic Recorder (PGR) were used (Knott & Hersey, 1956). The PGR is a continuous recording instrument which converts electrical signals into the form of a graphic trace on a long strip of paper. This paper moves slowly beneath a stylus which sweeps repeatedly across it, from one edge to the other. Electrical signals fed onto the stylus cause electroplating of the paper with a ferrous dye, the darkness of which is roughly proportional to the intensity of the signal. As used in echo sounding, the start of each stylus sweep also triggers a sound pulse in the water. This pulse and its echoes from the sea floor and from any underwater objects are received by a transducer which converts sound waves to the electrical signals which are fed onto the stylus. In this manner all of the sound, including unwanted noise, received during the sweep of the stylus is recorded. The stylus sweeps on a regular schedule; thus in a series of sweeps the echoes from fixed objects will generate bands of darkening parallel to the edge of the paper, while echoes from moving objects will give bands that migrate in accordance with the motion. The distance of the dark band from the position of the stylus at the moment it triggers the pulse (normally close to the edge of the paper) is proportional to the distance between the transducer and the object causing the echo. True noise appears as random speckles. Fig. 4–1 shows two typical recordings taken where many objects, probably individual fish which are being passed over by the ship carrying the transducer — at a speed much higher than their own — have caused echoes. Thus the darkened band corresponding to a sequence of echoes from each individual fish has a crescentic shape caused by the first decreasing and then increasing distance from transducer to fish, and also by the corresponding increase and then decrease in the intensity of the received echo. Recordings of this type have been made many times in both shallow and deep water. They can be used to learn much about the population densities, distribution, and

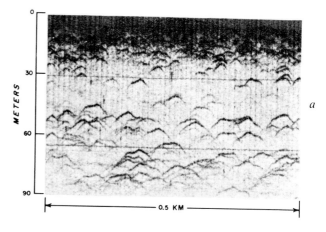

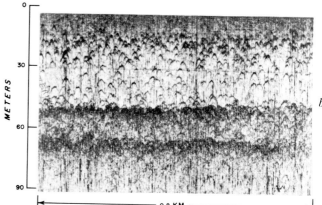

Figure 4–1. Echo sequences produced by sound scatterers, either individual fishes, or groups of fishes, as a ship carrying echo-location apparatus passes over them. The crescentic shape of the echoes is due to the decrease, and subsequent increase, in the sound travel time as the ship approaches and passes over each scatterer. The crescents in (a) are more elongated than in (b) since the ship is traveling more slowly in (a).

habits of free-swimming animals, and were enticing invitations to would-be photographers many years ago. The crescentic echo sequences are crude acoustical images of the fish, but the fish cannot be identified thereby; they cannot even be counted with certainty because such an echo might be caused by several individuals swimming close together. Nevertheless, if the apex of the crescent, the point of closest approach, occurred within the depth of focus of a camera pointing in the same direction as the echo sounder, a good photograph of the object causing the echo should be obtainable. This was the basis for the combinations of suspended echo sounders and cameras, assembled in order to identify the deep scattering layers.

Apparatus and observations

The echo-location apparatus has proceeded through two stages of design. In the first the PGR was employed as recorder and programmer. Two separate transducers, both connected by electrical cable to the ship, were used for radiating sound pulses and for receiving. Two UQN–1b transducers of the Edo Corporation (1951) were used, and they were mounted on a rigid frame with the camera between them, all three pointing in the same direction. A strobe light (see chap. 3) was mounted in various positions to provide side lighting (fig. 4–2).

Sound pulse generation and receiving apparatus was adapted from the Edo UQN–1b echo sounder, suitably interconnected with a PGR. In order to provide good resolution the PGR sweep rate was 8 sweeps per sec (the 50-fm scale), and a sound pulse 0.5×10^{-3} sec was used. This sweep rate and pulse length permits a resolution of 30 to 40 cm along the sound beam. The angular resolution, of the order of 15° to 30°, is not powerful enough to eliminate, but can merely reduce, echoes from objects which are outside the field of view of the camera.

The camera controls were adapted from those described in chap. 3 in order to place the control of the actuation of the camera in the hands of the observer. He would keep continuous watch of the recording, causing the camera to actuate whenever an echo was judged appropriately placed. Fig. 4–3 illustrates one of the better PGR records and photographs obtained.

Instruments based on this design were used during several expeditions of Woods Hole ships between about 1953 and 1958. They were successful in photographing individual fish and small schools of fish at depths down to about 60 m (Johnson, Backus, Hersey, & Owen, 1956). When this work was done, suitable electrical cables were not available for work at depths greater than about 60 m. After several frustrating attempts to photograph some of the deeper scattering layers, using inadequate cables, the apparatus was abandoned for a second design. This instrument, designed by Thomas Gifft, consisted of a very high-frequency pinger (30 kc/sec) which operated continually for weeks on end. The receiver was a pair of Atlantic* BC–10 hydrophones mounted about 2.5 cm apart (half wavelength at 30 kc/sec) in a box formed of SOAB (a sound absorbing material), 2.54 cm thick and open only at the side corresponding to the direction of the main lobe of the pinger. The two hydrophones were connected to independent amplifiers in series-aiding and, separately by bridging, in series-opposing. Without the SOAB box the two connections provided mutually perpendicular figure-8 directional patterns of reception. The first connection permitted the reception of echoes from objects on or near the acoustic axis of the pinger, reception from the opposite lobe being suppressed by the SOAB box. The second connection provided reception out to the side. The expected generous overlap of the two directional patterns means that echoes from the same object off-axis may be received by both combinations but presumably with different intensities. An analog computer formed the ratio of the two signals and fed the result to an oscilloscope. The camera was triggered when the ratio of series-aiding to series-opposing signals exceeded a certain adjustable value for an echo being received within an adjustable interval of distance from the apparatus. Thus the triggering was completely automatic. This instrument was tested during a cruise of *Chain* to the Mediterranean in 1959 but it was not successful in photographing objects that had caused echoes. Aside from trivial problems of minor equipment breakdown, the most serious shortcoming, to some observers at least, was the lack of the automatic and continuous record provided by the PGR in the previous design. The power of such a device for detecting inadequacies and for providing permanent record of the unexpected should not be underestimated.

Future possibilities

It is very likely that free-swimming animals, scattering layers, and other mid-water objects will be observed mostly from deep submersibles in the next several years.

Figure 4–2. Echo-location apparatus for mid-water photography. Left to right on the iron frame are the electronic flash, transmitting transducer, camera, and receiving transducer. The orienting fin is to the left. The suspension bridle allows either vertically or obliquely downward orientation of the camera and echo sounder.

* Atlantic Research Corporation, Alexandria, Virginia.

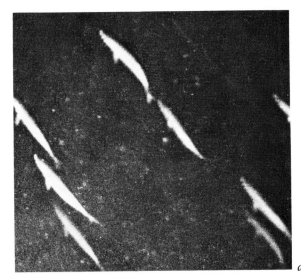

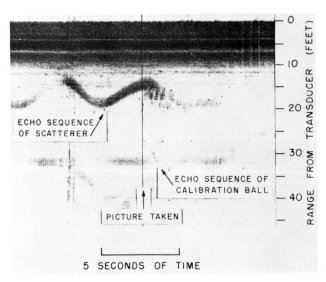

a

b

Figure 4–3. Photograph (*a*) and corresponding acoustic record (*b*) of eight unidentified fish at a depth of 30 m in a water depth of about 1,830 m. The echo-sounder record shows the scatterer (in this case, a school) at a range of 12.5 ft (3.8 m). The echo at about 30 feet is from a brass sphere used as a means of calibrating echoes from other objects. The 50 mm-focal-length lens was focussed at a distance of 8 ft (2.44 m) in water (depth of focus from about 7 ft [2.14 m] to 9 ft [2.75 m] at f/6.3, the aperture used); thus this picture is somewhat out of focus. By computing the length of field of the photograph at the 12.5-ft range and multiplying by the proportionate length of the image of the fish, the length of the fish was estimated to be about 7 in (17.8 cm).

Already Barham (1963) and others at the Naval Electronics Laboratory working in the bathyscaph *Trieste*, and Cousteau's group using the diving saucer, have made significant observations. Future observations must be documented by means of photography and capture (see chap. 19) but as yet the development of effective systems to this end has scarcely begun. It is worth observing that by now many ocean-floor photographs in which animals appear have shown that even very small animals can be identified, especially with the aid of stereography and color (see chap. 19). Although photography from submersibles appears promising, there seems also to remain an important role for suspended cameras with acoustic echo location. In part this is because the identification of the scattering layers still depends largely on circumstantial evidence. Another compelling influence is the ease with which fishes can be detected by echo location well beyond the limits of visibility. When identification has been made deliberate, then many other ecological and physiological problems can be attacked. Such an instrument might be suspended and controlled from a manned submersible, be made part of a robot submersible, or be suspended from a ship. Presumably both acoustical and photographic instruments can be adapted, but the main question to be decided is how to control the instruments in the ocean.

4–3. Echo location in photography of the sea floor

Historical notes

The ability to monitor the height of the instruments above the bottom of the ocean by means of an echo sounder has been widely appreciated ever since the late 1930's. Ewing, Vine, & Worzel included experiments with such a technique in the early work with cameras in shallow water (chap. 1). Edgerton & Cousteau (1959) adopted an echo-sounding method as a principal means of measuring the height of the camera above bottom.

A complete echo sounder might be mounted on the camera and feed its information to the ship via electrical cable as was done in the photography of free-swimming animals. However, as in the case of the latter, lack of suitable long cables in the 1940's and early 1950's removed such design from consideration. Ewing, Vine, & Worzel, as detailed in chap. 1, had placed a receiving transducer on the camera to receive echoes of sound pulses from an echo sounder in the ship for some of their experiments in the shallow water of the continental shelf.

Edgerton (1955) constructed a camera incorporating a sound-pulse generator, a wartime magnetostriction transducer, which was shock-excited to radiate pulses approximately once a second during an entire lowering, except when a bottom-sensing switch turned it off. The sound was monitored by means of a hydrophone, amplifier, and earphone combination, and the observer adjusted the

Figure 4-4. Bottom photograph taken near Pantelleria Island, off the coast of Sicily. Correct focussing was accomplished by an acoustic method in which a sonar Pinger was turned off by a bottom-sensing device when the camera was the correct distance from the bottom. The sound was received by a combination of hydrophone, amplifier, and earphones, and the camera triggered only when the sound signal was switched off.

length of suspension cable so that the sound pulse was just between operating and being turned off. Using this procedure bottom photographs in excellent focus were obtained (fig. 4-4).

Later, Edgerton & Cousteau (1959), again using a self-contained energy source, mounted the sending transducer on the camera and relied on the ship's echo sounder for receiving echoes. Their work was in the deep sea at depths generally greater than 500 m. It was possible to obtain the height of camera above bottom simply by measuring the time interval between the arrival at the ship of the sound pulse, or ping, and its bottom echo, because both paths are nearly vertical (fig. 4-5). For this use Edgerton adapted a transducer based on the design of the UQN-1b of the Edo Corporation, discussed below, for the sound source. This new transducer with its energy package and programmer he called the Pinger. The timing of the interval between the direct and bottom-reflected sound pulses or pings was first done by visual inspection of a display on the A-scan of a cathode ray oscilloscope. This method, though often successful, provides no record, and is frequently unusable because of high noise levels.

The requirements of the data display have led to the development of a considerably more complicated system

of instruments, both for the measurement of camera height and for the interpretation of the acoustic data. The acoustic reflectivity of the bottom varies within wide limits (well over a factor of 10) and the attenuation of the sound with distance of travel varies at a somewhat greater rate than the inverse square of the distance, due to scattering and absorption. The unwanted noise at the

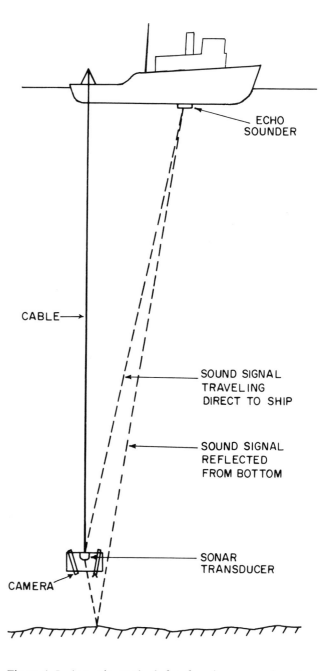

Figure 4-5. Acoustic method for focusing an underwater camera. The height of the camera above the bottom is measured by the difference in travel times of the direct and reflected sound signals.

echo sounder is another variable to be considered. It is affected by the local weather, conditions on the ship, and sometimes by sound-producing animals in the water. Many of these last produce clicks that sound quite similar to the pings discussed above, so that some signal processing has proven essential even to detect the pings. For this purpose recorders like the PGR, as described above, have been effective. The instrument system to be described below was evolved to meet practical requirements at sea, partly during Edgerton's early experiments with Cousteau and partly with the author and others at Woods Hole during the late 1950's.

The apparatus and operating procedures

The block diagram of fig. 4–6 shows the functional arrangement of the instruments.

The combination of pulse programmer, generator, and the associated transducer, as mentioned before, has been called the Pinger. Its programmer may or may not be interconnected with the camera and light source, but neither has been electrically coupled to the ship except for special experiments. The echo-sounding transducer and receiving amplifier have been standard research equipment on the ship, frequently the Edo UQN–1b or other similar design. The recording system has included a recorder similar to the PGR and may also include an oscilloscope, magnetic tape recorder, or other means of visual display or permanent record.

The Pinger. The Pinger transducer was adapted from the Edo UQN–1b echo-sounding transducer in that the same design of sensitive element was used. The UQN–1b

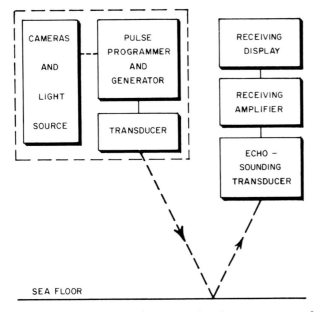

Figure 4–6. Functional diagram showing the arrangement of parts for acoustic monitoring of an underwater camera.

transducer is a piston consisting of a mosaic, about 23 cm across, of anhydrous dihydrogen phosphate (ADP) crystals. The Pinger transducer, illustrated in fig. 4–7, differs from the UQN–1b by lacking its air-filled baffle and by using aluminum, instead of cast steel, for the housing. The front of the mosaic which is directed toward the reflecting object is immersed in oil behind a neoprene covering; the back is mounted on an acoustically compliant sheet glued to an aluminum mounting plate. Radiation through the neoprene window is roughly 10 db higher than through the aluminum back plate. This means that, since the acoustic reflectivity of the sea bottom at nearly normal incidence is commonly .05 to 0.3, the sound radiated directly to the receiving transducer through the back plate will be of the same order of intensity as sound radiated through the neoprene window and reflected from the bottom.

The Pinger is driven by the circuit shown diagrammatically in fig. 4–8. When switch S_1 is open, capacitor C_1 is charged to 900 v and the trigger grid of the discharge tube is at ground potential so that no current flows through the main gap of the tube. Current flowing through the high resistance R_4 supplies a small "keep-alive" discharge. When S_1 closes, a voltage of 450 v is momentarily applied to the trigger grid as capacitor C_2 begins to charge. This voltage causes electrons from the small "keep-alive" discharge to initiate a discharge in the main gap. The charge on capacitor C_1 then flows through the transformer primary, inducing a high voltage on the secondary and thus on the Pinger. When C_1 is discharged, the tube discharge terminates because resistor R_1 is too large to permit the flow of minimum current necessary for a discharge in the main gap. When the discharge ceases, C_1 can then recharge to 900 v. S_1 opens, C_2 discharges through R_2, and the cycle is ready to repeat.

The electrical energy of discharge for each ping is about 1.5 watt sec. The peak level of a ping at a range of 1 m exceeds 0.3×10^6 dynes/cm², or about 110 dbμb, and has proved strong enough for use in the Puerto Rico Trench where the water depth is about 8,000 m.

The Pinger's programming circuits provide for a 1-ping-per-sec repetition rate, held accurate to about a part in a thousand, or better, so that pings will be recorded roughly in synchronism with PGR sweeps (see fig. 4–9). A mechanical escapement clock is provided in some commercial models while, for special applications, very accurate crystal oscillators have been used (Breslau *et al.*, 1962). If the depth of the water exceeds about 750 m, a Pinger operating without interruption can read only differences in depth, a difficulty which has been avoided by having the Pinger omit one ping in, say, every ten or twelve. This is done by programming S_2 to be open for each tenth or twelfth closure of S_1. The silent interval

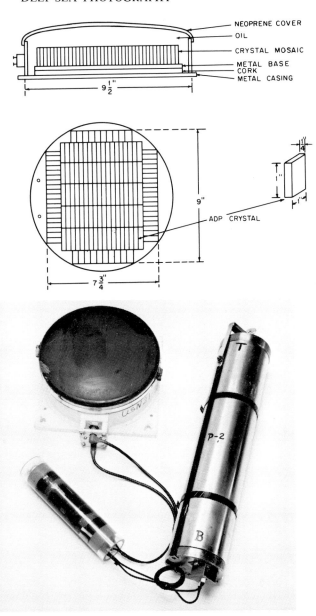

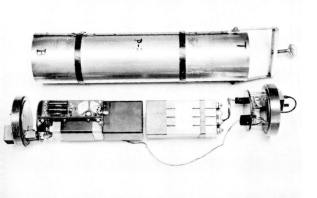

Figure 4–7. Sonar Pinger. (a) Schematic drawing showing cross section of pinger transducer. (b) Schematic drawing showing open view from above, arrangement of piezoelectric crystals, and a single crystal. Application of voltage to the terminals causes oscillations of the individual crystals. The sound wave thus generated passes through the neoprene cover to the ocean bottom, and through the metal base and casing to the surface. The transducer is oil-filled to equalize the internal pressure to that of the sea water. (c) Sonar Pinger, showing power-supply unit (enclosed in a cylindrical aluminum pressure case), transformer, and transducer. (d) Open view of Pinger power supply, showing batteries and electronic circuitry.

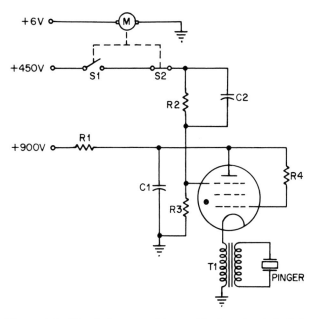

Figure 4–8. Circuit for sonar Pinger. Closing of switch S_1 by the motor M triggers the discharge tube, so that current flows through the primary of transformer T_1. This induces the high voltage necessary to shock excite the Pinger. Switch S_2 is programmed to cause a missed ping at regular intervals.

serves as a reference by means of which travel times, up to the 10 or 12 sec between omissions, can be read. In many photographic expeditions from Woods Hole the omitted-ping program has been made to correspond with the actuation of the camera. As will be illustrated below (figs. 4–10 and 4–11), this design has the convenience that particular photographs can be matched exactly with the acoustic record, thus enhancing the value of both for interpretation.

The Pinger is started by completing an electrical connection external to the pressure case, after which it operates until the camera is on deck again, or until its batteries are exhausted. Eight to twelve hours' continuous operation can be expected from a fully charged battery.

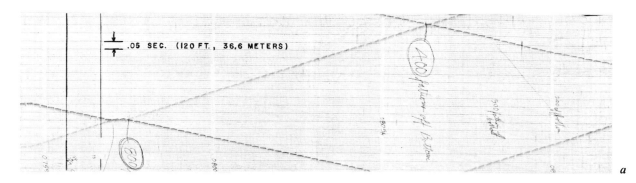

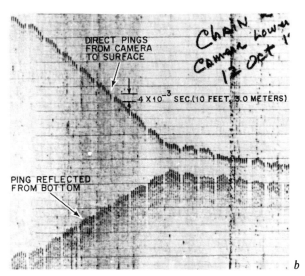

Figure 4–9. Portions of a PGR record of a camera lowering to about 4 fm (7 m) above bottom. In (a) the PGR is being operated on its 400-fm sweep, so that the traces of the direct and reflected signals cross when the camera is at 800 and 400 fm (1,500 m and 750 m) above bottom, forming a criss-cross pattern. In (b) the camera can be tracked as it approaches close to bottom and is maneuvered just above the sea floor. To obtain good resolution at close distances the sweep is changed to 100 fm or 50 fm. On this record every ninth ping is missed, thus enabling a particular ping to be identified with its own echo.

This battery life is adequate for a lowering and picture-taking phase in any depth of water, but in the deepest trenches the batteries might well fail during the raising of the camera. Because camera lowerings in deep water require over two hours, the film in the camera would be exhausted if the operation of the camera were not delayed until it reached the bottom. As discussed in chap. 3, this is done by means of a clock-driven timer switch whose delay is set for the estimated time of lowering. If the camera circuit causes the ping to be omitted, as discussed above, then the appearance of a periodic silence in the ping sequence indicates that the camera has begun to operate.

The receiver. Deep-sea photographers have used a variety of receiving transducers and amplifiers to record pings and echoes. The receiving system must be capable of receiving pings having peak pressures no higher than those characteristic of the ambient noise of the ocean in calm weather (see Vigoreux & Hersey, 1961; or Albers, 1965), in the frequency band matching that of the ping. For the Edgerton Pinger the appropriate band is about 5 kc/sec centered at 12 kc/sec. Other pulse radiators have been built to operate in a variety of frequency bands from about 0.8 to 30 kc/sec, and possibly higher. To the writer's knowledge only those with center frequencies from about 7 to 15 kc/sec have been used for bottom photography. Many echo sounders are designed to filter their signals in a band far narrower than the spectrum of a ping. It is highly desirable for good resolution, and it may be a simple matter, to broaden the filter of the receiver of an otherwise suitable echo sounder. One manu-facturer* has produced receivers with filters adjustable to match pings of different lengths.

Operations at sea. The PGR and many other graphic recorders in current use in oceanographic research are designed to permit a choice of several stylus-sweep periods, one of which is 1 sec. The PGR and a few others also have a choice of several shorter periods, the shortest practical ones of the PGR being $\frac{1}{4}$, $\frac{1}{8}$, or $\frac{1}{10}$ sec.

The combination of the 1-sec period and one of the shorter periods forms the basis for operations at sea, from Woods Hole, of the Edgerton Pinger and the PGR. While not identical to all other practice, the method of operation from Woods Hole ships is at least representa-tive. At the commencement of a lowering, and after the camera assembly has been attached to the steel cable, the pinger is started and the assembly lowered to a depth of 20 ft (6 m). There it is held while a series of pings and bottom echoes are recorded on the PGR in order to establish the rate of the Pinger's timing clock relative to that of the PGR. The cable is next paid out at a rate not exceeding 70 m/min, the free-fall terminal velocity of

* Thomas Gifft Associates, El Segundo, California.

the assembly. (In early experience with cameras, this speed was sometimes exceeded, with resulting cable tangles of fantastic complexity!) The PGR record, as shown in fig. 4–9, allows the observer to keep track of the lowering rate as well as the height above bottom, so

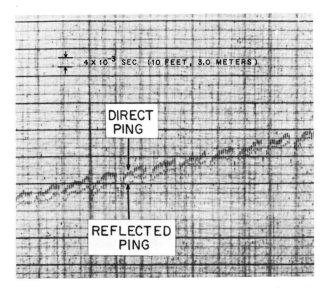

Figure 4–10a. Portion of the record of a camera lowering over a smooth, flat bottom. Since the sound signal is reflected from one surface only, the echoes are short and clear.

long as the depth of water is known at the start. The received pings are seen to drift to later times as the Pinger goes deeper into the water, while the bottom echo migrates to earlier times. Whenever the depth of water exceeds a distance corresponding to 1 sec travel of the sound pulse toward and away from the sea floor (this distance is 400 fm based on an assumed sound speed of 800 fm/sec; mean sound speeds between surface and bottom of the open ocean are generally 2 to 5 per cent higher) the two sequences of signals cross on the record at times when the camera is exactly an integral multiple of this depth interval off bottom; hence the notations on fig. 4–9a at the several crossings. After the crossing at 400 fm (750 m) from bottom has been passed, the lowering rate may be reduced, but, in any event, the assembly is held well off bottom until the camera and strobe start operating. As mentioned above, the start of camera operation may be signaled by the regular omission of one ping whenever the strobe light flashes. The operator may then assure himself that the camera is in fact near bottom by checking to see that the silent intervals corresponding to the missed ping and its echo occur on the same sweep of the PGR. At this point most observers prefer to increase the sweep speed of the PGR to $\frac{1}{4}$, $\frac{1}{8}$, or $\frac{1}{10}$ sec periods (the 100-50-, or 40-fm scales) for higher resolution. Fig. 4–9b illustrates the start of such a record.

Figure 4–10b. This photograph was taken on the lowering corresponding to the record of fig. 4a, and shows the monotonous, sediment-covered bottom. The lowering was made about 80 miles (110 km) west of Madeira on *Chain* cruise 7, July, 1959, and the photograph was taken at a depth of 2,917 m.

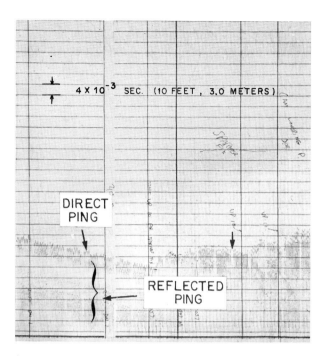

Figure 4–11*a*. Portion of the record of a camera lowering over a rough bottom. The sound is reflected from many facets of the rocky surface, causing the complex echo sequence shown above, and it is often difficult to detect which represents the true bottom directly below the camera.

Representative results

In figs. 4–10 and 4–11 acoustic observations are compared with corresponding bottom photographs over a smooth bottom, and over a rugged area of the mountainous slopes of Plantagenet Bank south of Bermuda. The PGR record suggests where a smooth bottom is to be anticipated (fig. 4–10) and the corresponding photographs usually show a flat mud or sand bottom. The bottom may be cut by trails of animals, and bottom-dwelling animals may be photographed, but there is little relief. Other records show more complex acoustic echoes, and, correspondingly, the photographs may show more striking relief (figs. 4–11*a* and *b*).

During early experiences with deep-sea photography in areas of high relief, cameras which were lowered by means of $5/16$- or $1/4$-in steel cable were sometimes lost, apparently due to their becoming snagged on rocks. Later on, similar cameras were recovered after rather severe snagging when they were lowered on $1/2$-in cable. Since that time, we at Woods Hole have preferred the stronger cable and have lost very few cameras even when photographing a rugged bottom. Nevertheless, it is prudent to take account of the local topography when conducting such observations; for example, by moving the camera downslope during a lowering.

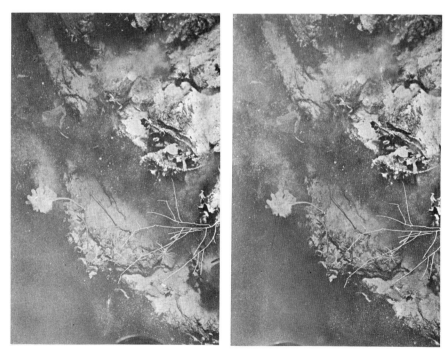

Figure 4–11*b*. This photograph was taken at the instant indicated by the arrow in fig. 4–11*a*. The lowering was made over the western slope of Plantagenet Bank, near Bermuda, on *Chain* cruise 9, October, 1959.

Figure 4–12. Cousteau's Troika during early tests on *Calypso* in the Mediterranean Sea.

Continuous strip photography

Suspended cameras.   During the early use of cameras, which were capable of taking several hundreds of photographs per lowering, little care was taken to achieve continuity of coverage. The motion of the camera over the bottom was controlled only by the drift rate of the ship, thus the distance between exposures was irregular and successive photographs seldom overlapped. As ships which could be closely controlled at slow speed came into use, continuous strip photography became practicable. By 1963 a few successful photomontages, made by piecing together successive photographs, had been made, and in the *Thresher* search a Woods Hole team, headed by S. T. Knott and relying on continuous strip coverage as a search technique, was the first to find debris from the submarine. The same general technique was greatly extended by Brundage, Buchanan, & Patterson (chap. 6) during the investigation of the wreck itself. Since that time a number of successful photomontages

of areas in the North Atlantic and Indian oceans, and in the Mediterranean Sea, have been constructed. These are discussed in chap. 11. They, too, depended on careful maneuvering of the ship, but, unlike parts of the *Thresher* search operations, did not have the benefit of references such as taut-wire buoys, pingers resting on the bottom, or other precise navigational aids. Several such systems are at present under active development. To take full advantage of this method, very precise locationing of the ship is necessary. Navigation by satellite or, for local control, acoustical systems basically dependent on pingers and receivers, which may be fixed relative to the bottom, are the most promising techniques for accurate positioning.

The Troika.   Cousteau has towed cameras mounted on a sled (the Troika) which provides a second method of obtaining continuous photographic coverage of a strip of ocean floor. The Troika (fig. 4–12) was designed to be asymmetric so as to assume the correct attitude for towing at the end of a long cable from the research ship. The

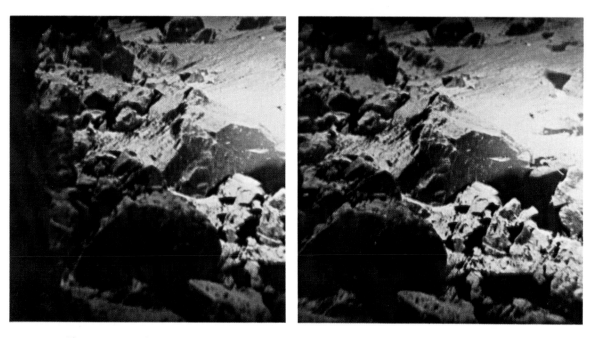

Figure 4-13. Taken in the rift of the Mid-Atlantic Ridge, from Troika towed by *Calypso*.

tow cable is attached near the stern of the body, but towing is done by means of a weaker cable attached at its bow and onto the tow cable. If the Troika is snagged on the bottom then the weaker cable should part first, permitting the Troika to tumble and free itself. When thus snagged the Troika was to be hauled in immediately, to avoid further damage or loss.

Cousteau has towed the Troika along the Mediterranean sea floor, in the median rift of the Mid-Atlantic Ridge, and in other regions of great interest (figs. 4-13 and 4-14). Motion pictures, time-lapse, and repeated single photographs have been made with great success from the Troika. One remarkable feature of this work is the freedom from interference by particulate matter stirred into the water by the vehicle itself. Because of the continual forward motion the cameras, pointed ahead of the vehicle, are commonly inspecting clear water.

Future developments

The manipulation of suspended deep-sea cameras is in an elementary state. The art is capable of developing by improved view-finding techniques, such as television and high-resolution sonar, by greatly improved navigation, and by remote sensors and controllers made possible by the use of long steel-clad electrical cables. For more efficient operation, cameras should be made to sink faster (for example, four to five hours of lowering is required to achieve a two-hour period of photographing the bottom of the Puerto Rico Trench). Manipulation devices should allow for extreme possibilities in side-lighting and

should enable the observer to leave a camera near the bottom for an extended period to take time-lapse photographs for such purposes as measuring the time rate of change of the sea floor. Many other possibilities can be imagined.

It is obvious that photography from manned and unmanned submersibles will be developed and used intensively in the next few years. Some scientists have felt that the great potential of submersibles would mean an end to deep-sea photography by means of suspended cameras. At present the use of a submersible is difficult and very costly compared with the corresponding use of a suspended deep-sea camera. Others have suggested that, for work on the sea floor, extended coverage of an area by suspended camera should precede the use of a submersible as the best means of avoiding serious difficulties and otherwise insuring the success of the submersible's program. From experience of the recent past, both approaches to the manipulation of cameras for deep-sea photography have lasting advantages, and each will be worthwhile using and improving for many years to come.

References

Albers, Vernon M., 1965: *Underwater acoustics handbook*. Penn. State Univ. Press, 356 p.

Barham, Eric G., 1963: Siphonophores and the deep scattering layer. *Science*, **140,** no. 3568, 826–28.

Breslau, Lloyd R., J. B. Hersey, Harold E. Edgerton, & Francis S. Birch, 1962: A precisely timed submersible pinger

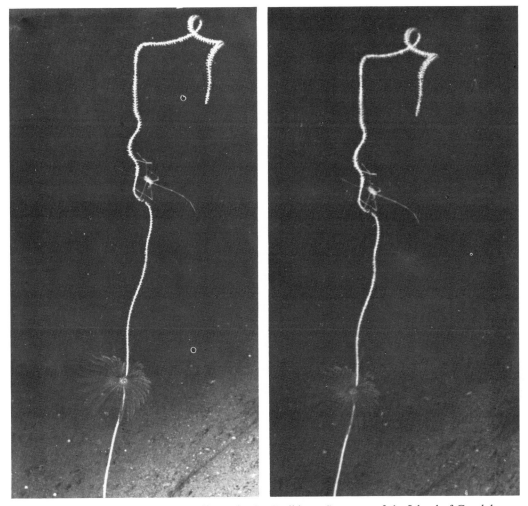

Figure 4–14. Troika meets abyssal epifauna in the Caribbean Sea, west of the Island of Guadaloupe.

for tracking instruments in the sea. *Deep-Sea Res.*, **9**, no. 2. 137–44.

Edgerton, Harold E., 1955: Photographing the sea's dark underworld. *Nat. Geog. Mag.*, **107**, no. 4, 523–37.

———, & Jacques Y. Cousteau, 1959: Underwater camera positioning by sonar. *Rev. Sci. Instr.*, **30**, no. 12, 1125–26.

Edo Corporation, 1951: Instruction book for sonar sounding set AN/UQN–1b. Department of the Navy, Bureau of Ships.

Harvey, E. Newton, 1939: Deep-sea photography. *Science*, **90**, no. 2330, 187.

Johnson, Henry R., Richard H. Backus, J. B. Hersey, & David

M. Owen, 1956: Suspended echo-sounder and camera studies of midwater sound scatterers. *Deep-Sea Res.*, **3**, 266–72.

Kanwisher, John, & Gordon Volkman, 1955: A scattering layer observation. *Science*, **121**, no. 3134, 108–9.

Knott, S. T., & J. B. Hersey, 1956: High-resolution echo-sounding techniques and their use in bathymetry, marine geophysics, and biology. *Deep-Sea Res.*, **4**, 36–44.

Vigoreux, P., & J. B. Hersey, 1962: Sound in the sea. In *The sea; ideas and observations on progress in the study of the seas.* Vol. 1, *Physical Oceanography*, M. N. Hill (ed.), Interscience Publ., New York, 476–97.

5. Photogrammetry applied to photography at the bottom

M. D. Schuldt, C. E. Cook, and B. W. Hale   *U.S. Coast and Geodetic Survey, Washington, D.C.*

Abstract

Photogrammetric techniques used in stereographic aerial surveys are applied to deep-sea stereophotography, permitting the photoanalyst to describe the subject quantitatively. The technique and conditional requirements are discussed and a few contoured and cross-section examples are shown. Internal accuracy is approximately one part in 2,000.

5–1. Introduction

It is the purpose of this chapter to describe photogrammetric measurements of the relief and the size of objects appearing in deep-sea photographs. This technique requires stereographic pairs of photographs and is basically similar to that of aerial photogrammetry. Its use eliminates the uncertainties arising from comparison with an object of known dimensions appearing in the field of view and/or reliance on shadow effects. The application of this method does, however, impose a few conditions and requires some specialized equipment.

5–2. Conditions

Varying degrees of success in photogrammetric contouring can be achieved with varying picture quality, but for best results the image must be sharp and evenly lighted, with good contrast. These conditions are basically a function of equipment, system configuration, and film; however, deficiencies in lighting and contrast can, within reason, be improved upon in the photo laboratory. For deep-sea work, the conditions of stereophotography must be met by a synchronized system of two cameras and a light source suitably mounted on a supporting vehicle. This procedure cannot be replaced by the forming of a stereo pair from two successive photographs taken by the same camera, since the motion of the camera cannot be controlled as precisely as is done in aerial photographic mapping.

The following condition is imposed on the ratio of base-line length to lens-subject distance. Let b be the distance between the lens centers of the two cameras, and h be the distance from the subject to the lens plane. Then, with all values expressed in the same units,

$$k_1 \leq \frac{b}{h} \leq k_2 \tag{1}$$

where k_1 is the lower limit imposed by the stereoscopic photograph measuring instrument and k_2 the upper limit imposed by the angular field of the camera lens; h is limited by the transparency of the water and its actual value can be derived from the system geometry and the amount of overlap in the stereopair; b is preset — assuming a value for h — so that the ratio is within the prescribed limits. The value of h then becomes the working distance of the cameras from the subject. (Typical values are $b/h = \dfrac{1}{3}$, $k_1 = 0.25$ and $k_2 = 0.4$.)

5–3. Procedure

To enable the dimensioning of any subject appearing in underwater stereophotographs, each frame of a stereopair is projected and viewed so that points common to each frame are superimposed, and the left and right frames are seen with the left and right eye respectively. There are four ways of doing this, one of which is illustrated in fig. 5–1 and will be discussed; the others will be mentioned briefly.

The dichromatic method

The method illustrated in fig. 5–1 requires that each frame of a stereopair, whether in color or black and white, be printed on black-and-white, glass dipositive plates for use in a dichromatic multiplexing projection stereoplotter. One frame is then projected in red, the other in blue-green, and, by wearing spectacles with matching red and blue-green lenses, the operator can view the image in three

69

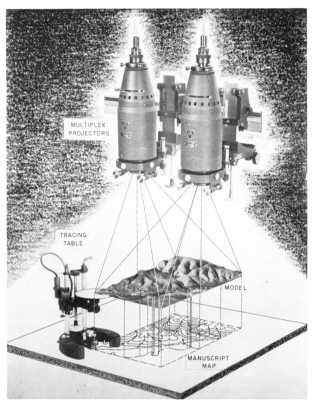

Figure 5–1. Principle of the Multiplexing Stereoplotter. (Photo courtesy of the U.S. Geological Survey.)

dimensions. This projection-viewing system permits, through filtering, the viewing of one frame with one eye and the other frame with the other eye. A table serves as the x-y plane and is perpendicular to the axis of projection, which is the z axis. The plane of focus is parallel to and above this table. The scales of both the x-y plane and the z axis are functions of the ratio b/h. However, the z scale is also a function of the focal lengths of the photographing and projecting cameras, and hence the two scales can be, but are not necessarily, the same. Fig. 5–1 shows a small circular projection screen mounted on a suitable fixture and aligned parallel to the image plane. The screen's center serves as the reference point for either tracking the image contours or measuring elevations. Directly below this point, and in contact with the table, is a pencil which traces the horizontal path of the reference point. The fixture can be moved freely in the x-y plane and the screen can be moved vertically, through gearing, by a graduated dial. The dial readings are convertible to true height differences from an arbitrary zero reference plane. Since the vertical scale can be adjusted to fit the vertical range of the screen, the reference point can be made to coincide with any desired point in the image. By moving the reference point along the curves of coincidence, those image elevations are sketched, and thus the entire relief can be contoured at any desired interval.

Figure 5–2a. Pair of stereoscopic photographs, showing sand ripples. Depth, 18.3 m. Location, 35°26.7′N 75°24.0′W. July, 1963.

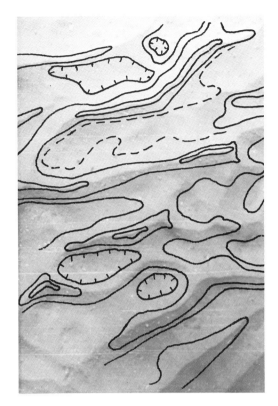

Figure 5–2*b*. Contours drawn on one of the stereopair in fig. 5–2*a*. Contour interval, 15 mm.

Also, by moving the reference point along any horizontal line and adjusting it vertically — noting dial readings and *x-y* coordinates — any desired cross section of the image can be drawn. Fig. 5–2 shows contour map and section profiles constructed from the respective stereoscopic pairs using the above method. Fig. 5–3*a* shows a stereoscopic pair from which the contours of fig. 5–3*b* were constructed.

The size of the photographs and illustrations was altered to fit these pages. Actually, contouring is done on a much larger scale than shown here; somewhere in the neighborhood of one square meter. The contour intervals and cross sections shown were chosen for no other reason than to demonstrate the technique. Intervals could just as well have been larger, smaller, or irregular depending on the purpose. Likewise, the cross sections could have been located anywhere and oriented in any direction. Furthermore, any object, regardless of its orientation with respect to the coordinate system, can be dimensioned provided that it has sufficient definition in the projected image.

The absolute accuracy of measurement of the instrumentation used for preparing the accompanying drawings is better than one part in 2,000. Level datum or coordinate axis orientation is dependent on the assumption that the plane of the photograph is perpendicular to the gravity vector. Level data could be incorporated into the

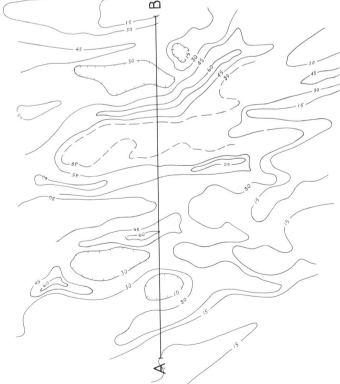

Figure 5–2*c*. Enlarged contour map of the sand ripples in fig. 5–2*a*. Contour interval, 15 mm.

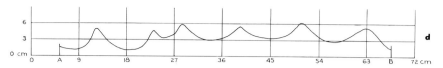

Figure 5–2d. Cross section along the line A–B in fig. 5–2c.

system to verify this assumption or provide the necessary data to compute the angular correction.

Other methods

The projection-viewing method discussed above is referred to as the dichromatic method; there are also the flicker, optical, and polarized-light methods. The dichromatic method requires black-and-white transparencies whereas either color or black-and-white positive transparencies may be used in the other three. Glass transparencies are preferred because of their stability, but ordinary film may be used.

The flicker method is a synchronized projection-viewing system which utilizes the image retention capabilities of the eyes. Alternately, the left frame is seen with the left eye and the right frame with the right eye, and the images alternate or flicker at such a rate that both frames appear to be viewed simultaneously, thus giving depth to the image. In the optical method the frames are viewed through an optical system to achieve image depth. The polarized-light method projects one frame with light polarized in one direction and the other frame with light polarized in a direction at right angles to the first. Then, when viewed with spectacles fitted with appropriate lenses, the image is seen to have depth. Contouring and

profiling with these methods is analogous to the procedure described above, but with modification according to the particular method.

5–4. Applications

Although any sampling system has limitations, the techniques described can be applied to the measurement of: ripple characteristics for interpretation of the water movements they reflect; bottom roughness for sonar studies; grading of sediments into fine, medium, and coarse; and the linear extent of such grades. Geological features can be measured, such as width and height of outcrops, size of fracture patterns, inclusions in conglomerates, size of minor structures like nodules for mineral-resource evaluation studies, pillow lava, etc. The dimensioning of organisms for quantitative description and taxonomic studies, and the measurement of burrows, mounds, tracks, and trails for biological studies are also subject to this technique. A single stereopair or a mosaic with stereo between adjacent pairs can be analyzed. The mosaic allows examination of large features not definable in a single pair. Further refinement of the system and development of operating techniques will extend the applications.

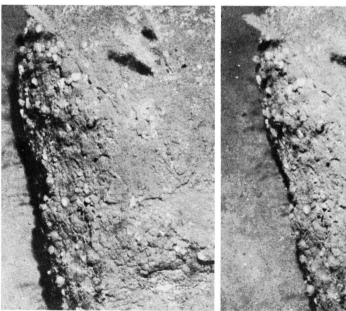

Figure 5–3a. Pair of stereoscopic photographs, showing a rock. Depth 2,750 m. Location, 38°15.0′N 71°21.0′W. October, 1961.

Figure 5-3b. Contours drawn on one of the stereopair in fig. 5-3a. Contour interval, 10 mm.

5-5. Summary

A method has been described to utilize stereophotography beyond the usual visual sensation of three dimensions. Within its limitations the system permits the scientist to measure what he sees without being there himself or taking samples. The method elevates deep-sea stereophotography to the status of a sampling device and can be used alone or correlated with other sampling means, permitting the extension of "point" information to "area" information. This technique allows an accurate quantitative description of objects and features that otherwise could only be described in nondimensional terms.

To quote Lord Kelvin (1824–1907): "I often say that when you can measure what you are speaking about, and express it in numbers, you know something about it; but when you cannot express it in numbers, your knowledge is of a meagre and unsatisfactory kind; it may be the beginning of knowledge, but you have scarcely, in your thoughts, advanced to the stage of science, whatever the matter may be."

6. Search and Serendipity*

W. L. Brundage, Jr., C. L. Buchanan, and R. B. Patterson *U.S. Naval Research Laboratory, Washington, D.C.*

Abstract

Towed underwater-camera systems were used by the U.S. Naval Research Laboratory during 1963 and 1964 to photograph major parts of the *Thresher* (SSN–593) hulk bottomed at 2,500 meters. The equipment and techniques were tailored for the search mission and produced many useful bottom photographs. The photographs contain abundant serendipitous (unsought-for) information about the natural environment in a relatively small area beyond the base of the continental slope. Patterns of pelagic sediment dispersal indicate the presence of gentle bottom currents. Remarkably uniform linear patterns of rocks suggest the past occurrence of a high-speed turbidity current. Photographs reveal the relative abundance and habits of the larger benthos, including some rarely photographed bathypelagic animals. An approximate index of faunal activity may be expressed in the alteration of sediment marks separated in time by one year.

6–1. Introduction

During the period from June, 1963, to September, 1964, personnel from the U.S. Naval Research Laboratory (NRL) took more than 100,000 ocean-bottom photographs in the *Thresher* search area. In the process of looking for the sunken submarine, we, like the Princes of Serendip, reaped an unsought wealth . . . in ocean-bottom information. Preliminary studies of the physical, biological, and geological phenomena observable in these photographs are now in progress. While the analyses are far from complete, a number of interesting aspects can be reported at this time. This paper will discuss the equipment and techniques used to take the photographs and some of the information derived from them.

* The authors are indebted to members of the staff at the Woods Hole Oceanographic Institution who made helpful comments and suggestions. Particular thanks are due R. H. Backus and D. W. Bourne for their reviews of the typescript and advice on the biological observations. Additional thanks are expressed to the many members of the Sonar Systems Branch at the NRL who made constructive criticisms of the manuscript.

6–2. Equipment and techniques

At the time of the loss of *Thresher*, the NRL had recently acquired a "Edgerton, Germeshausen, and Grier (EG&G) stereo-camera system," but had no experience in its use. With the decision to participate fully in the ensuing search, it became necessary to develop a deep-ocean photographic capability during actual operations in the search area. The urgency and uniqueness of this project, combined with a general lack of communication in the field of underwater photography, made it necessary to evolve this capability somewhat independently of more experienced groups. As a result, our equipment and techniques differ to some extent from those used by other organizations.

Conventional stereo-camera systems mounted on towed "Unistrut" racks were used during most of 1963. On some occasions a slow-scan television camera, built by Lear Siegler, Inc., and modified for deep-submergence use by the NRL, was used in conjunction with the photographic equipment. This unit scanned once every four seconds and the shipboard monitor presentation was photographed with "Polaroid" equipment. The underwater photographic cameras were operated only when objects of interest were observed on the television presentation. The video signals as well as control signals were transmitted through an insulated center conductor in the tow cable. The telemetry system is described more fully by Flato (1965).

An improved camera system was used during the final cruise in 1963. The strobe light was mounted on the nose of a streamlined body and two cameras were mounted on the tail, tilted out 17° from the vertical to each side in order to obtain wider coverage. The advantage of this technique in a search of this kind was demonstrated when small pieces of debris were recognized in a 1964 photograph. A thorough study revealed that a photograph of *Thresher*'s sail would have been taken on June 3, 1963, if tilted cameras had been used instead of a stereo pair.

Upon completion of the first year's exercise in the search area, it became possible to review the work and make plans for increasing the effectiveness of the equipment for future operations. A number of experiments were conducted with filters and processing systems as well as camera and light arrangements.

On the basis of these experiments and the review of previous work two instrument packages, like the one shown in fig. 6–1, were built. Two 200-watt-sec strobe lights are mounted on the nose, tilted 14° from the vertical to each side. The reflectors are mounted with adjustable brackets which allow a further tilt of the light pattern. Three EG&G cameras occupy the space beneath the tail fins. One camera is mounted vertically with the long dimension of its film format parallel to the longitudinal axis of the towed body. The other two are mounted at tilt angles which vary between 25° and 37° from the vertical, with the long dimension of their film formats normal to the body's longitudinal axis. An example of one typical field of view, covered by the cameras, is illustrated in fig. 6–2. The cameras were towed 7 to 8 m above the bottom. This is a greater operating distance than is generally used and results in some loss of resolution. In searching for a large object, the resolution could be sacrificed to obtain greater coverage.

The starboard camera is displaced transversely from the vertical camera, with the result that the percentage overlap of the two fields of view is a function of height above the bottom. The overlap may thus be used to determine the scale of the photographs. Though the field of view of an oblique camera is trapezoidal its image is rectangular. The resulting distortion can be seen by comparing fig. 6–3, which represents an unrectified photograph, with the field of view of the port camera shown in fig. 6–2. An example of the camera coverage is given in fig. 6–4, which is an unrectified photomontage of bottom pictures taken simultaneously by the three cameras.

The battery, telemetry, and other systems are mounted between the cameras and lights. The camera system is operated from the ship at a rate which is dependent upon the speed of travel, so that overlapping coverage is obtained. For example, at towing speeds of about two knots the cameras are triggered at five-second intervals.

Several different films were used successfully during the two years. Kodak Plus-X and Tri-X were both used and for the longer ranges the higher-speed Tri-X is preferable. During most of 1964 Dupont 331B was used. It is a thin-base film which allows as many as 700 exposures in a single lowering. Rough tests indicate that 331B is similar to Tri-X in both speed and grain size.

A variety of film-processing systems was used during the period. In 1963 all film was processed aboard ship in Smith-type rewind tanks, using Acufine or Ethol UFG developers. For best results, this system requires close time and temperature controls which are difficult to main-

Figure 6–1. The towed instrument vehicle, containing two 200-watt-sec strobe lights (left) and three EG&G cameras (right), was built of aluminum to minimize the effect on the magnetometer (end of boom, extreme right). Just forward of the tow point is a transponder (tall housing) and the battery container. The long black tube at the bottom contains a lateral echo-sounding transducer. The EG&G pinger electronics and transducer are forward of the cameras and the horizontal housing under the tow point contains controls. The other two horizontal cylinders contain sonar electronic components. Lead ballast is placed in a center space and the vehicle is lowered on an armored coaxial cable which has a breaking strength of 13.6 metric tons.

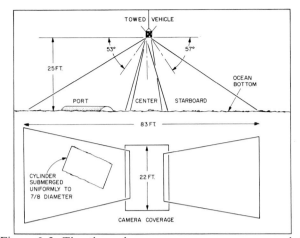

Figure 6–2. The three deep-sea cameras were arranged to provide optimum coverage of the sea floor when searching for a large object. The percentage overlap in the field of the star-board camera, compared with that of the center camera, is a measure of the distance above the bottom. Note the shape of the buried cylinder to the left.

tain in small shipboard installations. These difficulties can be overcome by the use of a two-solution developer such as Diafine. Within wide limits, this type of developer is insensitive to variations in time and temperature. Diafine was used with success in the relatively inexpensive Morse G–3 hand-operated tank. The tank can be loaded in a changing bag, eliminating the need for a dark room. This system is probably an ideal solution to the processing problems of any low-budget photographic survey.

The techniques described here are treated more fully by Patterson & Brundage (1965). The search tactics employed during 1964 are reviewed by Andrews (1965).

6–3. The search area

The most thoroughly photographed part of the deep-sea floor is the site of the *Thresher* search, centered at approximately 45°45′N, 65°00′W. The search area, shown in fig. 6–5, extends from the base of the continental slope seaward to the upper part of the continental rise, as defined by Heezen, Tharp, & Ewing (1959). Details of the bathymetry in the area are discussed by Hurley (1964). The greatest number of photographs were taken in the east central portion of the area, at an average depth of about 2,500 m. This portion slopes less than one degree toward the south.

A layer of suspended particles near the bottom compounded the problems of the photographic search. The presence of the layer was suspected when vast differences in quality were noted between photographs taken at 8 and 10 m above the bottom. Observers aboard the bathyscaph *Trieste* also noted the presence of a turbid zone

(Keatch, 1964). Recently Ewing & Thorndike (1965) reported a similar decrease in water transparency near the bottom of the continental slope, using the term "nepheloid zone" to describe the layer. Film studies indicate that the nepheloid layer in the search area varies both in time and place. Precisely how is not yet known.

6–4. Serendipity

The photographs taken by the NRL span the periods June 3 to September 3, 1963, and June 13 to September 12, 1964. The large numbers of pictures covering the relatively long periods permit studies which are not normally possible in ocean bottom photography. For example, the relative abundance of visible animals can be computed and any changes in the population which occur within the sampled time span can be noted.

"Urchin scale"

During the early part of the *Thresher* search films were scoured rapidly for clues. Distance above the bottom could not always be readily determined and many dubious estimates were made of the size of objects in the photographs. In June, 1963, a number of light-colored sea urchins were photographed on and near familiar objects such as a quart milk container and a "Winston" cigarette carton. From their size relative to the known objects, (excluding spines) these urchins were determined to be 4 to 5 cm across. Since the majority of the population did not appear to vary greatly in size (see fig. 6–6), the regular sea urchins served as a convenient scale. The scale helped to establish the identity of the object photographed in fig. 6–7. This curiosity, photographed in August of 1963, was at first thought to be a railroad-engine wheel used to anchor a surface buoy in the search area. Later it was

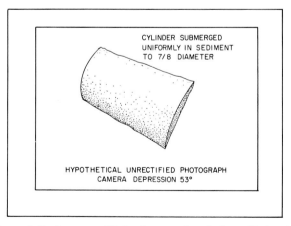

Figure 6–3. An unrectified photograph of the cylinder in fig. 2 looks like this. Rectification can be applied to oblique photographs of the sea floor but it does little to improve the shape of objects with relatively large vertical relief.

Figure 6–4. This is an unrectified photomontage made up of one vertical and two oblique photographs taken simultaneously about 6 m above the bottom. The two lines of linear rocks are actually parallel and may mark the path of a turbidity current. North is toward the right. The lines of rocks are approximately 2.5 m apart.

identified as the two-ton mushroom anchor carried by *Thresher*. The anchor was probably the most massive part of the submarine photographed in 1963.

Further checks, made on larger groups of urchins in the area, confirmed earlier measurements. Approximately 85 per cent of one large sample falls within the range from

4 to 5 cm. Unless otherwise specified, the urchin scale was used to measure all sizes of objects and fauna mentioned in this paper. For example, to determine the scale in fig. 6–6 (a center photo) all the urchins but the two obvious juveniles were measured to determine an average "urchin unit." The long artifact was then measured in

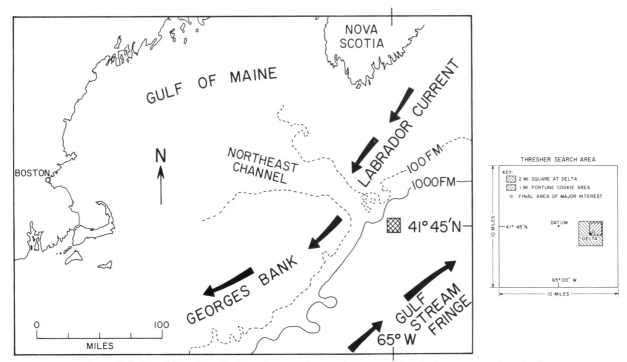

Figure 6–5. The 10-by-10 mile *Thresher* search area is the crosshatched square southeast of a deep reentrant in the continental shelf called the Northeast Channel. Detailed sections within this square are shown in larger scale to the right. All of the bottom photographs shown in this chapter were taken in the east central part of this area at an average depth of 2,500 m.

urchin units and this number was multiplied by 4.5 giving a length of approximately 110 cm. Unrectified oblique photographs are not used to make measurements unless at least three regular urchins can be found close to the object of interest.

Bottom sediment

The size distribution of bottom sediment at the site is bimodal, reflecting sediments derived from two entirely different sources. The dominant size range is formed by foraminiferal ooze of pelagic origin, while the secondary size range is made up of rocks from terrigenous sources. The modal size of the pelagic sediment falls somewhere in the silt range (Wentworth scale) and can be analyzed adequately from a small grab sample. An extensive photographic survey seems to provide the only practical means of measuring the rock sizes. Photographs indicate that the rocks, presumably ice-rafted into the area, make up about 10 per cent of the bottom sediment. Most of the rocks appear to be gravel to cobble-sized, although boulders larger than one meter are not uncommon. The boulder in fig. 6–8 exceeds three meters in diameter.

The majority of rocks seem to be randomly distributed, but after looking at thousands of pictures two patterns appeared often enough to be more than accidental. Hersey (1964) also noted these rock patterns in photo-

graphs taken in the same area. The more common arrangement can be described as a central cluster (fig. 6–9). It generally consists of one or more boulders surrounded by smaller rocks which sometimes appear to be fragments broken from the large boulders. Rocks abruptly discharged from floating ice might settle to the bottom in a partially frozen clump, forming a central cluster.

The second distributional pattern is linear and has evoked a number of theories concerning its origin. The three photographs of this pattern (fig. 4) were taken at a distance of about 6 m above the bottom. The next photograph taken with the port camera revealed that the upper line of rocks continues to the outer edge of the field of view. The three fields of view total more than 20 m, and the lineation is approximately north-south with north to the right in the montage. All such patterns that were measured by compass were within five degrees of the same alignment.

The linear pattern may be more common at the site than the photographs indicate. Keach (1964) noted for instance, "long ridges of loose broken rock jutting incongruously from the sand-silt floor," observed through a port in the bathyscaph. The results of a side-looking sonar survey (Clay, Ess, & Weisman, 1964) suggest that there may be numerous long lines of rocks throughout the area, most of which are oriented north-south. If these lines really are rocks, it is logical to assume that some single cause or event is responsible for the uniform pat-

Figure 6–6. The majority of these regular sea urchins measure about 4 to 5 cm across, not including the spines. Since they are visible in many photographs the urchins can be used as a rough approximation of scale. Here a group of echinoids have been attracted toward what appears to be a broken mop.

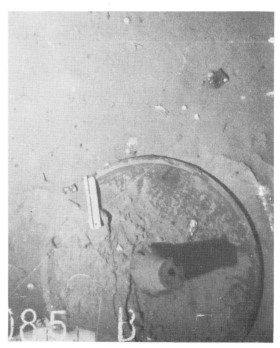

Figure 6–7. The urchin scale helped to identify this piece of wreckage. It is the 1,769 kg mushroom anchor carried by *Thresher* and photographed by NRL in August, 1963.

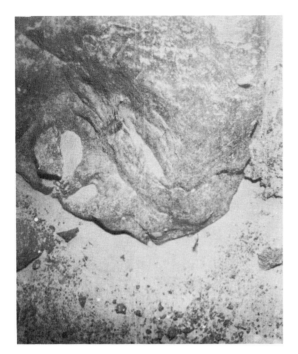

Figure 6–8. This huge boulder is more than 3 m across. Note the several sea pens growing on and near it, and the comparatively small size of the urchins.

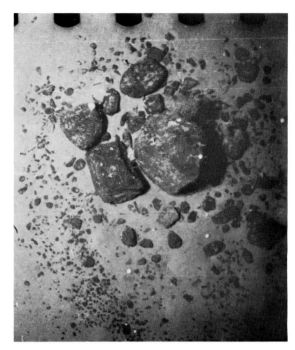

Figure 6–9. Some rocks on the sea floor near *Thresher* are distributed in a central cluster of large rocks surrounded by smaller ones. The pattern is probably the final disposition of a load of glacial debris dumped by a mass of ice.

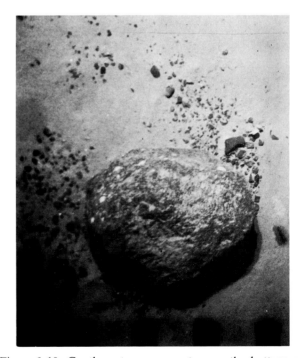

Figure 6–10. Gentle water movements near the bottom are believed to be responsible for the sediment-covered area off to one side of this boulder. The same apparent differential sedimentation was noted just beyond the west to northwest sides of many other large rocks.

tern. A large-scale turbidity current, similar to the 1929 Grand Banks current (Heezen, 1963), would reach a speed of better than 50 knots at the bottom of the continental slope. Such a high-speed flow would be capable of moving cobbles and even larger rocks, perhaps redepositing them in long lines marking its path. Many of the linear sonar reflections parallel the trend of a small submarine valley located near the 64°57′ meridian (Clay *et al.*, 1964, fig. 6–7). It is suggested that a turbidity current might have caused the linear pattern by redepositing rocks eroded from steeper portions of the continental slope north of the search area.

Bottom currents

Drogue current measurements down to 2,300 m were made in the search area during the period from May 29 to June 3, 1963 (Shonting & Cook, 1964). Northeasterly currents of about 15 cm/sec were measured at all depths greater than 100 m.

A paucity of low-cohesive (sand-sized) particles in the bottom sediment may account for the absence of ripples or scour marks noted in the bottom photographs. Other evidence in the photographs, however, suggests the existence of water movement next to the bottom. Many boulders like the one in fig. 6–10 are bordered on one side by what appears to be a current "shadow zone." This zone of entrapment for pelagic particles is typically found on the bottom just beyond the west to northwest side of each boulder. The same current direction is indicated by the attitude of the flexible sea pen to the right of the boulder in fig. 6–10.

A bottom navigation aid (the so-called "Fortune Cookie") implanted in July, 1963, was photographed in September, 1964 (fig. 6–11). The plastic placard, tethered to a sash-weight anchor, has apparently been moved by currents. The most conspicuous marks which it made, roughly define a 100-degree arc from 240° T through 340° T (variation 20° west). The EG&G compass is suspended from the towed camera vehicle.

A slow-moving westerly flow might characterize a "nepheloid current." Ichiye (1964) suggests that a nepheloid current may exist over a slope having a near-bottom density gradient formed by a nepheloid layer. The motion would be to the right of the downslope direction. It is also conceivable that the indicated westerly flow is part of a more widespread, and probably transient, deep-current pattern. Volkmann (1962) reported deep westerly currents over the continental slope and rise south of Cape Cod, 400 to 600 km to the southwest.

Returning to fig. 6–11 note that the outer periphery of concentric marks is apparently a series of radial depres-

Figure 6–12. Brittle stars appear to be one of the most common animals in the area. They are evident in almost every photograph taken accidentally at close range.

sions. This indicates that sediment has been removed from the bottom after having been stirred into motion by the action of the placard. The same motions of the placard and its tether probably also agitated the sediment which has been redeposited within the fold of the Fortune Cookie. It is concluded therefore that bottom currents in the area are easily capable of transporting the pelagic ooze after it has been stirred into suspension but that the threshold velocity necessary to initiate particle movement is probably not reached.

Bottom fauna

Among the benthos large enough to be identified in the photographs are representatives of at least six phyla. The echinoderms greatly outnumber the other bottom dwellers. Photographs also reveal coelenterates, poriferans, chordates, arthropods, and mollusks in order of decreasing abundance. The interpretation is probably biased by the relatively great distance from subject to camera.

Members of all five living classes of the echinodermata are visible at the site. Ophiuroids (brittle stars) probably are the most numerous because they appear in practically every photograph taken when the cameras come unusually close to the bottom (fig. 6–12). The light-colored, regular echinoids (sea urchins) are evidently next in abundance. These urchins appear to be among the more active com-

Figure 6–11. This bottom marker, attached to a sash weight, was dropped to guide *Trieste* in 1963. The marker, known as a "Fortune Cookie," was down for more than one year when this photograph was taken. The marks in the sediment suggest that the placard was moved by bottom currents predominantly in a westerly quadrant. The compass is suspended from the towed instrument package.

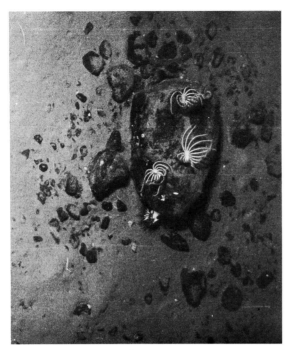

Figure 6–13. Comatulids (feather stars) were photographed on many large rocks. Two sea anemones are attached to the lower end of the big boulder.

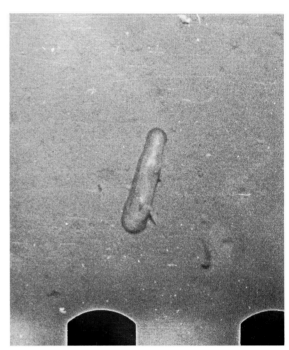

Figure 6–14. The elasipod sea cucumber typical in the area has conical papillae growing from its dorsal surface.

petitors for food. A concentration is always noted on and near any organic material (see fig. 6–6). The *Trieste* photographed a similar group on a large piece of shallow-water kelp which had probably drifted out to the area before sinking.

Other echinoderms, less common in the locality, were crinoids (feather stars), holothurians (sea cucumbers), and asteroids (sea stars). Many a large rock serves as the perch for a twelve- to fifteen-armed crinoid. Sometimes as many as three or more individuals share the same boulder (fig. 6–13). Sea cucumbers were most often noted on parts of the bottom devoid of rocks. The holothurian in fig. 6–14 is about 15 cm long. Several six-armed sea stars may be seen on and near the rocks in fig. 6–9.

At least three types of coelenterates are relatively common in the search area. Some sea pens were often found on and near large rocks. (See figs. 6–8 and 6–10). Another sea pen-like coelenterate, which resembles a bent piece of wire (fig. 6–15), was usually seen in rock-free areas. Two sea anemones may be seen on the lower end of the boulder in fig. 6–13.

A leaf-like sponge is found on many rocks. The shape of the colony varies greatly but one of the more common patterns consists of two to three lobes (fig. 6–16). The white form on the rock in fig. 6–17 has been tentatively identified as an incrusting sponge. Irregular shapes are not uncommon for such colonies in this area. A few

bottom-dwelling octopuses were photographed partly hidden by rocks. The specimen in fig. 6–18, however, was caught out in the open.

Bathypelagic fauna

The animal in fig. 6–19 is swimming or drifting just above the bottom. This 80 cm specimen could be a large member of the bathypelagic octopoda. Another specimen about 60 cm long can be seen forward of the *Thresher*'s upper rudder in fig. 6–20. The same animal was seen in a photograph taken 10 sec later. Its rate of progress was calculated at about 13 cm/sec.

Estimates of the relative abundance of bathypelagic fishes can be made from bottom photographs. Obviously the conclusions reached must be qualified by the possible influence which the sampling method has upon the population sampled. Due to a number of factors about which we have little information, a fish may be attracted, repelled, or unaffected by the audible, visible, and other emissions from the towed instrument package. In some cases the same fish is seen in a number of successive photographs apparently following the camera rig. The number of fish repulsed and hence not photographed must be left to speculation.

The smallest and most commonly photographed type of fish in the area is the macrourid (rat-tail) shown in fig. 6–21. Twenty macrourids, measured by the urchin scale, ranged in length from about 19 to 32 cm. The average length in this sample was approximately 26 cm. The fish in fig. 6–22 resembles a member of the family Moridae. These dark fish were usually seen alone or in pairs, except toward the late summer of 1964, when as many as eight were counted in a single frame. The five morids in fig. 6–23 are fairly representative of the population, ranging in size from roughly 26 to 45 cm. The Moridae, or deep-sea cod, are the second most abundant fishes seen in the area. Another deep-sea fish, third in

Figure 6–16. The three-lobed structure on top of the boulder is a deep-sea sponge.

Figure 6–15. This long, wire-like structure is probably a sea pen. The specimen in the photograph is over 30 cm long.

Figure 6–17. The white form on the boulder is probably another kind of sponge. Note the irregular shape.

Figure 6–18. The benthic octopus in this photograph is estimated to have a tentacle span of about 50 cm. The large eyes are probably a distinct advantage to a carnivore which may depend upon dim bioluminescence to detect its prey.

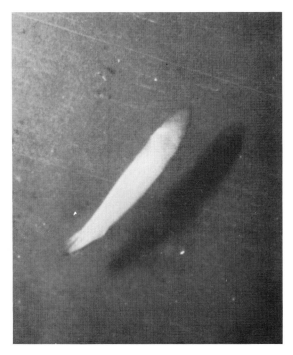

Figure 6–19. This animal resembles certain nektonic octopods. Some are covered with a gelatinous membrane which could obscure morphological details in a photograph.

Figure 6–20. The towed camera vehicle caught a similar swimming animal just forward of *Thresher*'s upper rudder. A later photograph permitted calculation of its rate of progress at approximately ¼ kn.

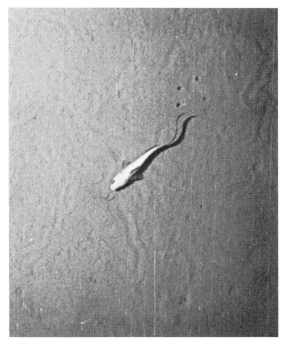

Figure 6–21. This fish is the type most often photographed in the search area. Macrourids (rat-tails) like this are among the most numerous of fishes on the continental slope. The slim tapering tail is a feature shared by many diverse groups of deep-sea benthic fishes.

Figure 6–22. Second in relative abundance at the locale is the dark fish in this photograph. It resembles a morid or deep-sea cod.

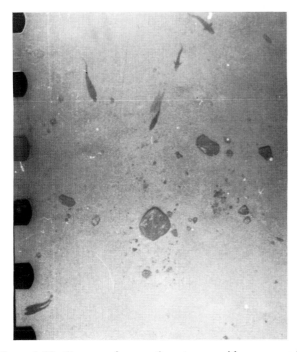

Figure 6–23. Groups of more than two morids were rarely seen in the area except in September, 1964. The five depicted here range in length from about 26 to 45 cm.

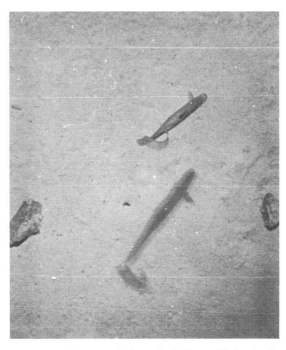

Figure 6–24. This unusual view of a halosaurid clearly shows the translucent and flexible nature of its tail. It is another typical benthic fish frequently caught by cameras at the site.

relative abundance in the vicinity, is the halosaurid shown in fig. 6–24. The ribbon-like caudal fin is often flexed in several bends of more than 90°. The thin pectoral fins are usually held in a high "dihedral" of about 45°. Halosaurs are also usually seen alone in the photographs.

Several other kinds of fishes were photographed less frequently throughout the two four-month periods. Skates appeared less than a dozen times. The specimen in fig. 6–25 is longer than 100 cm and was photographed on July 23, 1963. It closely resembles the three species: *Raja hyperborea*, *R. jenseni*, and *R. radiata* (Bigelow & Schroeder, 1953, pp. 206, 213, 255). The geographic location and depth tend to rule out all but *R. jenseni* which has been captured in this general vicinity (Templeman, 1965).

In late August, 1964, a chimaera was seen for the first time in the bottom photographs. Fig. 6–26 is an oblique photo of what appears to be the deep-water Chimaera, *Hydrolagus affinis* (Bigelow & Schroeder, 1953, pp. 539–545). Compared with the average size of the nearest sea urchins, the chimaera is approximately 95 cm long. The location is within the recorded geographic range of *H. affinis* although the depth would be slightly greater than that previously recorded (Bigelow & Schroeder, pp. 543–44).

The only picture in which a notocanthid was recognized is fig. 6–27. One distinguishing characteristic of the family Notocanthidae is the series of short separated

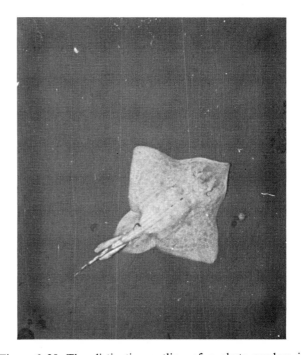

Figure 6–25. The distinctive outline of a skate renders it perceptible in photographs taken at the limit of the cameras' underwater range. This short-range photograph of a male skate shows details like the coloration and a dark parasite (near the base of the tail).

Figure 6–26. The chimaera is a fish rarely photographed in the area. The large triangular pectoral fins and the short caudal filament are distinctive features of the deep-water chimaera, *Hydrolagus affinis.*

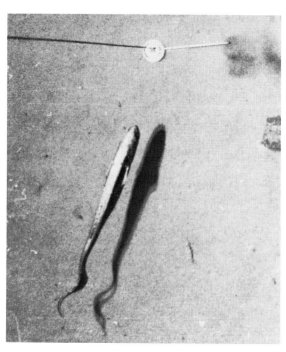

Figure 6–27. Note the series of short disjointed dorsal spines on this fish. The spines are characteristic of the family Notocanthidae. This individual is approximately 90 cm long.

Figure 6–28. Activities of the benthos are evident from trails like these. Other photographs indicate that several types of animals produce trails. These were made by spatangoid urchins, two of which appear in the picture (Bourne, 1965).

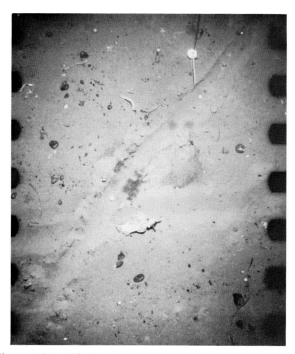

Figure 6–29. This September, 1964, photograph reveals two bathyscaph trails that were made by dragging a spherical lead weight of about 110 kg. The more distinct trail was ploughed across the barely perceptible trail made about a year earlier. The vagile benthos tend to obliterate such marks.

dorsal spines perceptible in the photograph. Comparison with the 7.6 cm head of the EG&G compass indicates that this specimen is about 90 cm long.

Benthic faunal activity

Several kinds of animals in the area leave distinctive marks in the bottom sediment. One of the typical trails is shown in fig. 6–28. The spatangoid urchins (Bourne, 1965) responsible for these trails may be seen partly buried near the bottom of the photograph. Many other pictures revealed the kinds of microtopographic features usually linked with the activities of bottom dwelling animals.

The bathyscaphs (*Trieste*, in 1963 and *Trieste II*, in 1964) made several dives in the search area. During maneuvers on the bottom the deep submersibles each dragged a cable terminated by the same lead sphere. This device helps the craft to maintain neutral buoyancy near the bottom, but it leaves a very distinctive trail about 25 cm wide. The double mounds ploughed up on each side of a freshly made trail have crisp, definite outlines in contrast to trails of an earlier date which are gently rounded and beginning to blend smoothly into the surrounding sediment. A photograph taken on September 10, 1964 (fig. 6–29), contains both a 1963 and a 1964 bathyscaph trail. The new trail crosses from upper right to lower left. The older trail is nearly horizontal in the picture. Note that the fresh trail has attracted two sea urchins.

In the discussion on bottom currents it was inferred that the currents were too weak to erode the bottom sediment. If this is true then bottom currents probably do not contribute directly to the alteration of the bathyscaph trails. Instead, it seems more likely that the activities of the benthos tend to destroy the trails. Near-bottom water movements contribute to the obliteration of the marks in a secondary manner by redepositing any sediment thrown into suspension by the animals. The differences noted in the bathyscaph trails may therefore represent an approximate index of bottom faunal activity in the search area.

6–5. Conclusions

The NRL bottom photographs from the *Thresher* search area were made for the specific purpose of obtaining photographs of the sunken submarine. The observations and interpretations presented here are the preliminary results of an effort to make fuller use of the unsought-for information inherent in the photographs. Although some of this information may not be new, it is felt that the effort at least demonstrates the value of taking large numbers of photographs within a relatively small area of the ocean bottom. Because of the large amount of available information on this area, it may well serve as a proving ground for future deep-search techniques. While testing new skills additional photographs will be taken. The photographs should be examined for information which will augment the studies presented in this chapter.

References

Andrews, F. A., 1965: Search operations in the *Thresher* area —1964, Section I., *Naval Engineers J.*, **77** (4), 549–61.

Bigelow, H. B., & W. C. Schroeder, 1953: *Fishes of the western north Atlantic*, part II, sawfishes, guitar-fishes, skates and rays, and chimaeroids. Sears Foundation, Yale University, New Haven, Conn.

Bourne, D. W., 1965: Personal Communication.

Clay, C. S., J. Ess, & I. Weisman, 1964: Lateral echo-sounding of the ocean bottom on the Continental Rise. *J. of Geophys. Res.*, **69** (18), 3823–35.

Ewing, M., & E. M. Thorndike, 1965: Suspended matter in deep ocean water. *Science*, **147** (3663), 1291–94.

Flato, M., 1965: Long cable telemetry and command system. *Proc. Nat. Telemetering Confer.*, Session X, 176–79.

Heezen, B. C., 1963: *The Sea*. (M. N. Hill, ed.) Chap. 27, Vol III, Turbidity Currents (742–75) Interscience Publishers, N.Y.

———, M. Tharp, & M. Ewing, 1959: The floors of the oceans, I. The North Atlantic. *Geol. Soc. Am.*, Spec. Paper 65, 122 p.

Hersey, J. B., 1964: Search for *Thresher* by photomosaic. *USN J. Underwater Acoustics*, **14** (2), 311–18. (Unclassified report in Confidential Journal.)

Hurley, R. J., 1964: Bathymetric data from the search for USS *Thresher*. *Int. Hydro. Rev.*, **41** (2), 43–52.

Ichiye, T., 1964: Some hydrodynamic problems for a nepheloid zone. Unpublished Tech. Rept. Lamont Geological Observatory, Columbia U., N.Y. 24 p.

Keach, D. L., 1964: Down to *Thresher* by bathyscaph. *National Geographic Mag.*, **125** (6), 764–77.

Patterson, R. B., & W. L. Brundage, Jr., 1965: Deep sea photographic search techniques. *Trans. Ocean Sci. and Engin. Confer.* M.T.S. and A.S.L.O., **2**, 1178–97.

Shonting, D. H., & G. S. Cook, 1964: Drogue current observations in the *Thresher* search area. Unpublished Rept. TM. No. 307, Naval Underwater Ordnance Station, Newport, R.I., 37 p.

Templeman, W., 1965: Rare skates of the Newfoundland and neighboring areas. *J. Fish. Res. Bd. Canada*, **22** (2), 259–79.

Volkmann, G., 1962: Deep current observations in the western North Atlantic. *Deep-Sea Res.*, **9** (6), 493–500.

7. Deep-sea photography in support of underwater acoustic research

Carl J. Shipek   *U.S. Navy Electronics Laboratory, San Diego, California*

Abstract

In order to fulfill a need for more information about the microrelief of the sea floor, the U.S. Navy Electronics Laboratory (NEL) has developed a new type of lowered oceanographic equipment. This consists of a compact combination of proven underwater instruments which enables the simultaneous and efficient collection of related data, including stereoscopic bottom photographs for photogrammetric measurements, sampling of sediments and bottom water, and measurements of currents. The system has been used over a range of depths and geographic locations and the knowledge gained from the study of the collected data is making substantial contributions to the solution of problems involving underwater sound.

7–1. Introduction

Interest in the observation and measurement of sea-floor microrelief has been stimulated by studies of sound waves reflected from the sea floor, since the distortion of those waves in the reflection process depends upon the nature and roughness of the reflecting surface. Most major and many intermediate features of submarine topography have been located by soundings and mapped, but this method is not of sufficient resolution to observe the fine details of the relief. The U.S. Navy Electronics Laboratory has developed an oceanographic survey system which enables the simultaneous collection of various specific data for detailed study of the sea floor.

7–2. The new deep-sea oceanographic system

Design (*NDSOS*)

The component parts of this system, shown in figs. 7–1a and b, are unique neither in design nor function, but as a combination they constitute an especially useful approach to the collection of oceanographic data for correlation with acoustic data. Also, since the NDSOS enables simultaneous collection of related data on deep-sea lowerings, its use has saved considerable ships' time, and thus has permitted the inclusion of further measurements needed in underwater acoustical investigations but sometimes necessarily omitted because of time shortage.

The components are as follows:

1) a stereoscopic pair of 70-mm cameras,
2) a repeating flashlight for illumination,
3) a water sampler with reversing thermometer rack,
4) a sediment grab sampler,
5) two short plastic corers,
6) a water-current velocity meter and recorder,
7) a magnetic compass in the field of photographic view,
8) a tip and tilt recorder,
9) a 12-kc/sec sonar transducer for communications to the surface vessel.

The volume of the combined system is made as small as possible by fitting the components into a strong, light-weight, hoop-type frame which resists fouling and reduces water resistance during lowerings. Parts that must be protected from sea water and external pressure are enclosed in cylindrical stainless steel cases. The system is balanced so that the camera axes are vertical.

Operation

Control of lowerings.   Until closed-circuit television is added for visual control, sonar continues to be the only practical means for positioning the system so that the sea floor is in correct focus for the dual cameras. An Edgerton, Germershausen, and Grier 12-kc/sec pinger, mounted on the frame, monitors the lowering of the NDSOS and prevents unwanted touchdowns, as described in chap. 4. The cameras are held at a height H, which may vary from 6 to 12 ft (18.3–36.6 cm) above the bottom. The 600-lb (270 kg) weight of the NDSOS and the weight and water resistance of the $\frac{1}{2}$-in diameter lowering wire act as a sea anchor, tending to control the drift rate of the system above the sea floor.

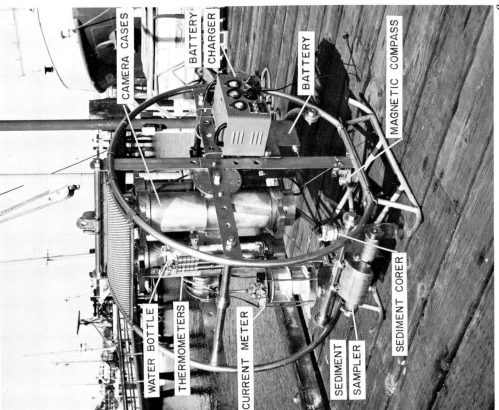

Figure 7–1. The new deep-sea oceanographic survey system, designed and constructed by the Navy Electronics Laboratory in San Diego, California. (a) Front view. (b) Back view.

Recording of photo orientation. Knowledge of the orientation of sea-floor photographs is often important, particularly in areas where bottom currents and ripple marks are present. The image of a magnetic compass held in the field of photographic view is recorded on each exposure.

Stereo photography of the sea floor. Since the primary function of the NDSOS is to collect information on the micro-roughness of the sea floor, photogrammetry plays an important part in the analysis of the data. As pointed out in chap. 4, the control of underwater cameras is not fine enough to permit the use of successive overlapping photographs as stereographic pairs for scaling relief. Hence it is necessary to take stereo pairs simultaneously, by using two matched cameras whose fields of view overlap in the target area. Fig. 7–2 illustrates the optics of stereoscopic photography.

The cameras utilize 70-mm Tri-X film and are equipped with f/6.3, 80-mm, wide-view Ektar lenses. They are fitted into cylindrical stainless steel cases capable of descending to maximum depths. A cone of clear Plexiglas is fitted into the end of each case to provide a watertight window. The optic axes of the dual cameras are separated by a distance of 12 in (28 cm). When measured from a stereo pair this represents the distance between photo centers and serves as a handy scale to measure photo width and camera height. The operation of the cameras

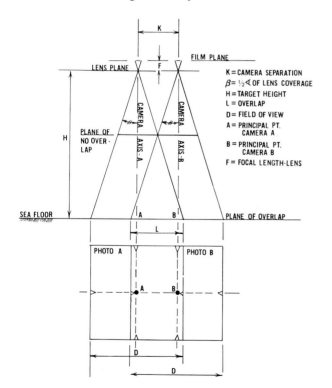

Figure 7–2. Illustrating the optics of stereoscopic photography.

and flash unit are electrically controlled through a special programmer, and the film advances automatically after each exposure. Photos are taken at 15-sec intervals for a period of two hours.

After return to the surface and sufficient time for the component parts to reach temperature and moisture equilibrium the cameras are removed from the cases and unloaded for processing. After processing the photographic pairs are mounted for stereoscopic viewing in a plotter. Any distortion due to skew, tip, and tilt between the cameras can be detected in the photographs, enabling the necessary mechanical adjustments to be made for correction.

The photographs are examined for the interpretation and identification of bottom targets, sediments, benthonic animals, and rock outcrops. After this the standard photogrammetric procedures described in chap. 5 are employed to record heights of bottom features in the selected microrelief sampling pattern.

Sampling of stereo overlap for acoustic purposes is accomplished using rectangular coordinates. X and Y denote horizontal position and Z determines vertical heights above this position. To reduce the number of variables for computer readouts, sampling is done in the X direction only on a pre-selected interval, I. This interval is determined by the magnitude of the micro-roughness under study. On a smooth surface having negligible relief, such as an abyssal clay surface, the interval selected would be rather large (± 2 cm). On an extremely rough surface containing many irregular targets (relief up to 20 cm) a small sampling interval (1 cm) would be selected to obtain a reliable interpretation of the micro-roughness. The pattern of sampling follows a definite procedure. Sampling starts in the upper left-hand corner of the stereo overlap and proceeds along numbered successive horizontal lines. The sampling interval is the same for both X and Y directions. The intersections of all X and Y lines, or X points of sampling, represent points of vertical relief which are measured with the stereo plotter. The accuracy is $\pm \frac{1}{2}$ cm. The resulting point-in-space locations can then be plotted as microprofiles along the X-direction sampling line or transformed as information to cards or tape for computer readout.

Sampling of bottom sediment. Bottom samples taken concurrently with the photographs are enclosed in a specially designed grab sampler (see fig. 7–1*a*). This encompasses an area of 0.44 ft² (0.04 m²), penetrates the bottom to a depth of 4 in (10 cm), and encloses 0.1 ft³ (0.003 m³) of sediment. Also two short corers, each 9 in (23 cm) in length and 2.5 in (6.4 cm) in diameter are attached to the frame near the grab sampler, to insure recovery of bottom material in case the grab fails to operate.

The grab operates with a slicing action, rather than by scraping or pushing, so that the sample collected is practically undisturbed. Small cores are extracted for sound velocity determination, and measurements of sediment shear strength are made with a portable shear vane. Also samples of the sediment are bottled for color, grain size, and weight analyses.

Sampling of bottom water. Measurements of temperature and salinity are used for the accurate determination of sound velocity in water. Both quantities are measured directly from water samples collected at about 24 in (60 cm) above the bottom in Fjarle bottles equipped with reversing thermometers.

Transparency measurements are likewise important, since they indicate the quantity of organic matter suspended in the water and the effect which this turbidity has on the transmission of sound waves. Transparency is measured in two ways: directly, by comparison of the Fjarle-bottle sample with a sample of double-distilled water, and indirectly from study of close-range bottom photographs. By considering such factors as camera-to-bottom distance, exposure time, type of film, contrast, color, and the nature of the bottom a qualitative idea of the water transparency can be obtained.

Measurement of bottom currents. Bottom currents are effective in forming, changing, and destroying sea-floor features in a variety of environments. A Savonius current meter is mounted on the frame; bottom currents cause rotation of a set of vanes actuating a pulse counter that prints out a record on a Rustrack recorder. Fig. 7–3 shows a portion of such a recording. The measurements are reliable only after the system has touched bottom and is resting firmly on the sea floor. This is accomplished during the period of sediment and water sampling, which lasts about 10 min, depending on weather and sea conditions.

7–3. Scientific observations, results, and conclusions

A number of lowerings have been made in the Pacific Ocean, in a variety of depths and bottom environments. Fig. 7–4 shows the positions of oceanographic stations occupied by the author with the NDSOS. Figs. 7–5*a* and *b* are a typical pair of overlapping sea-floor photographs taken at a station in the western Pacific Ocean and show manganese nodules scattered on the surface of an abyssal red clay.

Figs. 7–6 and 7–7 illustrate the construction of a typical microprofile made in the same area from photogrammetric sampling. The large hump is a manganese nodule that sampling line no. 3 has passed over. The sampling interval was 1 cm. In this profile the maximum

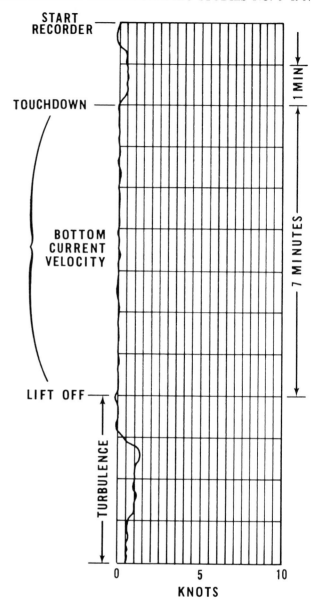

Figure 7–3. Portion of a current velocity meter record. Two horizontal divisions represent a current velocity of 1 knot. The recording paper moves at a speed of 15 in/hour (0.1 cm/sec).

relief did not exceed 3.5 cm, which would not be detectable by present-day echo sounders.

Distribution of low-order relief is highly irregular and observations indicate that animal burrowing, nodule formation, frequency of rock outcrops, and sediment distribution may follow certain trends which will enable the prediction of microrelief over great distances. At the present time insufficient numbers of profiles have been constructed to draw conclusions as to the exact nature and distribution of the microrelief, but representative samplings taken at a number of stations are pointing the way to solutions of this problem.

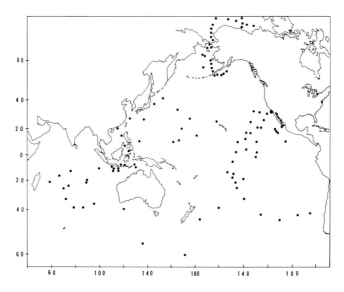

Figure 7–4. Positions of stations occupied by the NDSOS in the Pacific Ocean.

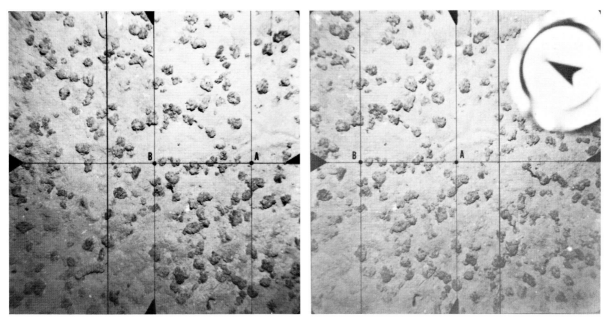

Figure 7–5. A typical pair of overlapping sea-floor photographs taken in the western Pacific Ocean, and showing manganese nodules resting on red clay. The depth of water is 3,240 fm (5,930 m), the camera height is 5.75 ft (1.75 m) and the width of the field of view of each photograph is 3 ft (0.9 m). The stereo overlap, indicated by broken lines, is 66 2/3 per cent. A and B are the optical centers of the left and right members of the pair. The distance between these two points is equal to the camera separation, 2 7/16 in (6.2 cm). The vertical heights of the visible manganese nodules range from 1 to 8 cm. The total vertical relief is 10+ cm (4 in).

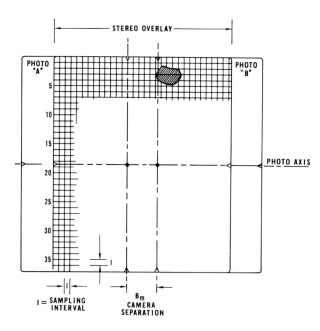

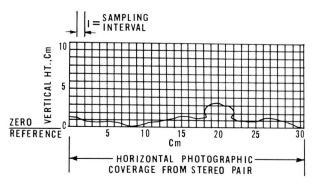

Figure 7–7. Profile of sea-floor microrelief. The profile is constructed from sampling line no. 3 in fig. 7–6. The large hump is a manganese nodule.

Figure 7–6. Microprofile sampling pattern of stereo overlap. The sampling interval, *I*, is 1 cm. Sampling line no. 3 is profiled in fig. 7–7.

References

Sharp, H. O., 1951: *Practical photogrammetry*. Macmillan Co., New York.

Shipek, C. J., 1961: Microrelief on the sea floor. *Science Teacher*, **28,** no. 6 (October).

———, 1965: Photoanalysis of sea-floor microrelief. NEL report no. 1374.

8. A multi-shot stereoscopic camera for close-up ocean-bottom photography[*]

David M. Owen *Woods Hole Oceanographic Institution, Woods Hole, Massachusetts*

Abstract

Highlights of a deep-sea stereoscopic camera that was assembled by the writer in 1960 are given in this article. The prototype camera, particularly when adjusted for close-up bottom photography, has contributed to several research activities at the Woods Hole Oceanographic Institution. Representative bottom photographs accompany the camera description. Some detailed interpretations of the camera's photographic results are found in chaps. 14, 15, and 21 in this volume; titles of additional papers or reports involving the use of this camera are found in the references, together with other pertinent writings of a background nature.

8–1. Introduction

My active interest in deep-sea photography began in 1947, while participating in a seven-month cruise of the R/V *Atlantis* to the Mediterranean Sea. During this expedition I operated some single-shot cameras that had been loaned by W. Maurice Ewing of the Woods Hole Oceanographic Institution and Columbia University for use in the scientific program (Owen, 1949). Ewing, Vine, & Worzel (1946) and others, of course, preceded the writer in the use of deep-sea stereoscopic photography.

8–2. Technical description

The earlier stereoscopic cameras that I used were either single-shot, or had serious depth limitations. Originally, two separate camera housings were utilized for the stereo effect (Owen, 1951). Later, a single conventional stereo camera using No. 120 film was installed in a single housing which had two windows (Owen, 1952; Owen, 1954; and Owen, 1957).

In 1960 two separate 35-mm spring-driven, semi-auto-

matic Robot cameras were mounted together in one housing with two windows (figs. 8–1 and 8–2). The Robot camera has been used from time to time by a number of workers (Ewing, Vine, & Worzel, 1946; Shumway, Dill & Kierstead, 1954; and Shipek, 1957) due to its compact size and spring motor drive. In addition, the use of two or more individual cameras naturally offers a choice between stereo photography and/or simultaneous exposure of two different types of film, i.e., black-and-white and color.

Each contact of a pilot weight with the ocean floor actuates the two cameras simultaneously — exposing up to twenty-four stereoscopic pairs of photographs in a series of lowerings with each full winding of the spring motors. The pilot weight is used in conjunction with a spring-loaded bottom switch similar to an original design by Worzel (Thorndike, 1958). (In practice, after the moment of exposure in the lowering, the camera supporting frame itself continues on to contact and momentarily rest on the ocean floor before its arrival there can be detected or "lift-off" begun by the winch operator.) The two camera solenoids are also from an earlier Ewing project. One camera shutter closes the trigger circuit in an electronic flash unit of 50-watt-sec input (fig. 8–3). The complete camera assembly is shown in fig. 8–4.

When the bottom switch operates, four operations take place:

1. A pinger device (if incorporated with the camera) ceases to operate, thereby "signaling" arrival on the bottom, when the normally open contacts in a relay become closed — shorting out the pinger.

2. Circuit is closed between a 6-v battery and the two camera solenoids which are connected in parallel.

3. Current is applied to the heater terminals of a thermal delay relay. The normally closed switch contacts on this relay temporarily maintain the circuit for the current going to the camera solenoids in the second operation. When the thermal element has heated for 3 sec the switch opens, disconnecting the camera solenoids. This operation eliminates the relatively high drain incurred by the solenoids if the camera remains on the bottom longer than intended; the thermal delay relay uses only a

[*] WHOI Contribution Number 1751.

The development of the camera described here was supported by the Office of Naval Research, under contract Nonr-2196(00), NR 083–004.

Figure 8–1. Frontal view of two Robot cameras mounted end-to-end, with individual sole-noids; this unit is quickly detachable from the camera case cover. Four leads from the Jones plug are attached to Joy High-Pressure Terminals (X8372-103), which are protected externally by the short tube at far right. Corresponding view of deep-sea housing for camera assembly.

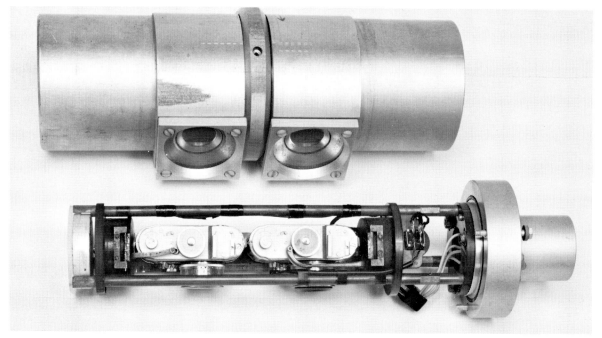

Figure 8–2. Top view of camera assembly, withdrawn from deep-sea housing. The housing was machined from a solid rod of aluminum alloy 6060-T6, giving a 3½-in (8.9-cm) bore (the bore was drilled off-center in the rod in order to leave sufficient bearing surface for the window inserts). The two windows are of Plexiglas, 1¼ in (3.18 cm) thick, truncated cones; O-rings provide the seals. In order to machine the window openings, this case was cut in two, and the two halves handled separately; they subsequently joined with a screw thread and O-ring seal. Myron P. Howland of the Woods Hole Oceanographic Institution Instrument Shop performed the work. A later case was handled in one piece—being shorter.

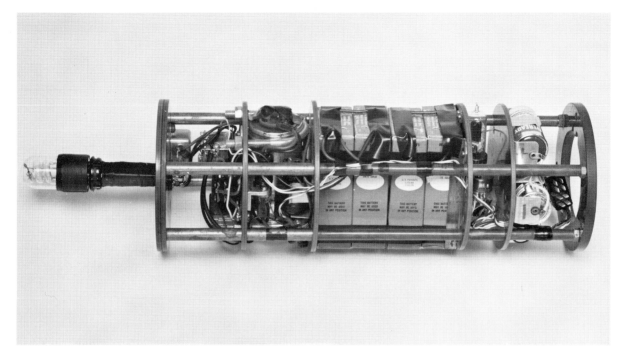

Figure 8–3. Electronic flash unit of 50-watt-sec input. It is a direct battery system with four Eveready No. 491, 240-v batteries connected in series-parallel, applied to a 525-Mfd, 450-vdc capacitor through a 1,000-ohm resistor. The flashtube is a Kemlite DX with a self-contained trigger coil. Charging is initiated by the bottom switch; full charge is reached generally within 5 to 10 sec. This unit also contains the control circuit for the cameras and pinger; a detailed explanation of the operation is found in the text.

fraction of the current required by the solenoids. The cameras usually are not triggered inadvertently while being lifted clear of the bottom by a rolling vessel, because of the short interval required for the heater element to cool before the relay contacts close again. The substitution of a normally open thermal delay relay, with a slight wiring change, would result in delaying camera operation until several seconds after both trigger weight and camera frame have come to rest on the sea bottom; this capability might be useful in some applications.

4. Actuation of a relay whose normally open contacts are in the electronic flash circuit between the negative side of the high voltage dry batteries used for this purpose and the 525 mfd, 450 vdc main capacitor. As a result, the flash unit is "on charge" only while tension is removed from the pilot weight line.

In practice, at the beginning of a station and before sending the camera on its way down, the camera flash unit is given a "topping off" charge, simply by lifting the pilot weight and allowing the tripping switch to act for a few seconds (also causing a test flash, if desired).

Although there can be a noticeable under-exposure, even the initial bottom contact following the lengthy descent time to about 5,500 m produces a flash and an ordinarily usable picture (of course, a low-leakage capacitor is required for this capability). While the pilot weight (and normally the camera frame also) rests momentarily on the bottom — usually 5 to 10 sec — the flash unit main capacitor is accumulating sufficient voltage for the next exposure.

The practical benefit of this circuitry, when used in a camera of this type, is that all batteries spend most of the time off duty. There is no need for a delayed-action clock timer to start the mechanism; use of such a timer requires the following of a close schedule in the camera lowering. By the same token the camera-flash components cease operation automatically when they are no longer needed. These features lend themselves well to situations where successive stations are made at close intervals in very shallow depths.

As in the case of all cameras of the bottom-trigger type, the focus, lens coverage (field of view), and exposure — in successive photographs — are inherently virtually identical, regardless of the roll of the ship above and of sudden changes in depth. Also inherent, it must be admitted, is the possibility of any instrument becoming hopelessly fouled on physical contact with the ocean bottom. At the same time, many camera operators would probably agree that more units have been lost through other mishaps which did not involve touching the bottom.

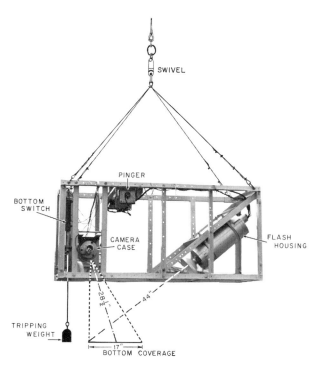

SWIVEL

PINGER

BOTTOM
SWITCH

CAMERA
CASE

FLASH
HOUSING

28½"

44"

TRIPPING
WEIGHT

17"
BOTTOM COVERAGE

Figure 8-4. Combination photo-drawing depicting camera assembly at moment of exposure on the ocean bottom. All components are shown enclosed in the frame, with the exception of the tripping weight drawn in the illustration; the frame has no protuberances. Note that the projected axis of the light source intersects the camera field on the bottom at the foreground edge of the picture, instead of in the center, producing a more even lighting. With the reflector surface painted flat white "hot spots" do not occur, and prints require little—if any—"dodging." The lens aperture required in this configuration, with Plus-X film and G filter, was f/10. The black-and-white films were developed normally—for fine-grain. Color film (High-Speed Ektachrome) required a setting of f/5.6, without filtration.

In perhaps its more unusual applications, this camera has been adjusted to photograph the deep-ocean bottom when only about 28 in (71 cm) from the camera windows. This method of attack differs from the approach used by some organizations, which have increased the camera-to-bottom distance to thirty or more feet (nine or more meters). Obviously, the application being described sacrifices extreme coverage for small-scale detail. The 30-mm focal length lenses on the Robots have a measured angle of view, in water, of about 30° (across the width of the negative). At the working distance of 28 in (71 cm), the actual field of view—assuming a flat bottom and a camera angle of 16° from the vertical—is about 43 cm from foreground to background; the area covered amounts to $\frac{1}{7}$ m² of the ocean floor.

Some results of this camera configuration—used at a number of benthic stations along a transect south of New England toward Bermuda—are shown in figs. 8-5 to

8-9. In this application, camera use was concurrent with the quantitative collections of bottom faunal samples. (See chap. 21.)

When a subject distance of some 28 in (71 cm) is involved, it is possible to make enlargements on a life-sized scale from the 1-in × 1-in (2.54 cm × 2.54 cm) negatives without objectionable grain appearing in the image (fig. 8-10). Such demand on a negative, however, requires critical focusing on the actual bottom distance (together with optimum exposure and processing of the film) for really satisfactory results. There will perhaps be future occasions when the planned focusing distance will be less than the current minimum of 28 in (71 cm), as the Robot lenses have appreciable near-focusing adjustment remaining. Possibly, exposure will be delayed, as explained earlier in this article, until after the camera frame itself has settled on the bottom.

The minimum working distance of 28 in (71 cm) used to date has also been of some benefit during suspended-camera work in shallow coastal areas where water turbidity was a factor (see Breslau, 1964).

In order to reduce the sources of doubt concerning the camera's focus and the actual field of view at such close working distances in water, the writer photographed targets (scales) through varying known distances of water in a laboratory setup with the same camera and type of window, and thus "constructed" the actual field of view —and lens angle—from the enlarged films. In these experiments, the lenses were manually focused for each target distance by means of a ground glass occupying the film plane. This setup, of course, enabled "direct" pre-focusing of the cameras for the expected working distances in the field. Confirmation of the camera's bottom coverage is provided on occasions when the trigger weight, of known size, is shown in the picture.

It should be noted here that the cloud of sediment which may be raised on impact of the trigger weight has occasionally been utilized in the estimation of relative grain size in the bottom sediment (see chap. 16, also Zeigler, Oostdam, & Owen, 1960; Athearn & Owen, 1962). A Phleger coring tube is sometimes attached to the camera frame, obtaining a sample of the bottom shown in the first photograph; subsequent contacts of the coring tube do not cause the initial core to extrude, they only contaminate the bottom end of the core sample (fig. 8-11). On one occasion, a heavy weight (separate from the tripping weight) was first to strike the bottom, providing a rough indication of bottom softness (fig. 8-12).

Concerning correction for the aberrations inherent in the use of conventional lenses and windows under water: when I have reason to use these lenses, with black-and-white films, I incorporate a yellow filter such as the Kodak Wratten K2 or G, in order to obtain at least a

Figure 8–5. Cruise A–264. Camera position 39°40.5′N, 70°38′W, in proximity of Woods Hole Oceanographic Institution Benthic Station G ⚹ 1. Photo depth 2,086 m. Sea urchin on clayey-silt bottom, with numerous tracks denoting animal activity on the surface; burrow at upper right. Note arm of brittle star at upper left.

partial correction through the use of more nearly mono-chromatic light. The angle of view in water is not altered by the use of a filter, and exposure must be increased to allow for the light-loss. Color film obviously is not applicable. But the narrower band of wavelengths involved after filtration reduces appreciably the elonga-tion of image points away from the center of the field that would be attributed to chromatic aberration. The

supposed disadvantage in loss of depth of field in the photograph (through opening the lens by perhaps 1½ f-stops in exposure compensation) is more than offset by the effect of the partial correction which is visibly obtained. Further details on this point, and on the design of correction lenses, are discussed in chap. 2 and by Thorndike (1950 and 1955), and Hopkins & Edgerton (1961).

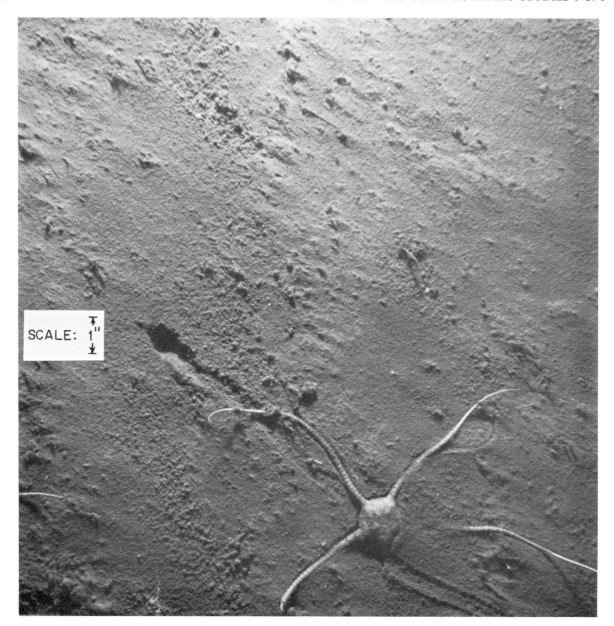

Figure 8–6. Cruise A-264. Camera position 39°45′N, 70°44′W, in proximity of Woods Hole Oceanographic Institution Benthic Station F ⚹ 1. Photo depth 1,500 m. Brittle star, probably *Ophiomusium lymani*, on a clayey-silt bottom which apparently has been recently marked by scour of currents that moved toward the lower right corner. Although this station was among those made during a benthic study, it received special attention by Owen & Emery (see chap. 15) because of evidence of discrete turbidity-current markings.

8–3. Camera mountings and stereoscopic formulas

Both cameras are mounted parallel on the same base-plate; thus, incorrect orientation of the two cameras in relation to each other is impossible. This is advantageous in the later stage of the mounting and viewing of stereo-scopic pairs of photographs.

The interocular distance involved is $4\,{}^{23}\!/_{64}$ in (11.073 cm). The distance from the outer window surfaces to the point in the water at which the overlap in stereo begins is $7\,{}^{5}\!/_{32}$ in (18.178 cm). When a window-to-subject "water distance" of $28\,{}^{1}\!/_{2}$ in (72.39 cm) occurs, the two photographs have about $72\,{}^{1}\!/_{2}$ per cent overlap across the principal point. It is ascertained that the following relationship applies to this camera,

$$D = \frac{7.15625 - \%\ \text{overlap}}{1.00 - \%\ \text{overlap}}$$

Figure 8–7. Cruise A-264. Camera position 38°45′N, 70°13.5′W, in proximity of Woods Hole Oceanographic Institution Benthic Station HH ⚹ 3. Photo depth 2,900 m. Burrows, tubes, and considerable animal activity on a silty-clay bottom. Dredge sample from this station yielded 748 animals per square meter. The larger holes are approximately 1 in (2.5 cm) in diameter.

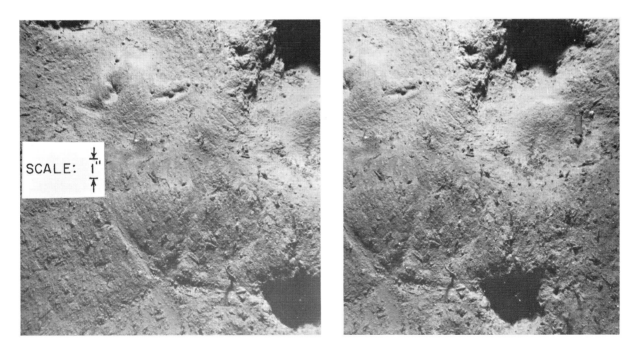

Figure 8–8. Cruise A-264. Camera position 39°29′N, 70°34′W, in proximity of Woods Hole Oceanographic Institution Benthic Station GH ⚹ 4. Photo depth 2,379 m. Tubes, fragments of tubes, and burrows on a silty-clay bottom. In left center, appearance of current effects on the raised knoll. Starfish indentation at lower left of picture. The large burrow at upper right has an opening of about 4-1/2 in (11.5 cm).

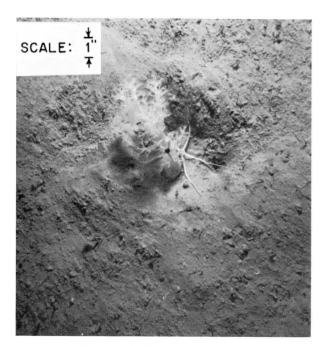

Figure 8–9. Cruise A-273. Camera position 39°29′N, 70°34′W, in proximity of Woods Hole Oceanographic Institution Benthic Station GH ⚓ 4. Photo depth 2,379 m. Branched organism in center depression is probably a coelenterate (order—*Pennatularia*), related to the common sea pens and sea whips; bottom is a silty-clay. Brittle stars are entwined with the organism. The diameter of the organism's body, where it meets the bottom, is approximately 1 in (2.5 cm); the depression in which the organism is found has a diameter of about 7 in (17.8 cm) at the rim.

where D equals distance from the outer window surface to the subject center of view in inches, and where $\%$ represents the fraction of overlap expressed in decimal form. It also follows that the actual amount of overlap W, measured in inches from left to right across the principal point, can be calculated by the formula

$$W = 0.538(D - 7.15625)$$

An example of a contour map of the bottom, which was made from a pair of stereophotographs taken with this camera, is found in chap. 15.

8–4. Camera lowerings and sonar devices

The weakest components of the prototype camera assembly limit the maximum operating depth to about 6,400 m. In shallow water, bottom contact with the camera frame is easily noted from the slackening of tension in the lowering cable. In fact, arrival of this camera on the bottom has been detected in about 5,490 m when the instrument was lowered on $3/16$-in hydrographic wire; loss of tension was displayed on accumulator springs.

However, these fortunate extremes in bottom-arrival detection were accomplished when accumulator-spring movements were not complicated by vessel motion in rough seas, and when excessive wire angles were not developed by the drift of the vessel or camera in wind or current. A contributing factor in the successful use of this method is in the design of the camera mounting frame, as shown in fig. 8–4. The frame is oriented horizontally, thus its total weight is removed from the lowering cable at first contact with the bottom. This frame is unusually stable when resting on the bottom, or on deck. The camera's longevity is further increased by the fact that all components are enclosed within the frame, with the exception of the tripping weight hanging below; moreover, there are no sharp protuberances in the frame itself to foul with the bottom (or with a loop in the cable).

My interest in a sonar device to signal bottom arrival or proximity was heightened in 1951 when, with the aid of Lloyd D. Hoadley, a continually recycling, 35-mm, electronic flash camera was assembled (Owen 1952, Owen 1954, and Owen 1957). In April, 1953, Edgerton — through the Development Fund of the Massachusetts Institute of Technology — received a purchase order from the Woods Hole Oceanographic Institution for the construction of a "bottom-contact indicator."

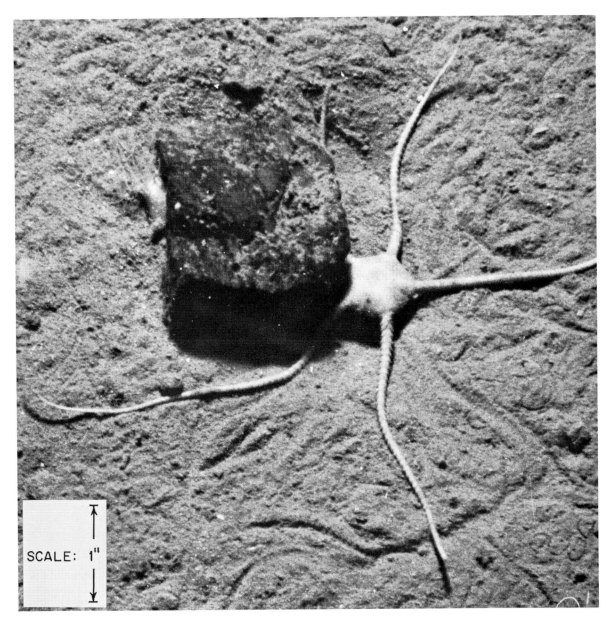

Figure 8–10. Cruise A-264. Camera position 39°40.5′N, 70°38′W, in proximity of Woods Hole Oceanographic Institution Benthic Station G ⚹ 1. Photo depth 2,086 m. Brittle star (probably *Ophiomusium lymani*) and cobble on a clayey-silt bottom. This photograph has been enlarged to approximately life-sized scale, and only a portion of the 1 × 1 in (2.54 × 2.54 cm) negative is shown here.

The device that was delivered to me in November, 1954, was of the same design which Edgerton had field-tested with Cousteau on the vessel *Calypso* earlier in 1954 (see Edgerton, 1955). Edgerton's prototype pinger was a nickel transducer emitting one ping per second during the camera lowering — audible with suitable listening equipment on the vessel above — until a mercury-tilt bottom switch shorted out the pulse.

This pinger was used for a time on an experimental deep-sea camera. During a loan of the pinger to A. R. Miller of the Woods Hole Oceanographic Institution, in the spring of 1959, the signals were picked up from a depth of 5,087 m. Later, the Edgerton device was incorporated with the present camera, until the pinger imploded at a probable depth of 5,124 m in 1961 (when the depth limit was exceeded).

A technical description of this early model pinger, and its uses by some other workers, was included with a discussion of a more advanced camera positioning sonar by Edgerton & Cousteau (1959), and by Edgerton,

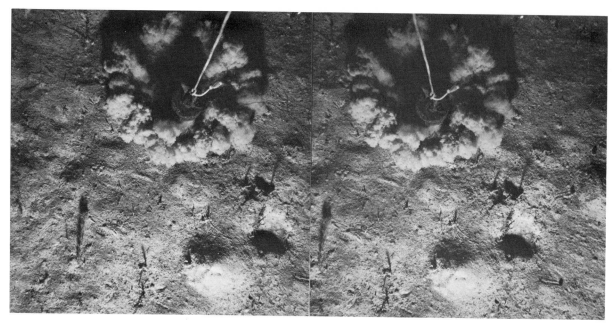

Figure 8–11. Stereo pair from the Tongue of the Ocean, Bahamas, showing the sediment cloud raised by the camera tripping weight landing on the sandy-silt bottom; depth 1,335 m. (Athearn & Owen, 1962). Tip of small coring tube attached to camera frame visible at the top right edge of the pictures. Field of view, foreground-to-background, about 23 in (58 cm), with a camera-bottom distance of 42 1/2 in (108 cm) along the lens axis, which was 3° from the vertical. Diameter of portion of tripping weight shown is 2 1/2 in (6.4 cm).

Cousteau, Hersey, & Backus (1960). A later model pinger assembly was obtained more recently for installation on the close-up camera described here.

References

Athearn, William D., & David M. Owen, 1962: Bathymetric and sediment survey of the Tongue of the Ocean, Bahamas, Part II: Bottom photographs. Woods Hole Oceanographic Institution, Ref. no. 62–27 (unpublished manuscript).

Breslau, L. R., 1964: Sound reflection from the sea floor and its geological significance. (Unpublished thesis submitted in partial fulfillment of the requirements for the degree of Doctor of Philosophy at the Massachusetts Institute of Technology).

Edgerton, Harold E., 1955: Photographing the sea's dark underworld. *Nat. Geog. Mag.*, **107**, no. 4, 523–37.

———, & Jacques Y. Cousteau, 1959: Underwater camera positioning by sonar. *Rev. Sci. Instrum.*, **30**, no. 12, 1125–26.

———, ———, J. B. Hersey, & Richard H. Backus, 1960: Underwater camera positioning by sonar. Woods Hole Oceanographic Institution, Ref. no. 60–17 (unpublished manuscript).

Ewing, Maurice, Allyn Vine, & J. L. Worzel, 1946: Photography of the ocean bottom. *J. Opt. Soc. Am.*, **36**, no. 6, 307–21.

Hahn, J., 1950: Some aspects of deep-sea underwater photography. *Photo. Soc. Am.*, Sect. B, **16**, no. 6, 27–29.

Hopkins, Robert E., & Harold E. Edgerton, 1961: Lenses for underwater photography. *Deep-Sea Res.*, **8**, no. 4, 312–17.

Johnson, Harry R., Richard H. Backus, J. B. Hersey, & David M. Owen, 1956: Suspended echo-sounder and camera studies of midwater sound scatterers. *Deep-Sea Res.* **3**, 266–72.

Ketchum, B. H., C. S. Yentsch, N. Corwin, & D. M. Owen, 1958: Some studies of the disposal of iron wastes at sea: summer 1958, Part 2: Bottom photography. Woods Hole Oceanographic Institution, Ref. no. 58–55 (unpublished manuscript).

Owen, D. M., 1949: *Atlantis* cruise 151 to Mediterranean area, Scientific Report No. 2: Bottom samples and underwater photography. Woods Hole Oceanographic Institution, Ref. no. 49–8 (unpublished manuscript).

———, 1951: Deep-sea underwater photography and some recent stereoscopic applications. *Photogrammetric Engineering*, **17**, no. 1, 13–19.

———, 1952: Two deep-sea cameras assembled for coastal, harbor survey, and mid-level photography. Woods Hole Oceanographic Institution, Ref. no. 52–62 (unpublished manuscript).

———, 1954: Recent developments in deep-sea photography at Woods Hole. In *Underwater Photography*, H. Schenck, Jr., & H. Kendall, Cornell Maritime Press, Cambridge, Maryland, 79–89.

———, 1957: Deep-sea photography at Woods Hole. In *Underwater Photography*, H. Schenck, Jr., & H. Kendall (2d ed., rev. and enlgd.), Cornell Maritime Press, Cambridge, Maryland, 57–68.

———, 1958: Photography under water. *Oceanus*, **6**, no. 1, 22–39.

Figure 8–12. Stereo pair from the Tongue of the Ocean, Bahamas, showing a 41-lb (18.6-kg) (in air) weight used in one attempt to determine the softness of the bottom; depth 1,322 m. From Athearn & Owen (1962). Camera tripping weight not shown. The white steel rod is 5 1/2 in (14 cm) in diameter, 6 in (15.2 cm) high, and is graduated with rings painted in black on the side at 1-in (2.54-cm) intervals; at least four rings can be discerned below the top of the weight, suggesting that the bottom was reasonably firm. The estimated speed of impact was 150 ft/min (46 m/min), and it is calculated that the weight had been resting on the bottom about 1.1 sec before the picture was made. Field of view, foreground-to-background, about 30 in (76 cm), with a camera-bottom distance of 47 in (119.5 cm) along the lens axis, which was 25° from the vertical.

Sanders, H. L., 1963: The deep-sea benthos. *A.I.B.S. Bulletin*, **13**, no. 5, 61–63.

Shipek, C. J., 1957: N.E.L. type III deep-sea camera. Report no. 768 U.S. Navy Electronics Laboratory, San Diego, California.

Shumway, G., R. F. Dill, & H. S. Kierstad, 1954: The U.S. N.E.L. deep-sea camera. Report no. 388, U.S. Navy Electronics Laboratory, San Diego, California.

Thorndike, E. M., 1950: A wide-angle, underwater camera lens. *J. Opt. Soc. Am.*, **40**, no. 12, 823–24.

———, 1955: Color-correcting lens for underwater photography. *J. Opt. Soc. Am.*, **45**, 584–85.

———, 1958: Deep-sea cameras of the Lamont Observatory. *Deep-Sea Res.*, **5**, 234–37.

Zeigler, John. M., Bernard Oostdam, & David M. Owen, 1960: A study of the bathymetry and sediments of the U.S. Navy Torpedo Range in Cape Cod Bay. Woods Hole Oceanographic Institution, Ref. no. 60–35 (unpublished manuscript).

9. Photographic measurements of bottom currents[*]

John G. Bruce, Jr. *Woods Hole Oceanographic Institution, Woods Hole, Massachusetts*

Edward M. Thorndike *Lamont Geological Observatory, Columbia University, Palisades, New York; and Queens College, Flushing, New York*

Abstract

Underwater cameras are used in several methods of making bottom-current measurements. Each method employs a tripod base that rests on the bottom, and the camera is used to photograph various current indicators such as an impeller, suspended balls or pendula of several weights, or neutrally buoyant drops of a liquid immiscible with water. Direction is given relative to a compass placed in the photographic field, and speeds ranging from less than 1 cm/sec to over 50 cm/sec can be measured by proper choice of the indicator. This article describes how such measurements are made, and indicates several other possible applications of photographic current meters.

9–1. Introduction

Underwater photography can be a useful tool in making deep bottom-current measurements if the camera remains fixed on the bottom, so that any water movement, turbulence, sediment transport, or movement of bottom animals can be related directly to the bottom to obtain absolute velocity measurements. Early work in this field, using weighted and unweighted ping-pong balls as current-sensing elements, is described by Ewing, Vine, & Worzel (1946).

A device developed at the Woods Hole Oceanographic Institution (Bruce, 1961) has been used successfully a number of times off the Bahamas, Bermuda, Cape Hatteras, the Strait of Sicily, and the Blake Plateau, and can be applied to measurements of:

1) bottom-current speed and direction;
2) bottom slope and characteristics;

3) estimation of ripple size and motion in a known bottom current;
4) sediment transport;
5) movements of bottom animals;
6) bottom shear, by using two or more meters;
7) bottom turbidity;
8) softness or penetrability of bottom.

9–2. Description of current meter

Fig. 9–1 shows the construction, the basic frame of which is an aluminum channel (later modified to a collapsible tubular tripod for ease in storage aboard ship). The camera, *A*, is mounted 6 ft (1.8 m) off the bottom and must be rigidly attached to the frame. A standard Edgerton camera was used with 35-mm film automatically advanced and exposed every 11–12 sec. Enclosed in the camera are a depth meter and chronometer which are photographed each time. The flash unit, *B*, and the power supply, *C*, are set clear of the field of view. The feet of the frame each have 50–100 lb of lead ballast, *D*, and an enlarged resting pad to prevent sinking in soft sediments. No large pieces of iron should be used which might deflect the compass (mounted on the platform, *F*). Within the field of the camera are an impeller-type current meter, *E*, with a revolution counter attached and a clear lucite platform, *F*. The rate at which the counter is photographed is obtained from the chronometer, and thus the turning rate of the calibrated impeller is known. Both impeller and vane are stiff polyethylene so that this unit is light in water. Plastic balls of various weights are suspended by light nylon strings from the platform. These may be weighted with lead shot to anticipate strong currents, or lightened to weigh a small fraction of a gram in sea-water by inserting pellets of buoyant polyethylene and thus become sensitive to weak currents (1 cm/sec). A liquid-filled compass is mounted at the center of the platform.

A calibration of the deflection of the balls with the current is made in a tow tank before use. It is necessary to

* WHOI Contribution Number 1752.

The development and use of the WHOI current meter was supported by the Office of Naval Research, under contract Nonr-2866(00) NR 287–004.

The work at Lamont was supported by the Office of Naval Research, under contract Nonr-266(48), with assistance from the Bureau of Ships, under contract Nobsr 85077.

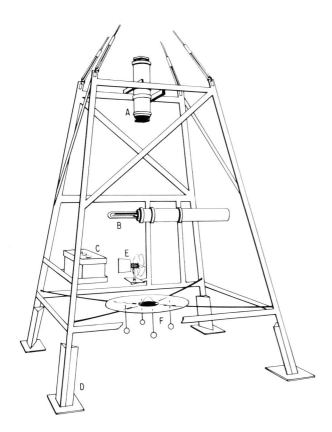

Figure 9–1. Bottom-current meter used by Woods Hole Oceanographic Institution. Basic parts are camera, *A*; flash unit, *B*; battery power supply, *C*; ballast, *D*; propeller current meter, *E*; deflecting balls and compass, *F*.

have two of the balls filled with lead shot to indicate the vertical, in case the device is lowered onto a sloping bottom. The strings of the heavy balls can then be used to obtain the vanishing point of parallel vertical lines in the photograph. A line then drawn from the vanishing point to the fixed point of the strings of the light, current-deflected balls can be used to determine their angle of deflection.

The line on which the frame is suspended is poly-propylene rope (⅜ in diameter) which is slightly buoyant in water, thus when the device rests on the bottom there is no load on the line, which floats clear of the frame without fouling (which might occur with a steel cable). The line may be buoyed off at the surface or tended from shipboard during the time on bottom.

9–3. Use of meter

Since the sea bottom is photographed at the same time as the counter, photographs may be compared to deter-

mine whether the frame has shifted relative to the bottom during the measurements. This method is quite sensitive, and very slow motions may be detected if the device is kept in position for several minutes. From the photo-graphs one can tell whether the frame has overturned by excessive shear, has penetrated too deeply into the bottom, or whether the instruments are in any way obstructed to prevent their obtaining a satisfactory record. The bottom roughness may often be determined by the photograph and thus give an estimate of the turbu-lence that might be expected close to the bottom.

Sediment transport and ripple formation may be cor-related with current strength. In a series of twenty-five lowerings on the Blake Plateau (Pratt, 1963) currents ranging from less than 1 cm/sec to 48 cm/sec were measured, indicating that the northerly flow of the Gulf Stream probably extends to the bottom on the Blake Plateau and also transports sediment across the bottom. Fig. 9–2 shows three photographs made on one of these lowerings during which sand transport was visible. The white circles drawn on the photographs are fixed relative to the frame and bottom. Changes in the quantity of sand in the rings may be seen in the sequence. On this cruise it was found difficult to make observations directly in the Gulf Stream although it was possible to work in areas of fairly strong surface currents (1–2 knots). As long as the line to the frame is kept slack, the frame will usually remain upright. Occasionally, in areas of strong currents, drag on the line becomes excessive and tipping results.

When the device was lowered off Cape Hatteras (Barrett, 1965) the bottom currents were generally less than 4 cm/sec, the stalling speed of the rotor used. In addition, the sediments were soft, and the frame sank up to the lucite platform, rendering the deflecting balls use-less for measuring current. Current estimates were possible from the moving mud cloud produced upon contact with the bottom. The cloud took from 35 sec to 3 min to drift clear.

A lowering made off Eleuthera, British West Indies, (Bruce, 1962) showed the movements of a brittle star at a depth of 1,200 m (cf. chap. 25). The speed of the animal (1 mm/sec), his direction, and the local slope of the bot-tom were measured. The plastic balls could be used as a reference to estimate the animal's size. Fig. 9–3 shows a mosaic of successive photos placed together to indicate the path traveled.

9–4. Current measurements at Lamont Geological Observatory

Developments similar to those described above have been carried on at the Lamont Geological Observatory,

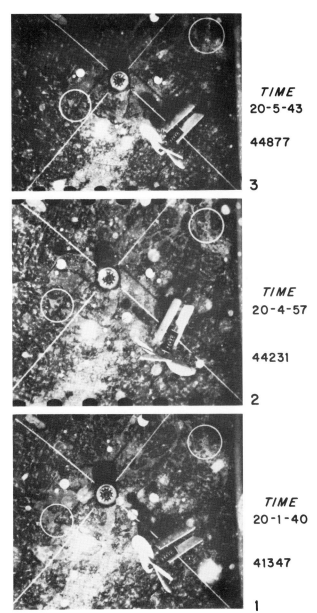

TIME
20-5-43

44877

3

TIME
20-4-57

44231

2

TIME
20-1-40

41347

1

Figure 9-2. Time sequence from a lowering on the Blake Plateau (Pratt, 1963) showing sand transport (under circles) over a manganese-nodule-encrusted bottom. The plastic balls are seen as white discs near the compass.

Currents greater than 1 cm/sec can be measured by merely photographing the displacements of thin aluminum sheets hanging as pendula from horizontal axes. Compact brass bars supported on the same axes give the vertical. The separation of the two is a measure of the current. Loose teflon bearings keep friction low. The deflection is rather insensitive to the direction of the velocity so that, with three sets of pendula having their axes set 120° apart, at least one gives a deflection very near to that for a current at 90° to the pendulum face. The light pendula can be weighted to give greater range, but at the expense of sensitivity. Fig. 9-5 gives a typical calibration obtained in a towing tank. Direction is obtained by a liquid-filled compass and a separate direction vane mounted on a vertical axis.

This type of current meter has been used intermittently since 1961. During the summers of 1963 and 1964 — *Conrad* cruises 7 and 8 — it was used in the Caribbean area at approximately twenty stations by Gerard & Sexton.

Currents below the range of the pendulum system described above can be measured by ejecting drops of colored liquid into the field of view of a camera and photographing them at a known time interval (Thorndike, 1963). The liquid should have approximately the same density as sea water and should be immiscible with it. Mixtures of ethyl benzoate and *n*–heptyl alcohol dyed with Sudan–3 are satisfactory. A white background with dark lines is helpful in obtaining displacements from the photographs. A time interval of approximately ½ sec has been found suitable.

Drops have been ejected from a storage tank in three ways. In the first two methods, a gear has a rod threaded through it. As the gear is rotated, the threaded rod, which is prevented from turning, advances and presses against the plunger of a syringe thus forcing out liquid. In the first method, the gear is a ratchet which is driven by a pawl and electromagnet. In the second, the gear, a spur gear in this case, is driven by a second gear mounted on the same shaft as the armature of a motor. The armature is caused to rotate a half-turn at a time forcing out liquid with each half-turn. In the third, the liquid is heated and caused to expand and flow out of a syringe needle. A system of valves allows the reservoir to be cooled and prepared for another expansion without drawing sea water into the needle. In the first method, the abrupt motion of the syringe plunger throws drops clear of the needle provided that it is clean. The other two methods require an electromagnet and arm to knock off the drops.

This current meter is obviously more complicated than the one using the displacements of pendula and has not received much use. However, it has the ability to measure very slow currents and it requires no calibration but

using somewhat different equipment. In the Lamont apparatus (fig. 9-4) the camera, light, and power supply are located in a single housing, 4.25 in (10.8 cm) in inside diameter and 10 to 20 in (25.4 to 50.8 cm) in length. The camera lens is at the center of the front plate of the housing and a small electronic flashtube is mounted in a separate aperture to one side of it. A single window covers them both. Current-sensing elements and a compass are attached to the housing. The entire unit hangs from a tripod in order to have its direction near to the vertical.

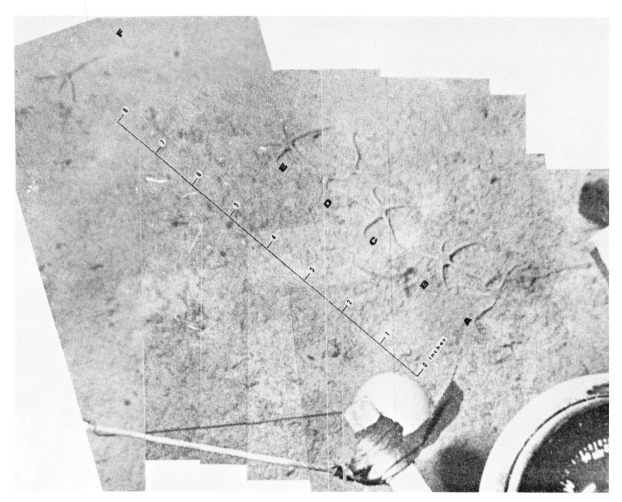

Figure 9–3. Mosaic of photographs of moving brittle star off Eleuthera, B.W.I. (Bruce, 1962). The first five positions of the animal, *A-E*, are 36 sec apart (representing every third photograph) whereas the final position, *F*, occurs about 5.5 min after *E*.

merely a knowledge of the geometry of the instrument and the speed of the timing motor. It is also well suited for studies of very small-scale turbulence. The two types of current-sensing elements, pendula and free drops, can readily be combined into one instrument as shown in fig. 9–4. A single camera now photographs both the pendula and the drops. The drops are used for very slow velocities, the pendula for high ones, and both are used in the mid range.

An alternative arrangement, which can use the same sensors as mentioned above or can merely supply a series of photographs of the cloud of sediment stirred up by the equipment has a camera and a strobelight mounted in the head of a coring apparatus. The equipment is started when the coring apparatus is triggered and photographs are taken until after the core pipe has been pulled clear of the bottom. The apparatus must, of course, be left stationary relative to the bottom long enough for the initial turbulence to die down and a series of photographs

to be taken. Equipment of this type can furnish bottom current information with the expenditure of very little time in addition to that used in normal coring operation. (An article on such equipment has been accepted by *Deep-Sea Research*, Ewing, Hayes, & Thorndike.)

References

Barrett, Joseph R., Jr., 1965: Subsurface currents off Cape Hatteras. *Deep-Sea Res.*, **12**, no. 2, 173–84.

Bruce, John G. Jr., 1961: Current studies off Plantagenet Bank. Woods Hole Oceanographic Institution, Ref. no. 61–17 (unpublished manuscript).

———, 1962: Photographic record of a moving brittle star. *Deep-Sea Res.*, **9**, 77.

Ewing, Maurice, Allyn Vine, & J. L. Worzel, 1946: Photography of the ocean bottom. *J. Opt. Soc. Am.*, **36**, 307–21.

Pratt, Richard M., 1963: Bottom currents on the Blake Plateau. *Deep-Sea Res.*, **10**, 245–49.

Thorndike, E. M., 1963: A suspended-drop current meter. *Deep-Sea Res.*, **10**, 263–67.

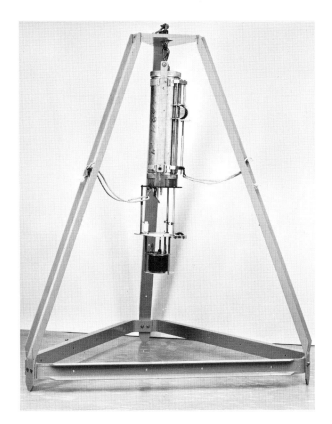

Figure 9-4. Photographic bottom-current meter used at Lamont Geological Observatory. The current-sensing elements and compass are in the field of view of a camera, contained in the cylindrical metal housing together with a light and power supply.

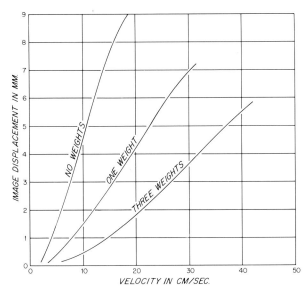

Figure 9-5. Calibration curves for thin aluminium pendula as current-sensing elements. The calibration tests are carried out in a towing tank.

10. Photographic nephelometers for the deep sea*

Edward M. Thorndike   *Lamont Geological Observatory, Palisades, New York; and Queens College, Flushing, New York*

Maurice Ewing *Lamont Geological Observatory, Palisades, New York*

Abstract

Photographic nephelometers have been designed and constructed for *in situ* measurements of light scattering in the deep sea. The method consists of a comparison of the amount of light which reaches a camera directly, after passing through an attenuator, with that reaching it indirectly by scattering in the sea water. The light is produced by either an electronic flashtube or a steady source, and the record consists of a series of photographs, or a continuous strip, showing the relative intensities of scattered light and direct, attenuated light. This gives a measure of the light scattering and thus an indication of the amount of suspended matter in the sea.

10–1. Introduction

The optical properties of sea water have been studied for many years. Particular attention has been directed toward measurements of the attenuation of sunlight in water near the surface, because of the biological importance of light for marine organisms. In recent years, detailed measurements of absorption and scattering have been made and theoretical treatments of the light field have been developed. The measurements have been made by collecting water samples and studying them in the laboratory and by lowering instruments and obtaining *in situ* observations.

Much less information has been gathered for the water well below the surface. Most of this has been obtained by collecting samples and bringing them to the surface for study. If closely spaced observations from the surface to the bottom are to be made, the advantages of *in situ* observation are enormous. In addition, the freedom from changed conditions and possible contamination is appealing. Thus a program for determining light absorption or light scattering at all depths and at many locations requires equipment that can make rapid *in situ* measurements. The value of light-absorption and light-scattering data for the identification of water masses, for the determination of their motions, and for the estimation of the material that they transport is clear. The photographic nephelometer described below was designed to provide information on light scattering in a preliminary survey of a large sample of the oceans.

10–2. Basic considerations

The nephelometer was designed to be sensitive to the presence of suspended matter in water and to be simple and rugged. It was to be lowered on a coring apparatus or on hydrographic wire. In essence, it consists of a light source, a baffle and calibrating attenuator, and a camera. The light source illuminates a body of water in front of the camera. The baffle prevents all direct light from reaching the camera window except that which has passed through the attenuator. The attenuator reduces the direct light and provides the camera with two regions having a known ratio between their brightnesses. These direct patches are used to monitor the light source and to calibrate the photographic film. The camera is a slight modification of our regular deep-sea camera (Thorndike, 1959). It provides a record of the amount of light entering the window from the different directions within its field of view. The record consists either of a series of photographs or of a continuous strip with marks from which the time can be obtained. After processing, the photographic film is run through a recording densitometer, and a comparison of the photographic densities of the regions receiving scattered light with those receiving light from the calibration attenuator furnishes a measure of the amount of light scattering and an indication of the amount of matter suspended in the water.

The arrangement of parts is shown schematically in fig. 10–1, and fig. 10–2 shows a nephelometer which is being used in conjunction with coring equipment. It is evident from fig. 10–1 that the blackening of any part of

* This work was supported by the Office of Naval Research, under contract Nonr 266(48), and by the National Science Foundation.

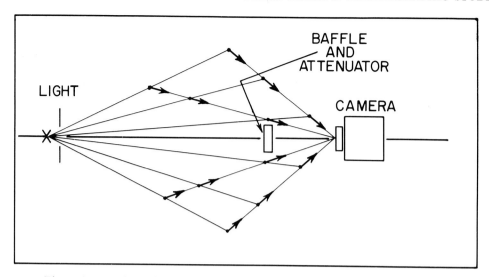

Figure 10–1. Schematic of a nephelometer which uses a continuous light source.

the film results from light that has been scattered through a range of angles. However, the relative importance of small-angle scattering compared with large is greatest in the inner region of the scattered-light patches. Thus, there is an indication of the angular distribution of the scattering.

10–3. Nephelometers using electronic flash-tubes as light sources

Two nephelometers have been constructed with electronic flashtubes as light sources. In these, the camera obtains a photograph in which the general light scattering shows as a gray background, and objects that are large enough to be resolved are shown in detail (fig. 10–3). One of these nephelometers had a single camera, and one had a pair of cameras with axes separated by 7.5 in (19 cm), giving a stereoscopic view of the water which enabled the linear size of the objects to be determined. In order to have a large volume of water illuminated, the distance from the light to the camera was made large — 3 m (10 ft). This, of course, required a large instrument, approximately 4 m (13.2 ft) long. No difficulty was encountered in mounting the equipment on the deck of the bathyscaph *Archimede* for dives in the Puerto Rico Trench. It was, however, somewhat awkward to handle from the research ship *Robert D. Conrad*.

The light sources were conventional in design, employing a 100-watt-sec flashtube powered by a 525-μF condenser charged to 480 v by two 240-v batteries connected in series. A small electric motor actuated the flashtube at a 10-sec interval. In the stereoscopic unit, each fifteenth flash was omitted to furnish a simple method of correlating the photographs taken with the two cameras.

Figure 10–2. A photographic nephelometer in combination with coring equipment. The light housing, attenuator, and baffle can be seen on the left of the picture in front of the camera, which is mounted in the core head.

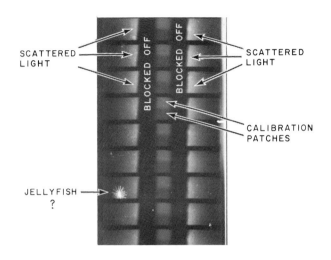

Figure 10–3. Record taken by a single-camera nephelometer using a strobe-light source. The center strip shows square calibration patches, and large scattering objects can be seen against the grey background of scattered light in the outer strips. In this instrument the flashtube is actuated once every ten seconds.

The attenuators consisted of two apertures covered by different amounts of opal glass.

The camera was extremely simple, consisting of a 75-mm, f/4.5 enlarging lens which formed an image on unperforated, 35-mm film transported at constant speed by a small electric motor. The short duration of the flashes makes it unnecessary to stop the film during an exposure, and the absence of sunlight during most of the descent makes a shutter unnecessary.

In use at sea these nephelometers were satisfactory in all ways except that they were a trifle large to handle.

10–4. Nephelometers using steady light sources

Several nephelometers have been constructed with light sources consisting of a small incandescent bulb (#67) powered by 13 or 14 nickel-cadmium, size D storage cells. The light remains on during an entire lowering. The camera film is transported at constant speed behind a slit aperture. Thus, the record consists of strips parallel to the length of the film (fig. 10–4). The two center strips are formed by light which has passed through the two apertures of the attenuator and the two outer strips by scattered light. Large particles in the scattering region produce streaks on the film rather than images corresponding to

their size and shape. With this arrangement, there is little or no advantage in illuminating a large volume of water, so the instrument is made smaller. A distance of 2 ft (61 cm) from the light to the camera has been used.

The attenuator consists of two apertures of unequal width covered by sheets of translucent teflon. The exposures for the two calibration strips are proportional to the widths of these apertures. The number of sheets of teflon covering the apertures is chosen to make the calibration strips of approximately the same optical density as the scattered-light strips.

The camera is the same as that used in the other nephelometer except that a 35-mm, f/2.8 lens is used, and a watch is added to give a timing trace. This is accomplished by allowing light from a very small bulb to reach the edge of the film after passing through a watch with a hole drilled for the purpose. A baffle on the second hand of the watch interrupts the light beam briefly each minute, producing a break in the trace.

Nephelometers of this type have been used with ease on hydrographic wire and have been built into coring equipment. In the latter arrangement, the camera is located in a tube passing through the core weight while the light and baffle are mounted on the pipe below the weight, as illustrated in fig. 10–2.

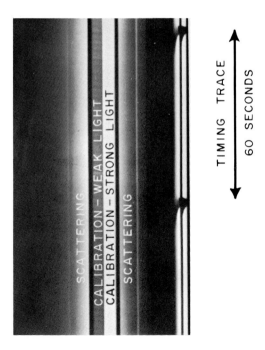

Figure 10–4. Record taken by a nephelometer using a steady light source. The two center strips are light and dark for calibration, and the outer strips are grey, recording the amount of light scattered. The strip on the left-hand side of the picture is a timing trace, the distance between two successive black marks representing a time interval of 60 sec.

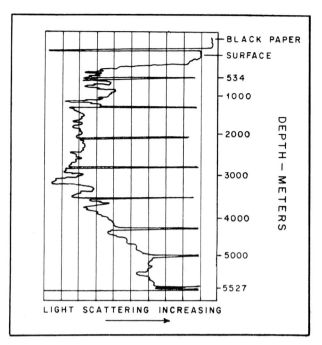

Figure 10–5. Variation of light scattering as a function of depth. A recording densitometer was used to measure the photographic density, or darkening, of the nephelometer film strip (negative). "Black paper" gives the reading for opaque film, and the spikes are positioning marks, given at 10-min intervals. The bottom was hit just after 80 min, at a depth of 5,527 fm. The end of the trace shows the increased light scattering near bottom; at the beginning there is some surface light, so that near the surface the true scattering is less than the curve indicates (taken at 20°22′N, 66°48′W).

10–5. Results

More than 500 nephelometer lowerings have been made from the ships of the Lamont Geological Observatory. Fig. 10–5 shows a representative graph of light scattering as a function of depth, based on a measurement of the photographic film by means of a densitometer.

The films show wide variations in light scattering with location. However definite trends and patterns are evident:

1) surface water normally has high and variable light scattering;

2) in the open ocean, mid water tends to have low scattering;

3) the variation of light scattering with depth is often smooth and gradual but occasionally it shows abrupt changes;

4) in some regions light scattering increases near bottom. The increase may be large or small, abrupt or gradual.

Clearly, light scattering must be studied in conjunction with other conditions for each location. A start has been made with the observations obtained off the eastern coast of North America (Ewing & Thorndike, 1965). Two possible explanations of the observed distribution of the nepheloid water were suggested: (1) that the cloudiness is associated with a water mass containing suspended matter that has remained more or less permanently in that layer, in effect serving as a marker of the water mass; or (2) that it results from sediment stirred up from the sea floor by turbulent flow of the bottom water, the flow being controlled in some cases by the influence of the suspended matter upon the density, and in other cases by other factors. Arguments were presented in favor of the second explanation.

10–6. Conclusions

Deep-sea photography clearly furnishes a useful tool for the *in situ* study of the optical properties of sea water at great depths. The instruments discussed here are intended for a rapid survey of light scattering. More detailed information can also be obtained photographically but, of course, the instrumentation loses some of its simplicity.

References

Duntley, Seibert Q., 1963: Light in the sea. *J. Opt. Soc. Am.,* **53,** 214–33.

DuPré, Elsie F., & Leo H. Dawson, 1961: *Transmission of light in water; an annotated bibliography.* U.S. Naval Research Laboratory bibliography no. 2, April, 1961.

Ewing, Maurice, & Edward M. Thorndike, 1965: Suspended matter in deep ocean water. *Science,* **147,** 1291–94.

Jerlov, N. G., 1953: Particle distribution in the ocean. Reports of the Swedish Deep-Sea Expedition, **3,** fasc. 2, Physics and Chemistry, no. 3, 71–97.

———, 1963: Optical oceanography. *Oceanography and Marine Biology, an Annual Review,* **1,** 89–114. George Allen and Unwin, Ltd.

Kalle, Kurt, 1945: *Der Stoffhaushalt des Meeres. Probleme der Komischen Physik,* C. Jensen *et al.* (ed.). Akademic Verlags, Leipzig.

Thorndike, E. M., 1959: Deep-sea cameras of the Lamont Observatory. *Deep-Sea Res.,* **7,** no. 1, 10–16.

11. Geological applications of sea-floor photography[*]

C. O. Bowin, R. L. Chase, and J. B. Hersey *Woods Hole Oceanographic Institution, Woods Hole,*
Massachusetts

Abstract

Photographs of the sea floor provide valuable geological information, and are especially useful when taken in conjunction with a program of echo sounding, seismic profiling, dredging, and coring. In this chapter aspects of the methods and role of deep-sea photography in submarine geology are illustrated by accounts of several investigations. The combined operation of photography and dredging is exemplified in a study of the Mid-Atlantic Ridge. Information from bottom photographs of the north wall of the Puerto Rico Trench has been combined with information from echo sounding, seismic profiling, and dredged samples to determine the structure and stratigraphy of an undersea area. A photographic survey of the Mona Canyon between Puerto Rico and Hispaniola has shown that outcrops of sedimentary rocks occur as deep as 3,700 m below sea level. A discussion of several camera lowerings in the Balearic Basin and in the Tyrrhenian Sea — western Mediterranean — indicate how bottom photographs may be useful in regional as well as detailed studies. An intensive photographic survey of a portion of the western and southern slopes of Plantagenet Bank, south of Bermuda, shows how bottom photographs can help reveal the nature of mountain slopes of undersea ranges. Finally, undersea photographs of the Seychelles-Mauritius Ridge, Indian Ocean, combined with dredged material, suggest that the ridge is almost completely covered by at least a veneer of calcareous rocks and sediments, and indicate the added value of continuous photomontages of the sea floor.

* WHOI Contribution Number 1753.

This work was supported by the Office of Naval Research under contracts Nonr-1367(00), NR 261–102, Nonr-2866(00), NR 287–004, Nonr-4029(00), NR 260–101, and the National Science Foundation under grant GP-2370. We are grateful to Henry B. Roberts of the Smithsonian Institution for the identification of the crabs in figs. 11–26 and 11–37; and to H. B. Fell of Harvard University, and other members of his group in the Museum of Comparative Zoology, for their identification of benthic fauna in figs. 11–19 to 11–38. We are grateful to K. O. Emery for his critical review of the manuscript.

11–1. Introduction

Since 1957, members of the geophysics group of the Woods Hole Oceanographic Institution have used deep-sea cameras for geological investigations, search for sunken ships, and survey of locations for cable runs and underwater instrument foundations. All, regardless of their original purpose, have produced valuable scientific information. In this chapter some aspects of the role of deep-sea photography in submarine geology are discussed, followed by accounts of several investigations in which photography has provided answers or posed problems which indicate where further research might prove most fruitful.

11–2. Methods

Photographs of the ocean floor are most useful to the marine geologist when taken as a complement to other methods of exploration, such as echo sounding and seismic profiling. They enable him to select the most significant locations for sampling, and should be used in any sampling program as a means of determining the geological setting of the samples. Photographs give visible form to his knowledge and frequently limit his speculations usefully and strikingly. The deduction, for instance, that currents do not exist in the deep sea was dealt a mortal blow when ripple marks were first observed in photographs of the deep-ocean bottom (chap. 1, this volume). Emery & Wigley (chap. 22, this volume) discuss their use of a combination of camera and grab sampler in biological studies on the continental slope. Our own work, in which photography has been used in combination with echo sounding or seismic profiling, has led mainly to sampling by means of dredging or coring.

Dredging is a crude but effective way of sampling the sea floor where seismic reflections suggest that rock may be recovered. However, it is not only very time-consuming in deep water (and thus costly), but results in the sampling of a very small and possibly unrepresentative area of the bottom. Even special dredges built to retain only coarse material such as cobbles and boulders frequently fail to bring up rock and, even if they succeed, yield little information about the manner in which the rocks occur on the bottom. Among the questions frequently left unanswered by dredging are the following: Are the rocks from outcrops, from a chaotic scree or talus slope, from gravel in the bed of a submarine canyon, or from deposits formed by the melting of icebergs? What is the nature and orientation of bedding or fracture surfaces, if any, in rock outcrops? Are there veins of intrusive material to be distinguished in them? Are the rocks fresh or weathered? What is the percentage of various bottom types in the area of interest (bedrock cliffs; rough, rocky bottom; smooth, sandy, or muddy bottom; etc.)? Many of these questions might well be answered by a series of photographs taken as part of the dredging operation. Investigations incorporating dredging and photography have been made, but in almost all of them the photographs were made separately from the dredging, so that one does not know with certainty how the samples occurred on the bottom. The correlation between photographs and dredged samples was improved by Hersey & Nalwalk (unpublished) when they attached a camera to the tow wire, 100 m above the dredge, and then lowered the combination to the sea bed while making way at a knot or two, towing the wire astern. The wire was payed out under tow until the camera and its pinger were in position to obtain photographs of the bottom. During the early part of the tow the dredge sampled the bottom and good photographs were obtained. It was necessary to pay out wire repeatedly in order to maintain the camera position near bottom. Many of these dredgings had to be terminated only because the camera-dredge combination, soon well astern of the ship, could no longer be detected by the ship's echo sounder, which is most sensitive to sound arriving from below it. In such a dredge haul by Nalwalk on the western slope of the Mid-Atlantic Ridge (at 28°53'N, 43°20'W) angular fragments of boulder-size rock, partially covered by sediment, were seen in three out of forty-one photographs (fig. 11-1). Other photographs showed only sediment. The camera mechanism failed after these photographs, but a few minutes later a sharp increase in wire tension signaled a strike. Upon recovery the dredge was found to contain about 300 kg of basalt and serpentinite in fragments measuring 10 to 50 cm across. Some of the larger fragments had the same general appearance as those in the photographs. Photographic evidence was lacking at the critical time during this dredging, and the full geological setting can only be surmised. One surmise is that the photographs correctly portray a large area of sediment sparsely dotted with fragments of rock. A second is that the dredge was towed

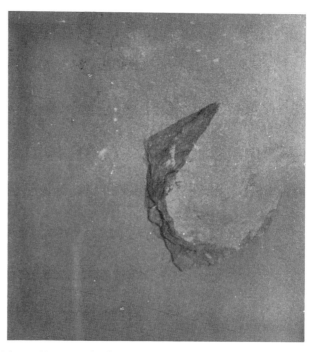

Figure 11–1. Rock fragment photographed by a camera attached to a dredge. Taken over the western slope of the Mid-Atlantic Ridge (28°53'N, 43°20'W, *Chain* cruise 21, December, 1961). The accompanying dredge collected about 300 kg of basalts and serpentinites in fragments measuring 10 to 50 cm across.

into a dense concentration of rock fragments, such as a talus slope, minutes after the camera had stopped photographing, and collected its entire haul at once.

Coring is well known to provide evidence about the local stratigraphic sequence to depths of 10–20 m. Many generalizations, for example Ericson *et al.* (1961), are based on correlating the cores with broad-scale morphologic features of the ocean floor, such as abyssal plains, continental rises, and sedimentary aprons surrounding oceanic islands. In such investigations, sea-floor photographs are seldom used directly as evidence, largely because core locations have been chosen more on the basis of the broad-scale geological setting, such as topography based on echo soundings, rather than the fine detail of a photograph. At present several investigators use a camera on the head of a core barrel in order to photograph the local setting of a core. The photographic technique described in chap. 4 for displaying long profiles of the bottom might well follow the discovery of an outcrop by means of the seismic profiler in order to obtain a truly detailed setting for subsequent cores. So far as the authors are aware, no such detailed survey has yet been attempted.

New sampling methods such as deep drilling, or deliberate sampling by means of free or tethered submersibles, are coming into use and will require the support of deep-sea photography for much the same reasons as our present limited sampling methods.

11–3. Applications

The north wall of the Puerto Rico Trench

The north wall of the Puerto Rico Trench (fig. 11–2) is a site where information from bottom photographs has been combined with information from echo sounding, continuous seismic profiling, and dredged samples to determine the structure and stratigraphy of an undersea area (Bowin, Nalwalk, & Hersey, 1966). The photographs of March 1, 1960, during *Chain* cruise 11, were the first to reveal rock outcrops on the north wall of the Puerto

Rico Trench, and they stimulated further investigations of the region. An outcrop of dark rock with wispy white veinlets (fig. 11–3) is believed to be serpentinized peridotite similar to the igneous rocks dredged from the main scarp of the north wall. Early photographs of the same series show abundant gravel with varying amounts of pebbles, cobbles, and boulders (fig. 11–4), while still later photographs (fig. 11–5) suggest outcrops of indurated pebble conglomerate that may be lithified or cemented older gravel. Gravel, commonly with plate-shaped fragments (fig. 11–6), was photographed probably on an upper portion of the principal scarp. Plate-shaped cobbles of thinly laminated silicified shale and chert, very similar in appearance to rocks in the photographs, were obtained in dredge hauls during a later cruise of *Chain* in 1962. A possible stratigraphy of the north wall

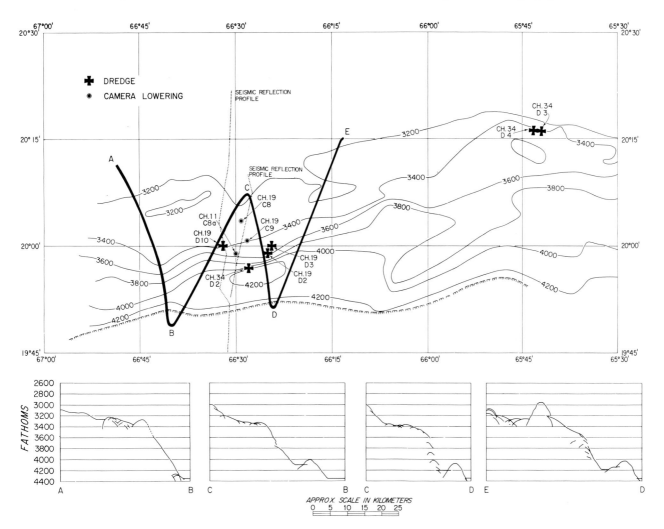

Figure 11–2. Bathymetric chart of the north wall of the Puerto Rico Trench showing locations of camera stations (*C*) and dredge hauls (*D*) made during *Chain* cruises 11, 19, and 34, individual echo sounding profiles (heavy lines), and two seismic reflection profiles (dotted lines). Depth contours are in fathoms based on an assumed sound velocity of 800 fm/sec.

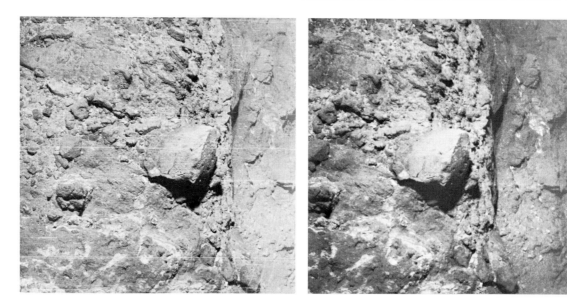

Figure 11–3. Dark rock outcrop with wispy white veinlets; probably serpentinized peridotite. Taken over the north wall of the Puerto Rico Trench (19°59′N, 66°30′W, *Chain* cruise 11, March, 1960).

of the Puerto Rico Trench was formulated from all the collected information (fig. 11–7).

Mona Canyon

Mona Canyon lies north of the Mona Passage between the islands of Puerto Rico and Hispaniola and is a steep-sided submarine re-entrant into the south slope of the Puerto Rico Trench. Earthquake centers beneath the canyon (Sykes & Ewing, 1965) suggest that it may be a tectonic feature. Dredging and photography of the slopes

of the canyon were carried out in 1965, from *Chain*, as part of a survey to ascertain whether the mass of igneous and metamorphic rocks which forms the spine of Puerto Rico extends northward beneath Tertiary sediments to crop out on the walls of the canyon. Dredged material includes semi-consolidated foraminiferal ooze from halfway up the eastern slope of the canyon, and slabs of gray friable mudstone from the lower part of the western slope. On the basis of foraminiferal content, both samples are considered to be Miocene in age (Ruth Todd, personal communication). Outcrops of blocky, apparently poorly

Figure 11–4. Boulder, with pebbles, cobbles, and gravel, on the north wall of the Puerto Rico Trench (19°59′N, 66°30′W, *Chain* cruise 11, March, 1960).

Figure 11–5. Outcrops of indurated pebble conglomerate on the north wall of the Puerto Rico Trench (19°59′N, 66°30′W, *Chain* cruise 11, March, 1960).

bedded, sedimentary rock characterized by regular joints spaced about a meter apart were photographed at a depth of about 3,500 m on the western slope of Mona Canyon (figs. 11–8 and 11–9). At somewhat greater depth on the western slope (3,900 m), there is an outcrop of rock having jointing that divides the rock into irregular polygons reminiscent of the surface of a pillow basalt (fig. 11–10). Rock outcrops and talus, from a depth of about 3,500 m on the east slope of Mona Canyon, are shown in fig. 11–11. The bedrock appears to contain bedding or shear planes which trend across the field of view from upper

right to lower left and appear to dip steeply toward the upper left. The surface of the outcrop is dark, but recently fallen fragments in the center of the photograph have white, broken sides, suggesting that the rock is a limestone possibly coated with manganiferous material. The square object near the center of the photograph is of unknown origin, but it may well be jetsam.

The exclusive contributions of the photographs in this survey are that whereas the dredging revealed only that Miocene sedimentary rocks occur on the sides of Mona Canyon, the photographs prove that there are *outcrops* of

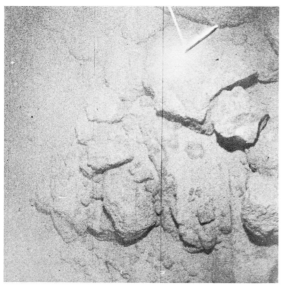
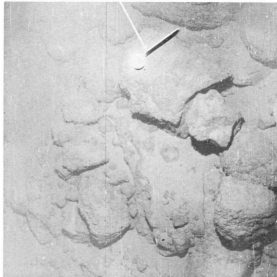

Figure 11–6. Gravel and plate-shaped fragments on the north wall of the Puerto Rico Trench (20°00′N, 66°28′W, *Chain* cruise 19, June, 1961).

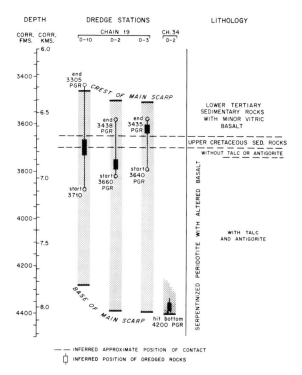

Figure 11–7. Possible stratigraphy of the north wall of the Puerto Rico Trench, constructed on the basis of photographs and dredge hauls.

sedimentary rocks as deep as 3,700 m below sea level, and keep open the possibility that igneous rocks crop out at greater depth.

Western Mediterranean

In the Mediterranean Sea many flat-floored basins are formed by ponded sediments transported largely from the surrounding land areas (Hersey, 1965). Sea-floor photographs have been made in two of the basins (fig. 11–12).

In the Balearic Basin, several tens of photographs were made at each of several locations while the ship was drifting. One location is near the continental slope off Algiers, three are on the abyssal plain — which appears to be a sediment pond — and the remainder are on the Rhone Delta. These photographs illustrate the sort of information sometimes found in photographs which appear from casual inspection to be devoid of interest. All those near Africa and over the sediment pond suggest a uniform mud bottom, without evidence of strong currents, exhibiting complex microrelief that appears to be mainly the result of burrowings by benthic fauna (fig. 11–13). In these same groups of photographs several (17 from a sample of 197 photographs) show small crater-like depressions or holes, each having a small mound of sediment beside it (fig. 11–14). Some

depressions (like fig. 11–14), appear to be freshly made, but others have been partly destroyed. The hole appears to have been dug, and the resulting sediment piled beside it, all in the same act. Fig. 11–15, taken over the Plantagenet Bank near Bermuda, suggests an explanation: a hole and sediment pile is formed as the small fish left of center plunges into the bottom, apparently seeking prey. Such diving fishes were also seen in one of Cousteau's motion-picture sequences exhibited at the Oceanographic Congress at the United Nations in 1959. Features corresponding to the depression and mound combination described above were not observed in any of the 210 photographs taken over the Rhone Delta. Is this an accident of inadequate sampling, or is there an association between the sediment ponds and this type of activity?

The contact between the abyssal plain of the Balearic Basin and the rather complex continental slope west of Italy in the Ligurian Sea was photographed in 1964 (Hersey, 1965), revealing rock outcrops. Some of these were dominantly light-colored, and others dominantly dark. The light-colored outcrop (fig. 11–16a) is covered in part by living animals, possibly coral. On the dark rock (fig. 11–16b) is a dark fossil coral, similar to some that were dredged soon after at a depth of 2,400 m, identified as *Desmophyllum cristagalli*, and dated as 31,800 ± 3,400 years old (Hersey, 1965). The particular value of photographs of *Desmophyllum cristagalli* is that they show the fossil to be in a life-like position, attached

Figure 11–8. Outcrop of bedded sedimentary rock on the west slope of Mona Canyon (18°49′N, 67°33′W; 3,500 m; *Chain* cruise 46, February, 1965).

to local rock, thus removing any question about its having lived where it was found.

In the Tyrrhenian Sea the sediment pond in the deepest basin is surrounded by rugged topography including apparent volcanic cones. Repeated attempts to dredge these slopes have produced only collections of fine sediment. Correspondingly, most of the sizable sample of photographs taken there suggest a complete cover of fine sediment. Nevertheless, Ryan *et al.* (1965) found evidence of outcrops (fig. 11–17) bordering the deepest sediment pond near the mouth of a submarine canyon there. This is one of the earliest examples of photomontage in deep-sea photography. Not only does the montage prove the presence of outcropping rock, but also striking evidence of current action is shown in rippled sediments and other flow patterns about the outcrop.

Plantagenet Bank

In 1959 an intensive survey was made of a portion of the western and southern slopes of Plantagenet Bank

Figure 11–9. Outcrop of bedded sedimentary rock near to that shown in fig. 11–8. (18°49′N, 67°33′W; 3,500 m; *Chain* cruise 46, February, 1965).

Figure 11–10. Outcrop of irregularly fractured rock, possibly pillow basalt, on the west slope of Mona Canyon (18°49′N, 67°33′W; 3,900 m; *Chain* cruise 46, February, 1965).

(near Bermuda), utilizing navigation by sound ranging, bottom photography, and dredging. Fragments of coral, limestone, and basalt (the latter identified by R. M. Pratt, personal communication) were dredged from several places on the steep slopes of the Bank. In all, some 5,000 bottom photographs were taken during lowerings, in depths of water ranging from 60 to 2,000 m. The photographs were taken in stereographic pairs, many of them in color. An impression of the scope of the geological information derived may be had from a selection of four camera profiles shown in fig. 11–18. All are presented without vertical exaggeration. The pattern along the profiles describes, in general terms, the corresponding photographs, examples of which are shown in figs. 11–19 to 11–38, located as indicated on fig. 11–18.

Many of these examples, in addition to illustrating the geological setting of the side of the sea mount, contain photographs of various animals which abound on these slopes. The larger, more prominent animals in figs. 11–19 to 11–38 are identified to some degree in the figure captions. The limitations and pitfalls in identifying animals from photographs alone are discussed in chap. 19.

The profiles suggest a steep slope at shallow depth, a more gentle slope starting at a depth of 500–700 m and terminating at about 1,200 m and, below 1,200 m, a very steep slope. The geological interpretation of the photographs themselves is greatly aided by the color photographs of profiles *D*, *H*, and *G*. Profile *C*, which consists

Figure 11–11. Rock outcrop and talus on the east slope of Mona Canyon (18°49'N, 67°20.5'W; 3,500 m; *Chain* cruise 46, February, 1965).

of black-and-white photographs, is mainly useful for distinguishing relief and features which are made obvious by their textures, such as the spheroidal concretions which abound at shallow depth, sand ripples, and fracture patterns in rock. In this geological setting very light-colored and very dark rocks, limestones and volcanics, and their clasts are the principal types. Thin coatings of sediment on either type of rock, or encrustation of manganese oxide, makes even tentative identification difficult. Nevertheless, color information, with the aid of such features as texture, evidence of layering, joints or fractures, or formational features like sand ripples, permits

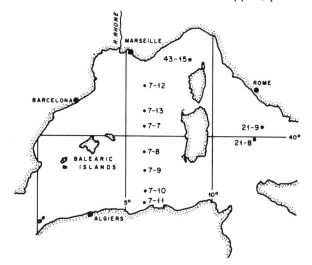

Figure 11–12. Locations of camera stations in the Balearic Basin (*Chain* cruise 7, in 1959), the Tyrrhenian Sea (*Chain* cruise 21, in 1961), and the Ligurian Sea (*Chain* cruise 43, in 1964).

quite reasonable distinctions to be made among types of rock and types of sediment. These classifications can be made with far greater confidence with the aid of color because with it one can generally distinguish between a light-colored rock and a dark rock, regardless of the effects of light and shadow. Some of the concretions and coral clasts appear to have a greenish or brownish hue (figs. 11–24 and 11–32) which may correctly indicate the coating on them, while some other concretions appear either lighter or darker. Otherwise, there is little striking color information from the rocks or sediments. Nevertheless, one can see from the color of many of the animals that the technique will provide color data if available in the subject.

The photographs of camera profile *D*, during which the camera rig was towed downslope from 200 to 1,000 m southeastward from Plantagenet Bank, reveal the following:

1. Numerous spheroidal concretions, 2 to 10 cm in diameter, are piled on the bottom (fig. 11–19) at depths less than 200 m. The concentration of them decreases with depth to 300 m, but similar objects are found here and there at greater depths. William Sutcliffe (personal communication) identified these objects as algal balls.

2. Light-colored rock and sediment form a long, steep slope between 200 and 650 m (figs. 11–20 to 11–23). The rock, nearly bare in places, dominates the upper (and steeper) slope, and the sediment covers much of the lower, less steep portion.

3. A steep cliff terminates the gentle slope of (2) at 650 m. No photographs of it were obtained, but it is evident from the echo-soundings.

Figure 11–13. Complex microrelief, probably caused by burrowing animals, on the floor of the Balearic Basin (42°00′N, 06°00′E; 2,450 m; *Chain* cruise 7, July, 1959).

4. Dark, rounded, spheroidal objects, finely pitted, are abundant in photographs taken below the cliff of (3) (fig 11–24), and massive dark rock surfaces, possibly basalt flows, were photographed in this region (figs 11–25 to 11–27). Patches of sand show by ripple formations (fig. 11–28) the presence of bottom currents at 800 m. More of the concretions intermixed with dark sub-angular rock fragments and abundant benthic life (fig. 11–29) are interspersed with dark rock outcrops (figs. 11–30 and 11–31) between 800 and 900 m. A gravel of coral fragments (fig. 11–32) was found still farther downslope (about 950 m) in an area where dark rocks are predominant.

5. Marked lineations in light and dark sediment (fig. 11–33), were found on a band over 100 m wide at about 850 m on profile *H*. Sand ripples of various forms in light-

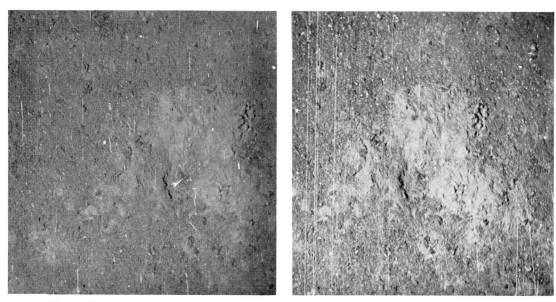

Figure 11–14. A depression and accompanying accrual of sediment which may have been made by a fish plunging into the bottom in search of food as shown in fig. 11–15. Features similar to this were observed in 17 out of 197 photographs taken over the abyssal plain of the Balearic Basin (39°30′N, 06°00′E; 2,370 m; *Chain* cruise 7, June, 1959).

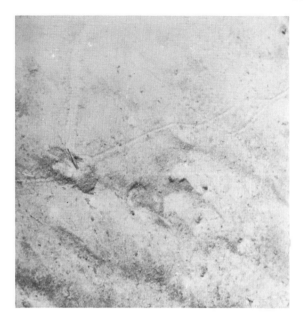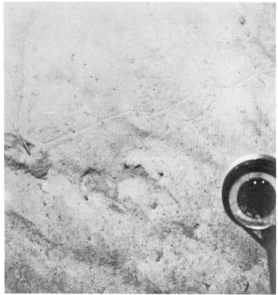

Figure 11–15. A small fish plunging into the sediment-covered bottom, presumably in search of food. Taken over Plantagenet Bank, south of Bermuda.

colored sediment characterize most of the photographs of profile *G* between depths of 950 and 1,200 m (figs. 11–37 and 11–38).

6. Rock outcrops (fig. 11–34) mark the change at 1,200 m from gentle to steep slope on profiles *G* and *H*, and light-colored rock and sediment intermingled with gravels of dark rock (fig. 11–35) show that these sediments both accumulate and are eroded on the steep slope below 1,200 m.

7. Markings having patterns somewhat similar to ripple marks were photographed several times at depths below 1,500 m on profile *H* (fig. 11–36).

These photographs may be interpreted as follows: The upper part of the slope is composed of bedded calcareous rock having varying degrees of compaction and cementation, probably constituting part of a limestone cap which extends over Plantagenet Bank and Bermuda itself. The cap overlies the main mass of the Bermuda-Plantagenet Bank rock, which is of volcanic origin. Outcrops and scree of the underlying volcanic flows, seen below 500 m, form the massive black rocks and fragments seen in photographs from the deeper part of the lowering. The alga which forms the balls noted on the upper part of the slope must have lived in shallower depths and rolled downslope as far as 300 m. The lineations and ripple marks in the sedimentary material are evidence of bottom currents (measured currents in the vicinity are as high as 1 knot to depths greater than these profiles). The dark spheroidal bodies at depth may be manganese nodules, which also grow in sites where currents are active, or they may be manganese encrusted algal concretions that have rolled downslope beyond the area of gentle slope. The light-colored sediment found in patches at all depths is probably calcareous sand or mud formed at shallow depths. A similar origin is suggested for the coral fragments such as those in fig. 11–32. The dark massive rock may well be the shallowest local exposure of the basaltic rock which presumably forms the pedestal of the Bermudas. The strange looking scene of fig. 11–36 may be a sand-covered area where currents were actively re-forming the ripples when photographed. This suggestion is supported by the fact that all photographs of profile *H* below that of fig. 11–36 had this appearance.

Seychelles-Mauritius Ridge

A complex arc of ridges, peaks, and shallow banks, flanked by water over 3,860 m deep, extends from the Seychelles Platform on the north to the Saya de Malha Bank, and southward to near the Island of Mauritius in the western Indian Ocean (fig. 11–39). Several of the Seychelles islands are underlain by Precambrian granite, a unique occurrence in the islands of the world remote from continental masses. Although a few other islands in the world, such as Madagascar and Japan, have granite exposed on them, these islands are closer to a continent and are separated from it by water no deeper than 1,830 m.

Many attempts were made (*Chain* cruise 43, in 1964) to dredge rock from the flanks of the Seychelles-Mauritius Ridge in order to obtain samples that might give clues concerning the age, structure, and history of this ridge. These attempts were largely unsuccessful, resulting in either the return of an empty dredge, or no dredge; a badly bent pinger, and a twisted tow cable. Several successful series of bottom photographs suggest that the dredging difficulties were due to a general scarcity of loose rock material, the bottom being usually composed

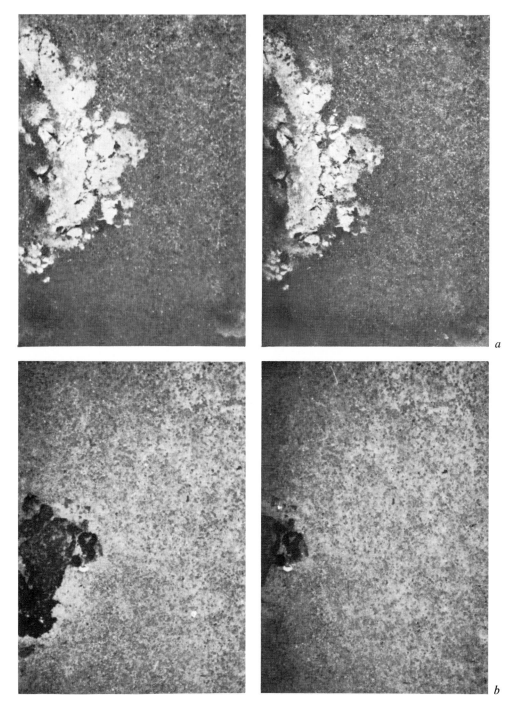

Figure 11–16. Photographs of the sea floor near 43°22′N, 8°40′E, on the edge of the Balearic Abyssal Plain. (*a*) shows a light-colored rock outcrop with unidentified living animals, probably coral, (*b*) shows fossil corals which have been identified in dredge samples as *Desmophyllum cristagalli* (*Chain* cruise 43, July, 1964).

of sand or bare outcrops of rock. This problem was realized aboard the ship, but the value of rock samples from these outcrops was considered such that dredging attempts were continued.

All the samples obtained on and along the flanks of the Seychelles-Mauritius Ridge with two exceptions (one small piece of granite southeast of Seychelles Platform and one possibly volcanic rock obtained northeast of Saya de Malha Bank) are limestone, coral fragments, or calcareous sand whose foraminifera have been identified as Recent by William A. Berggren (personal communication).

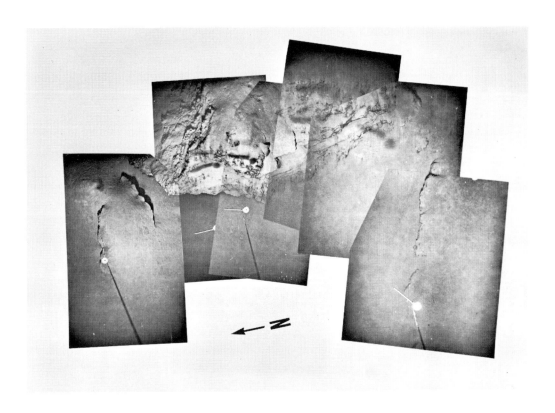

Figure 11–17. Photomontage of probable sedimentary rock outcrops at the mouth of a canyon leading onto the Tyrrhenian Abyssal Plain from the northeast (40°27′N, 12°49′E, *Chain* cruise 21, November, 1961) (plate 5 of Ryan, Workum, & Hersey, 1965).

The dredged material, combined with evidence from photomontages of the sea bottom, suggests that the Seychelles-Mauritius Ridge is almost completely covered by at least a veneer of calcareous rocks and sediments. Summaries of the preliminary results of the investigation of the Seychelles-Mauritius Ridge conducted from *Chain* have been reported by Bunce, Bowin, & Chase (1966) and Bowin, Bunce, & Chase (1965).

Figs. 11–40 and 11–41 reproduce portions of one of the photomontages of the sea bottom obtained during station number 46. The left edge of fig. 11–40 is but one photograph removed from the right edge of fig. 11–41. That the rocky portions of the slope are maintained bare of sediment by currents is evident from the complex pattern of ripple marks in sand which has been collected locally at a break in slope (fig. 11–40). The intricate pattern revealed in the ripples by means of the montage of overlapping pictures is stern warning against hurried interpretation of current direction from isolated photographs. The rock outcrop in fig. 11–41 shows horizontally bedded rock to underlie two ridges and two troughs. The rock outcrop is considered to be limestone on the basis of its appearance and the dredged material obtained in this area. The prominent oblong block about 3 m long shown in the photomontage lies in a closed depression along the axis of one of the troughs. It appears that currents have moved about the block so that it has abraded the depression in which it lies. This is similar to the formation of a pothole in a stream bed, but in the case of the oblong boulder the process is taking place 2,013 m below sea level.

Photomontage strips in stereographic coverage also provide the possibility of detailed charting of slopes by the simple extension of the photogrammetric techniques described in chap. 5.

References

Bowin, C. O., E. T. Bunce, & R. L. Chase, 1965: Structure of the Seychelles-Mauritius Ridge (Abstract). *Trans. Am. Geophys. Union*, **46**, (1), 102–3.

——, A. J. Nalwalk, & J. B. Hersey, 1966: Serpentinized peridotites from the north wall of the Puerto Rico Trench. *Bull. Geol. Soc. Am.*, **47**, 257–70.

Bunce, E. T., C. O. Bowin, & R. L. Chase, 1966: Preliminary results of the 1964 cruise of R/V *Chain* to the Indian Ocean. *Phil. Trans. Roy Soc.*, **259**, 218–26.

Ericson, David B., Maurice Ewing, Coesta Wollin, & Bruce C. Heezen, 1961: Atlantic deep-sea sediment cores. *Bull. Geol. Soc. Am.*, **72**, 193–286.

Hersey, J. B., 1965: Sedimentary basins of the Mediterranean Sea. *Proceedings of the 17th Symposium of the Colston Research Society*, held in the University of Bristol, April 5–9, 1965, 17, 75–91.

Ryan, William B. F., Fifield Workum, Jr., & J. B. Hersey, 1965: Sediments on the Tyrrhenian Abyssal Plain. *Bull. Geol. Soc. Am.*, **76**, 1261–82.

Sykes, Lynn R., & Maurice Ewing, 1965: The seismicity of the Caribbean region. *J. Geophys. Res.*, **70**, no. 20, 5065–74.

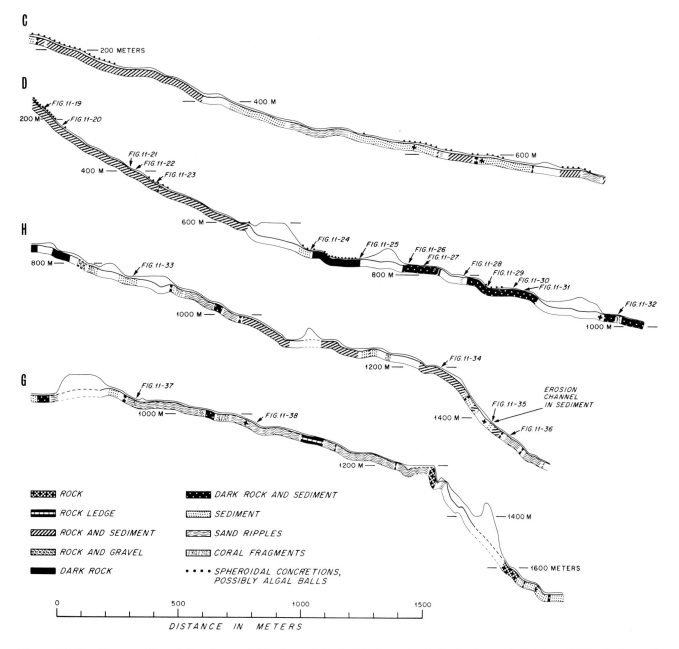

Figure 11–18. Four profiles of the slopes of Plantagenet Bank (*Chain* cruise 9, September and October, 1959). Each profile corresponds to the track of the ship during a camera station. About 500 exposures are taken during each station, while the ship proceeds at approximately 1 knot. The line above the profile of the slopes shows the position of the camera which, for best focus, was 8 to 12 ft above bottom. The profiles are constructed without vertical exaggeration and show the bottom types observed in each series of photographs. A selection of the photographs is presented in figs. 11–19 to 11–38.

Figure 11–19. Spheroidal concretions, probably algal balls, which are presumed to have been formed on Plantagenet Bank and subsequently migrated, in part by current action and in part by rolling down the slope (depth 155 m, average slope 32°).

Figure 11–20. Bare rock surface, with a crustacean and gorgonian (probably *Virgularia*) (depth 230 m, average slope 31°).

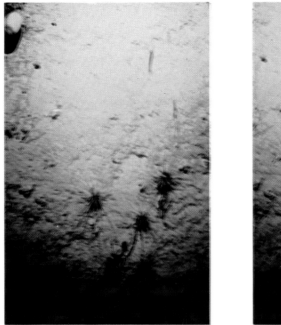

Figure 11–21. Sediment, probably as a thin cover over rock, with echinoderms (probably *Aspidodiadema*) (depth 390 m, average slope 31°).

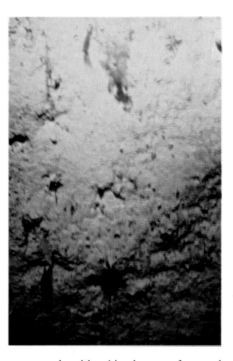

Figure 11–22. Sediment, probably as a thin cover over rock, with echinoderms, a fan coral (near bottom left) and probable Bryozoan (top center) (depth 400 m, average slope 20°).

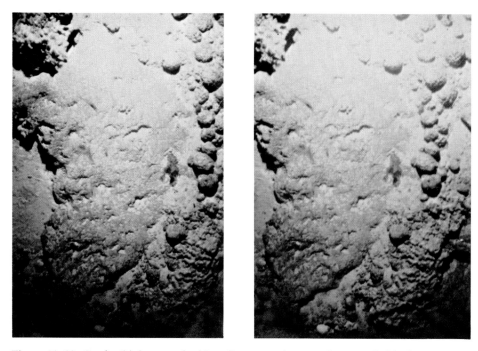

Figure 11–23. Rock, thinly coated with sediment, and concretions, probably from shallow depth. The prominent red animal is an unidentified shrimp (depth 450 m, average slope 16°).

Figure 11–24. Rock and shell fragments, with concretions (depth 710 m, average slope 17°).

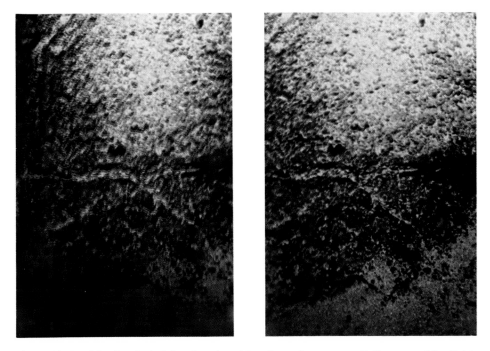

Figure 11–25. Massive dark jointed rock, with vein or fracture, partially covered by light-colored sediment. The animal is a siliceous sponge (depth 745 m, average slope 0°).

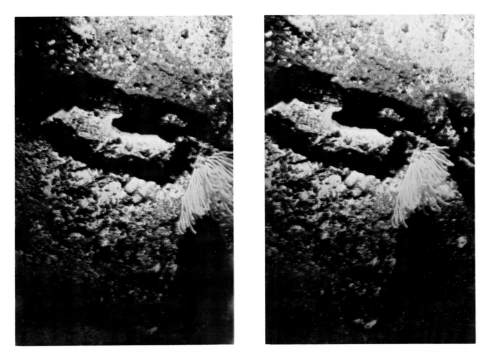

Figure 11–26. Dark rock ledge, massive and nodular, with a crab, gorgonian coral (probably *Calyptrophera*), and small, red crustacean. (depth 765 m, average slope 5°). The crab has been identified as Portunidae? *Bathynectes*, probably *Bathynectes superbus* (Costa) (H. B. Roberts, personal communication).

Figure 11–27. Dark rock, having a rough, ridged structure suggestive of lava flow (possibly this is aa). Here the camera rested on its side on bottom for nearly a minute. The orange gorgonian in the foreground assumed quite different positions in three successive photographs suggesting continual motion, presumably either in search of food or in response to current action. Other animals are a feather star, sponges, and an alcyonanian coelenterate, probably a gorgonian (far right) (depth 770 m, average slope 3°).

Figure 11–28. Sand-rippled surface, presumably calcareous, with an unknown animal (depth 810 m, average slope 15°).

Figure 11–29. Fragments of dark rock and various shells. The delicate, light-blue structures have been found in many photographs of this series, (as in fig. 11–30 below) and may possibly be *Cephelodiscus protocordata* (depth 855 m, average slope 5°).

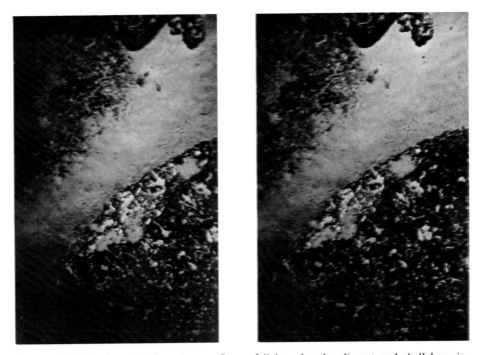

Figure 11–30. Dark rock ledge above a floor of light-colored sediment and shell breccia. A large antedonid feather star adheres to the rock face in the lower left corner (depth 855 m, average slope 3°).

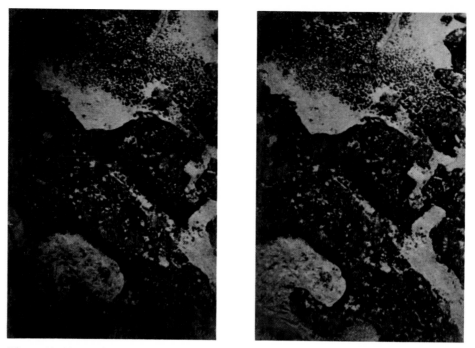

Figure 11–31. Dark rock, possibly volcanic, partially covered by light-colored sediment (depth 863 m, average slope 4°).

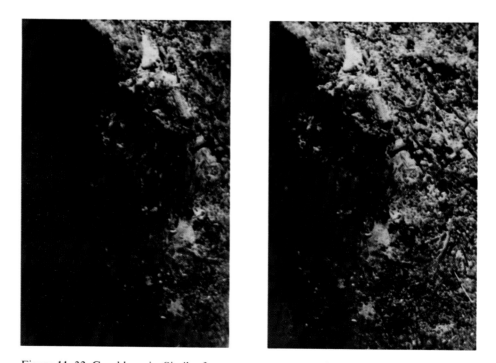

Figure 11–32. Coral breccia. Similar fragments were recovered by dredging near this location (depth 957 m, average slope 11°).

Figure 11–33. Striations in light-colored sediment, with pebbles and cobbles, suggesting mass movement of sediment as in a slump. This was selected from a group of thirteen similar consecutive photographs. The animal is a siliceous sponge (depth 858 m, average slope 6°).

Figure 11–34. Dark rock covered by light-colored sediment. The strong lineation could be either sedimentary or volcanic (depth 1,210 m, average slope 29°).

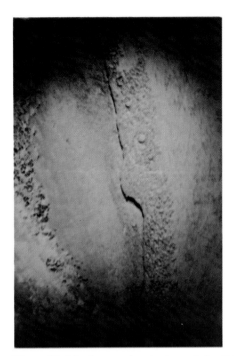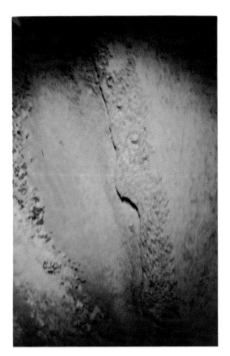

Figure 11–35. An erosional channel in light-colored sedimentary rock. Note the dark-colored pebbles on the left. The apparent slab of light-colored rock at right may be a layer of rock, or it may be an indurated crust forming the top of the sediment (depth 1,430 m, average slope 40°).

Figure 11–36. Many photographs of this type have been obtained. The scene is unfamiliar. The lineations suggest sand ripples, and yet the pictures do not look like sand ripples as do figs. 11–28 and 11–37. Possibly these are sand ripples photographed during a considerable motion of the sediment caused by bottom currents (depth 1,480 m, average slope 39°).

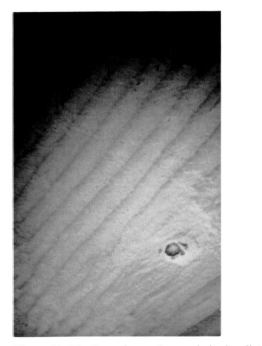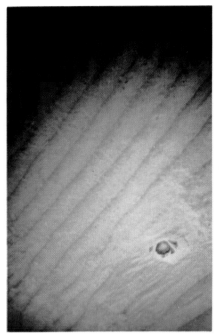

Figure 11-37. Severely regular sand ripples disturbed by a crab, identified as Geryonidae: *Geryon*, probably *Geryon quinquedens* Smith (H. B. Roberts, personal communication) (depth 950 m, average slope 22°).

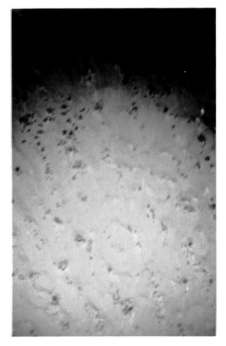

Figure 11-38. Irregular sand ripples over a gravel of dark rock (depth 1,048 m, average slope 33°).

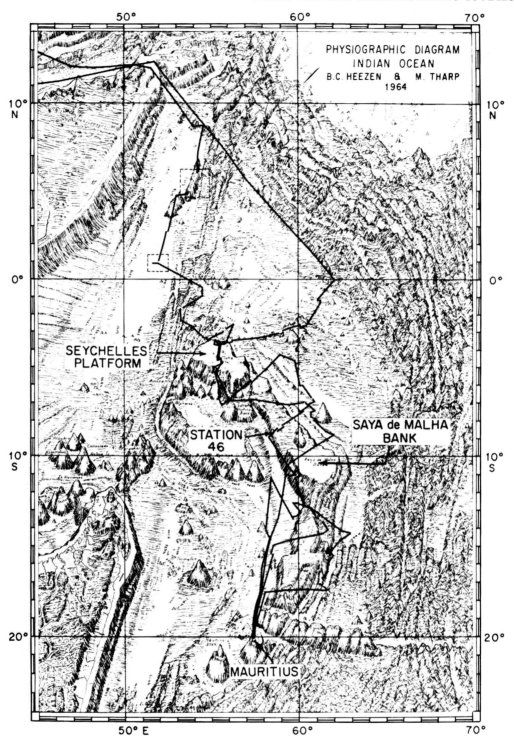

Figure 11–39. Portion of physiographic diagram of the Indian Ocean, showing the location of camera station 46. Photomontages obtained during this station are shown in figs. 11–40 and 11–41. (Published by the Geological society of America. Copyright, 1964, by B. C. Heezen and Marie Tharpe. Reproduced by permission.)

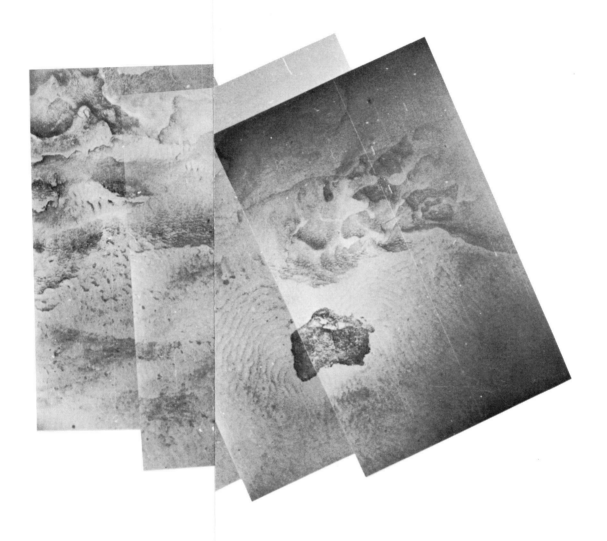

Figure 11–40

FEET

Figure 11–41

12. Advantages of color photography on the continental shelf[*]

J. V. A. Trumbull   *U.S. Geological Survey, Woods Hole, Massachusetts*

K. O. Emery   *Woods Hole Oceanographic Institution, Woods Hole, Massachusetts*

Abstract

Underwater color photographs of the continental shelf can provide information that is not readily obtainable from black-and-white photographs. Color improves differentiation between rocks of diverse types, between sands of contrasting mineral composition, and between encrusted and recently abraded gravels and rocks. Living and dead organisms may be distinguished by color, and the relative age of shell fragments can be estimated. Underwater color photographs of the continental slope and the deep-sea floor, as well as of shallow mud bottoms on the continental shelf, appear to present little advantage over black-and-white ones.

Fewer than one per cent of the tens of thousands of existing photographs of the sea floor were made with color film. This lesser use of color pictures is due to (1) difficulty in processing aboard ship in order to check equipment as work proceeds; (2) slow film speed prior to about 1960; (3) high cost of printing in journals and books; and (4) little or no improvement over black-and-white film for exposures at great depth.

A study of the geological history of the Atlantic continental margin of the United States was begun in 1962 as a joint effort of the U.S. Geological Survey and the Woods Hole Oceanographic Institution. About 2,000 bottom samples were obtained, mostly with a large clam-shell grab-sampler containing a light source and a camera designed to trip about a meter above the sea floor. Immediately afterward, the grab reaches the bottom and samples a 0.6-m area about equal to that of the photograph. A comparison of the photographs and corresponding samples was published (Emery, Merrill, & Trumbull, 1965); the present report is intended only to describe some of the color photographs independently of the samples.

About thirty-five acceptable color photographs were obtained in a few rolls of 35-mm color film (Kodak High-Speed Ektachrome) that were interspersed with many rolls of black-and-white film. The color film has the same sensitivity (ASA 160) as the black-and-white, so no change in the camera settings was required. The color photographs are from a depth range of 30 to 178 m, 75 per cent being between 30 and 60 m. The camera is a Robot Star having a 30-mm lens. It is enclosed in a brass Ewing-type pressure housing (Ewing, Woollard, Vine, & Worzel, 1946) that can safely be used to a depth of about 600 m. Exposures are made at $1/100$ sec and f/4. A fairly fast shutter speed is used with the strobotron light source because of the presence of effective amounts of ambient light atop the continental shelf. The spring-loaded, automatic-transport camera permits about fifteen stations to be occupied without need for opening the camera pressure housing.

Use of color rather than black-and-white film for bottom photographs on the continental shelf presents advantages for both geologic and biologic investigations. Examples are taken from the small collection at hand. The general lithology of gravels or rock outcrops can locally be determined from color photographs (fig. 12-1) in areas where some samples are available for calibration and where there is no concealing cover of attached organisms. Color may also reveal the presence of sands of differing mineral composition in the troughs and crests of sand ripples (fig. 12-2), where black-and-white pictures may give shades of gray suggestive only of variation in illumination. Similarly, color photographs can distinguish between gray detrital sand; yellow, iron-stained, relict sand; greenish-gray glauconitic sand; and white biogenic sand (Emery, 1965); and thus aid in determining the distribution of sediments of various origins on the continental shelves.

Encrustations of calcareous red algae and of other attached organisms on gravels indicate that the gravels

[*] WHOI Contribution Number 1754.
 Publication of this report was authorized by the Director, U.S. Geological Survey.

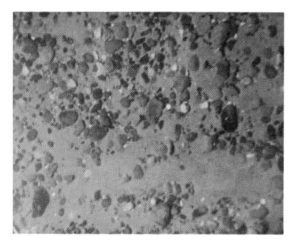

Figure 12–1. Pieces of quartzite and granite (tan and brown) and schist (gray) on a pebble-sand bottom at a depth of 30 m on Georges Bank. Station 1219, 41°31′N, 68°00′W. Scale is approximately the same for figs. 12–1 through 12–5, and is indicated by the 9-cm diameter of the lead trip-weight in 12–2, 12–4, and 12–5.

Figure 12–2. Current ripples in yellow, iron-stained, relict, quartz sand. Darker sand in the troughs probably contains an admixture of dark heavy minerals. Taken at a depth of 34 m on Nantucket Shoals. Station 1216, 41°12′N, 69°16′W.

Figure 12–3. Encrusting calcareous red algae on gravel at a depth of 35 m off Cape Sable, Nova Scotia. Station 1240b, 43°20′N, 65°47′W.

Figure 12–4. The number of light-brown, living, sand dollars far exceeds the number of white dead ones, as illustrated in this picture from a depth of 61 m on the central continental shelf south of Long Island, New York. Station 1283, 40°12′N, 73°45′W.

are moved about only rarely (fig. 12–3). Color film permits the encrustations to be identified far more reliably than does black-and-white. Another question of interest to both geologists and biologists is the relative number of dead and living shells on the bottom. Use of bottom photographs, particularly of colored ones (fig. 12–4), revealed that only about 4 per cent of the sand dollars (*Echinarachnius parma*) on the continental shelf off New England are dead; this implies that the fragile tests break up soon after the death of the animal. Similarly, the presence or absence of the chitinous covering of pelecypod shells (fig. 12–5), or of their leathery hinge, serves to

distinguish the shells of recently dead animals (Merrill & Posgay, 1964) from older ones which are usually white and have no large remaining areas of organic matter.

These examples indicate that color photographs of the sea floor are superior to black-and-white in certain environments; the superiority is not so evident in all environments. In turbid shallow water neither color nor black-and-white yields good photographs, but color is probably somewhat more revealing of organisms and bottom materials. Areas of the continental shelf that have the greatest color contrasts and are the most photogenic are, in decreasing order: rock, gravel, sand, and mud.

Figure 12–5. Chitinous cover on a pelecypod shell, as well as lack of separation of the valves, shows that the animal died only recently. The sandy bottom is at a depth of 53 m on the inner continental shelf south of Long Island, New York. Station 1277, 40°41′N, 72°16′W.

Figure 12–6. Color photograph, in air, of a bottom sample recovered in an almost undisturbed condition in a Smith-McIntyre sampler. The bright colors and sharp images obtained in air contrast with those of even the best underwater photographs. The sample was from a depth of 12 m on the inner continental shelf off South Carolina, 33°43′N, 78°45′W.

Only coral reefs and some other special environments have color contrasts that exceed those of *shallow* rock bottom. Mud bottom, on the other hand, is usually some shade of gray and it contains only grayish plants, worms, and other small animals. The continental slope and the deeper sea floor beyond exhibit the same lack of contrast as shallow mud bottoms, and color pictures present little advantage over black-and-white ones. The water above the deep-sea floor, however, does contain scarlet crustaceans, and water everywhere contains some fish whose recognition is enhanced by color pictures, especially where color patterns are significant. Without extreme care, expense, and time, underwater color photographs made remotely from aboard ship compare poorly with ones made in air in terms of sharpness of image and brilliance of color contrast (fig. 12–6).

References

Emery, K. O., 1965: Geology of the continental margin off eastern United States, in: *Submarine geology and geophysics.* W. F. Whittard and R. B. Bradshaw (eds.). Proc. 17th Symposium of the Colston Research Society, University of Bristol, April 5–9, 1965, 271–85.

———, Arthur S. Merrill, & James V. A. Trumbull, 1965: Geology and biology of the sea floor as deduced from simultaneous photographs and samples. *Limnology and Oceanography*, **10**, no. 1, 1–21.

Ewing, Maurice, G. P. Woollard, A. C. Vine, & J. L. Worzel, 1946: Recent results in submarine geophysics. *Bull. Geol. Soc. Am.*, **57**, 909–34.

Merrill, Arthur S., & J. A. Posgay, 1964: Estimating the natural mortality rate of the sea scallop (*Placopecten magellanicus*). Res. Bull. of the International Commission for the Northwest Atlantic Fisheries, no. 1, 88–106.

13. Photography of seamounts[*]

R. M. Pratt   *Woods Hole Oceanographic Institution, Woods Hole, Massachusetts*

Abstract

Seamounts are very effectively studied by deep-sea photography. Each photograph can be considered as a sample point when used in connection with bathymetric surveys, dredge samples, and cores, thus making it possible to delineate and map geologic and biologic zones on a seamount. Seamounts transcend through a great depth range and are characterized by minimal sedimentation which results in exciting and photogenic differences of environment over a short distance, as typified by our studies of Great Meteor and the New England Seamounts.

* WHOI Contribution Number 1755.

  This study was supported by the Office of Naval Research, under contract Nonr 2196(00) NR 083–004, and the U.S. Geological Survey. The assistance and instructions of H. E. Edgerton and J. B. Hersey in the art of bottom photography is deeply appreciated, as is the patience and understanding of the many crewmen and scientists who made the work at sea possible.

Seamounts are conical morphological features rising abruptly from the sea floor. They transcend vertically thousands of meters through the water column and a few of their crests project through the water surface as oceanic islands. Nearly all are volcanic in origin. Most are characterized by sharp irregular topography but a few, called guyots, have truncated, flat tops. The great contrast of geologic and biologic phenomena from crest to base makes them exciting subjects for underwater photography. Their precipitous sides and the erratic ocean currents around them make any work on them somewhat of a challenge.

In the past few years the Woods Hole Oceanographic Institution has given considerable attention to the photography of seamounts in the North Atlantic (fig. 13–1). Generally, photographs are taken in conjunction with dredging, coring, echo-sounding, and seismic profiling operations. In this way photographs can be related to the

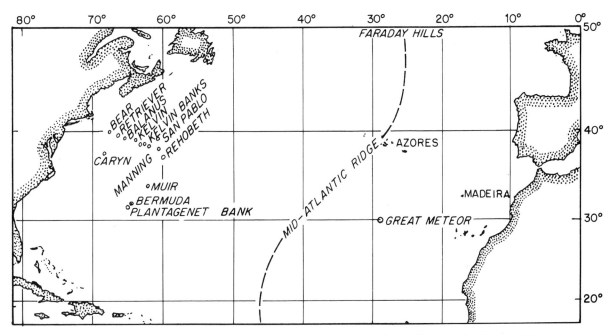

Figure 13–1. Index map showing the location of seamounts in the North Atlantic on which observations have been made by workers of the Woods Hole Oceanographic Institution. Seamounts discovered by early survey ships were generally called banks, i.e., Kelvin Bank and Plantagenet Bank. Others, like Great Meteor Seamount, are flat-topped and often referred to as guyots.

145

form and structure of the seamounts, and positive identification of the sediment and rock types seen in the photographs can be made. The Edgerton camera (Edgerton, 1963) used in the studies is capable of taking several hundred pictures in a continuous traverse. Each picture can be considered as a sample point when augmented by nearby dredge hauls or grab samples. Thus, the photography of seamounts is more than just a means of obtaining interesting pictures; it is a powerful tool for geological and biological mapping and environmental investigations.

Seamounts are uniquely defined topographic features and investigations of them must be planned in relation to their depth and topographic profile. Once the depth and location of the top are established, the relative location of any subsequent work can be easily related to the seamount, regardless of its absolute position. It is often feasible to plant a reference navigation buoy on the summit, and then locations on the sides of the seamount are uniquely established by an azimuth from the buoy and the depth. Even when the buoy is out of sight it is a positive check point to which the ship can return, as exemplified by recent work on Great Meteor Seamount (Pratt, 1965). More complex surveys have used several navigation buoys planted on the summit, and camera traverses were located by a combination of visual bearings, radar bearings, acoustic distances, and depth.

A continuous record of the depth of water is important information when working with cameras on a steep slope; not only does the depth aid in controlling and maneuvering the camera near bottom but it also establishes where the pictures were actually taken in relation to the profile of the seamount. Usually, the ship's echo sounder can be used for only short intermittent periods or it will interfere with the monitoring of the camera. One method used to obtain the true depth of the photographs while monitoring the camera is to plot the time that the pictures were taken along the depth profile of the camera traverse (fig. 13–2). As a safety measure, camera traverses are planned so that the ship slowly tows the camera downslope, either by drifting or under power. The interpretation of the bottom photographs on steep seamounts is complicated by the effect of slope scale changes (due to differences in camera-object distance) as well as by the usual problems of identification of objects in the photographs. A few representative samples of the objects photographed are an immense help in interpretation (figs. 13–3 and 13–4). Dredged objects are especially helpful in estimating the scale of the photographs, because it is difficult to judge how far off bottom suspended scales are, and objects dragged on bottom stir up suspended mud which clouds the picture. For the purpose of scale, certain species of brittle stars and coral with predictable size ranges are particularly useful, and, when available,

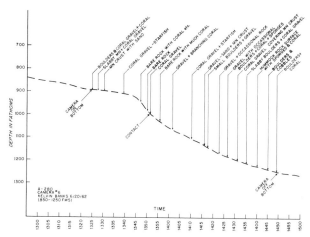

Figure 13–2. Camera lowerings with an Edgerton-type camera can last two hours and consequently photograph a considerable range of depths on a steep-sided seamount. This illustrates a simple graphic means of locating pictures along a depth profile. The depth profile is obtained from intermittent echo sounding and the time is photographically recorded in the camera. This depth profile can be fitted to a chart of the seamount to locate the position of the traverse. More complex systems using triangulating acoustic methods have been used to locate the exact position of the camera itself.

stereoscopic pairs of photographs aid greatly in the interpretation of scale and relief (see chaps. 5 and 7).

A typical seamount rises as a conical prominence from the ocean floor. If its summit reaches into the photic zone, reef coral and algae may cover its top (fig. 13–5). Even the crests of deeper seamounts have accumulations of coral and organisms of deepwater habitat (fig. 13–6). Globigerina and pteropod ooze commonly are the only sediments that can be deposited near the tops of seamounts because terrigenous, or land-derived, sediments are believed to be carried out primarily by turbidity currents and therefore confined to the abyssal floors. Often the carbonate remains of bottom-dwelling coral and hydroids augment the pelagic sand. Current-produced sand ripples are commonly observed in the summit sediment (fig. 13–7) because the projecting seamount obstructs the general water movement causing erratic currents that winnow out the fines. A few seamounts, especially guyots, seem to be capped with ooze that has been cemented into limestone (fig. 13–8 and fig. 13–9). The induration of soft carbonate sediment into limestone is probably connected with the subareal planation of the guyots.

The sides of all seamounts descend abruptly, resulting in the frequent exposure of hard rock. On the Great Meteor Seamount (fig. 13–10) and in the Mid-Atlantic Ridge area in general, fresh rock is usually exposed on the sides and numerous dredge hauls indicate that the

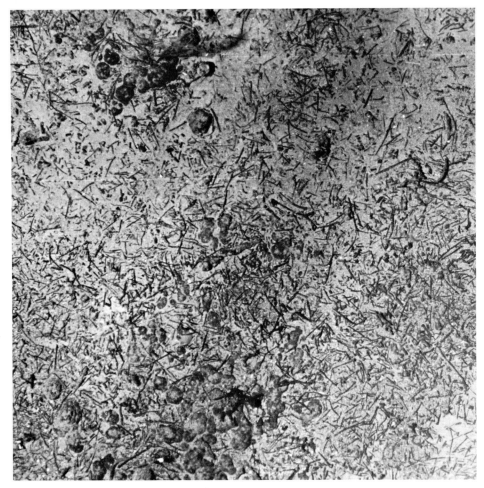

Figure 13–3. Coral debris and manganese nodules on the summit of Muir Seamount. The photograph should be compared with the dredge sample (fig. 13–4) taken at the same station. It often happens that much of the photographed area is covered with sediment that is virtually impossible to identify with certainty unless an actual sample is taken.

exposures are primarily basaltic. On others, as for example the New England Seamounts chain and Muir Seamount, most rock outcrops are so encrusted with manganese oxide that sampling of the bedrock is difficult. Other seamounts have yielded only altered rock. Rocky side slopes of seamounts are the best area for taking biological pictures because of the dominance of photogenic sessile organisms, and the apparently increased productivity around some seamounts when compared with the nearby ocean. However, where the exposed rock is covered with crusts of manganese oxide, not only is the sampling difficult but the substrate seems to be a relatively poorer support for sessile life. Toward the base of a typical seamount, sediment and debris from the sides tend to accumulate and photographs reveal a predominantly sedimentary ocean floor (fig. 13–11).

In our work on the New England Seamount chain (fig. 13–12) photographs have been the most instructive means of obtaining information. One of the objectives in the study of these seamounts was to find flat-topped peaks that might be related to relatively lower sea levels of the past. In this respect conclusive evidence was not obtained. Instead of vertical zonation, the photographs revealed a lateral environmental transition from deep water conditions characterized by pelagic ooze and manganese crusts on the outer seamounts (figs. 13–13 and 13–14) to slope conditions on the seamounts near shore (fig. 13–15). It is interesting that the seamounts can be used as environmental indicators of the surrounding and overlying water. In practice, they also tend to make old rocks available for study because seamount deposits are not subject to as rapid burial as the surrounding basin deposits.

This discussion on seamounts must conclude with the recurring thought that photographs have presented as many new problems as they have helped to solve old ones.

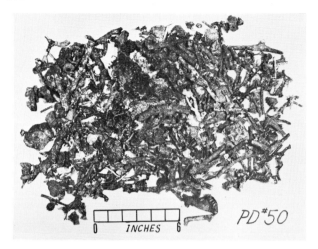

Figure 13–4. Manganese-encrusted coral and rounded manganese nodules dredged from the summit of Muir Seamount. From this type of sample we can judge reasonably well the scale and material photographed (see fig. 13–3) in the area, and conversely the photographs help us to judge the distribution of the dredged material and mode of occurrence on the bottom.

They have shown quite conclusively that on a favorable substrate, like rocky areas of seamounts, sessile organisms grow abundantly on the available surface. Photographs have demonstrated more than any other means the activity of currents at all depths in the ocean. Above all, photographs have shown the abundance of hard rock cropping out on the sides of the seamounts; the distribution and variety of hard crystalline rock is fundamental to our understanding of the recent tectonic and volcanic history in the ocean basins.

References

Edgerton, H. E., 1963: Underwater photography. In *The sea; ideas and observations on progress in the study of the seas.* Vol. 3, *The earth beneath the sea.* M. N. Hill (ed.), Interscience Publ., New York, 473–79.

Pratt, Richard M., 1963: Great Meteor Seamount. *Deep-Sea Res.*, **10**, 17–25.

Figure 13–5. The shallow zones where sunlight is important are characterized by algae and light-dependent, reef-coral growth as well as abundant animals. This reef environment on Plantagenet Bank, near Bermuda, is within the diving depth of skin divers. It should be compared with fig. 13–6 which is far below diving depth.

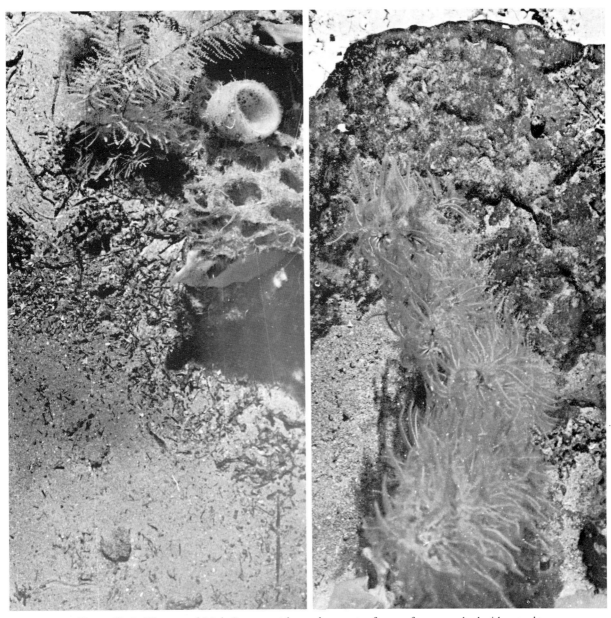

Figure 13–6. The top of Muir Seamount has a deep-water fauna of sponges, hydroids, coral, and crinoids analogous in many ways to shallow-water reefs. As on many seamounts, manganese encrusts most of the hard surfaces. Sponges, crinoids, and ophiuroids seem to be particularly abundant in deep-water environments in contrast to molluscan fauna which is scarce.

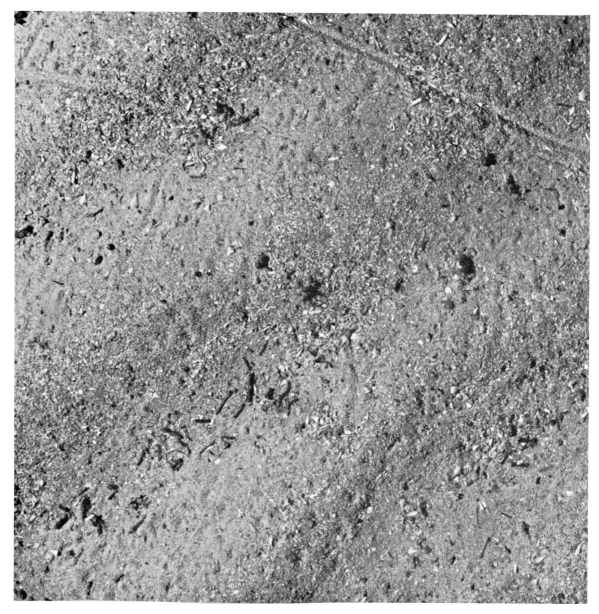

Figure 13-7. Rippled pelagic sand composed of the remains of Globigerina and pteropods. The pteropod shells are comparatively large and light and are somewhat differentially sorted into the troughs of the ripples. Seamounts, even near shore, have true pelagic sediments because they project above the ordinary range of terrigenous turbidites.

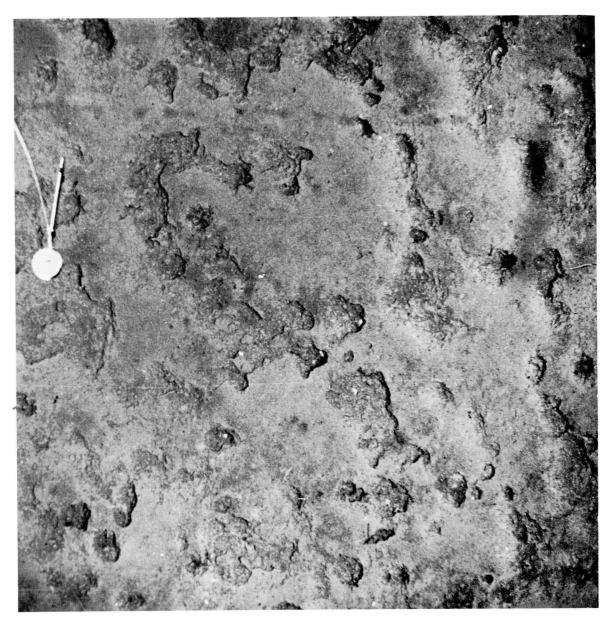

Figure 13–8. This photograph and the two following (figs. 13–9 and 13–10), were taken on the same camera lowering. They illustrate the manner in which zones of various rock types can be mapped from a photographic sequence down the sides of a seamount. The rock here is well-cemented, solution-pitted limestone on the upper terrace zone of Great Meteor Seamount.

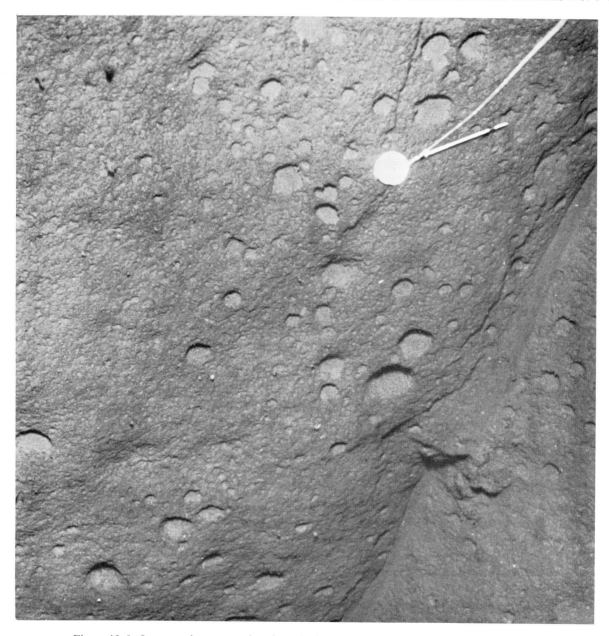

Figure 13–9. Loose carbonate sand and mud lying on a steep slope on the side of Great Meteor Seamount. The small pits and holes, probably made by animals, and the dammed sediment in the small gully indicate the steepness of the slope. In these photographs the 11-in compass vane gives some idea of the scale, but of course it is suspended somewhere short of the bottom.

Figure 13–10. Basalt, at the deeper end of the same lowering as the previous two pictures. Dredged samples and the characteristic ribbed surface indicate this is probably the glassy surface of an actual submarine volcanic flow. We have found that sessile organisms, like the sponge and coral shown here, are associated with rock outcrops at all depths.

Figure 13–11. An abyssal environment at the base of Rehoboth Seamount. Here the rounded nature of the boulders indicates that they may be talus from further up-slope, but the origin and distribution of large-sized sediment in the ocean is always somewhat of a problem. The pelagic ooze being deposited over the bottom is similar to that in the shallower pictures.

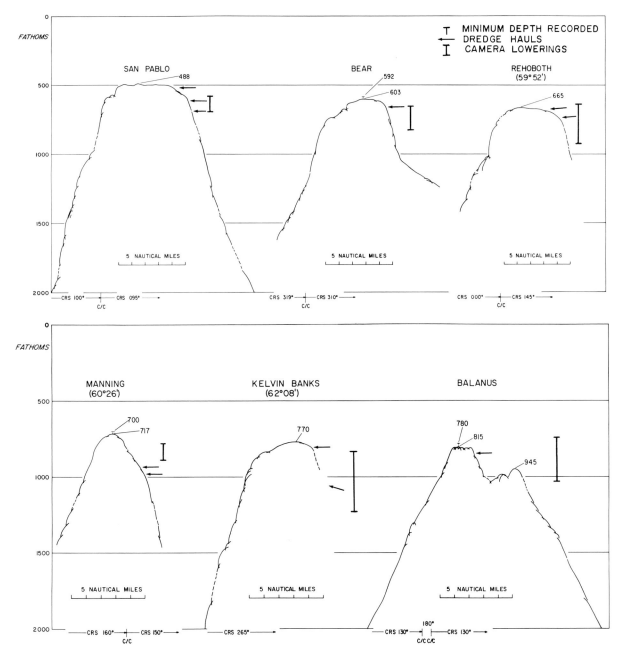

Figure 13–12. Cross sections traced from echo-sounding profiles across the six shallowest of the New England Seamounts on which are located the relative position of camera and dredge observations. Photographs have their greatest value when closely tied to sampling and bathymetric surveying. The next three figures (figs. 13–13 to 13–15) illustrate some of the types of bottom photographed during this survey.

Figure 13–13. Rock bottom similar to that photographed on numerous seamounts. In this case there is question about the nature of the rock since the only dredged material consisted of manganese slabs. The difficulty of sampling the rock is readily apparent from the photograph. The paucity of sessile organisms supports the hypothesis that this is largely a manganese-encrusted bottom.

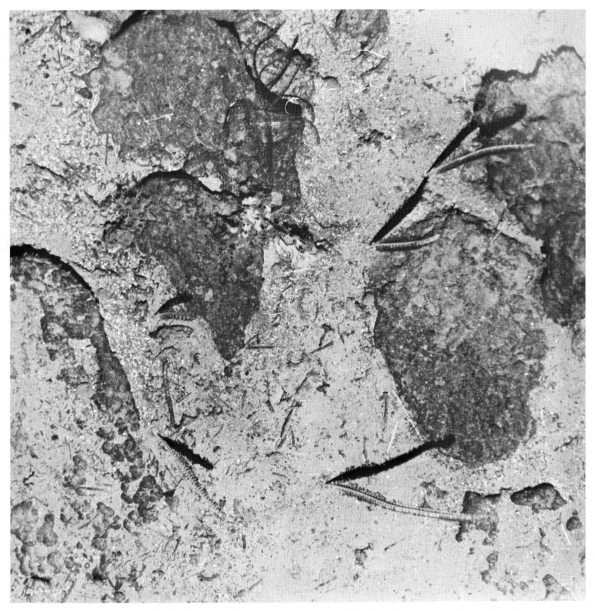

Figure 13–14. The hard rock of the seamount is usually draped with pelagic sediment as shown here. However, the numerous rock exposures indicate that the sediment must be largely removed from the tops and sides of the seamounts. A large crinoid, sponges, and sea pens make their home here.

Figure 13–15. The New England Seamounts extend into the slope environment of the New England coast. Unlike the typical offshore seamounts those near the coast are covered with glacially ice-rafted cobble and terrigenous sand. In some ways the contrast between the inshore and offshore environments is greater than between the various depths of the seamounts in this area.

14. Effects of tidal currents on the sea floor shown by underwater time-lapse photography[*]

David M. Owen, K. O Emery, and L. D. Hoadley† *Woods Hole Oceanographic Institution, Woods Hole, Massachusetts*

Abstract

A six-day series of photographs from a time-lapse camera positioned at a depth of 8.5 m in an area of fast tidal currents revealed movements of the sand bottom caused by the alternating currents. Ripple marks formed and reformed with their steep slopes facing opposite directions for flood and ebb currents, and they moved rapidly downcurrent at times of maximum current. It was deduced that the speed of the surface current was about twice as great when sand began to move as when sand movement ceased.

14-1. Introduction

Time-lapse movie cameras have long been used to depict the actions of slowly moving objects such as the opening of flowers, growth of plants, progression of seasons, and turbulence of clouds. They have also portrayed the movements of slicks on the ocean surface and the covering and uncovering of wide strand zones by the tide.

Perhaps the first application of time-lapse photography to studies of the sea floor occurred during the period 1953-55, when the first and third authors made three field tests in shallow Cape Cod waters at the suggestion of Columbus O'D. Iselin. The camera mechanism and underwater housings (fig. 14-1) were designed chiefly by Hoadley, while the accompanying flash lighting was developed by Harold E. Edgerton of the Massachusetts Institute of Technology. The supporting frame was designed by Owen, assisted by W. T. Hammond (formerly of the Woods Hole Oceanographic Institution).

The first and third field tests, made in Buzzards Bay during 1953 and in Cape Cod Bay during 1955, were mechanically successful but photographically unproductive due to excessive turbidities raised by storms. The second experiment, in 1954, was a success from these standpoints, and the results obtained from the favorable positioning of the camera form the basis for this report.

The choices of camera sites and some of the subsequent operations involved in the use of this camera were aided to a great extent by direct underwater observation. In fact, this project provided the impetus, and the basis, for the later and continued utilization of self-contained (SCUBA) diving in various institutional activities.

14-2. Equipment

In fig. 14-1 the fully assembled time-lapse camera is shown mounted atop its quadruped frame ready for lowering. Two electronic flash sources are attached low on two of the legs. The approximate weight of the entire assembly is 320 kg (700 lb). The frame is about 3 m square at the base and 2 m high (with camera housing). Depicted here is the arrangement used in the third experiment, when camera and lights were focused on an object on the sea bottom outside of the supporting frame; a current meter also appears in the field of view. In the second (Middle Ground) experiment, the camera and lights were aimed toward the bottom within the frame itself, and the current meter was not included. The lens-to-bottom distance was 2 m with the angle of sight approximately 15° from the vertical.

Basically, the mechanism consisted of a standard 35-mm still camera which had been coupled by Hoadley to a 35-mm motion picture magazine containing 1,000 ft of film. This unit was motorized and controlled by an adjustable timer which caused the film to advance nearly 8,000 frames at regular time intervals during a maximum of one month's operation, or a minimum of four days. The flash units, of 100-watt-sec input, were engineered and built by Dr. Edgerton to last the same length of time

* WHOI Contribution Number 1756.

The original camera development and early application described here was supported by the Bureau of Ships under contract NObsr-49058 and by the Office of Naval Research under contracts NGonr-27701, NR 083-004, and Nonr-796(00) NR 083-069. The report presented here was sponsored by contract Nonr-2196(00) NR 083-004 in the case of the first author, and by contract GS-8358 for the second author.
† L. D. Hoadley is now with Alpine Geophysical Associates, Inc., Norwood, New Jersey.

while continuously "on charge." At the conclusion of a maximum run the electronic flashtubes and the dry battery packs, which powered the camera mechanism and two light sources, were discarded and replaced with new equipment.

The pressure cases were designed to operate at depths of 1,800 ft (550 m) with a factor of safety of two.

14–3. Methods

The successful experiment was made in 8.5-m depth at the east end of a long sand shoal named Middle Ground, in Vineyard Sound near the Woods Hole Oceanographic Institution (fig. 14–2). The heavily weighted camera was left on the bottom for six days (0930 Eastern Standard Time on 10 August 1954 to 0900 on 16 August).

Figure 14–1. Time-lapse camera, mounting and lights.

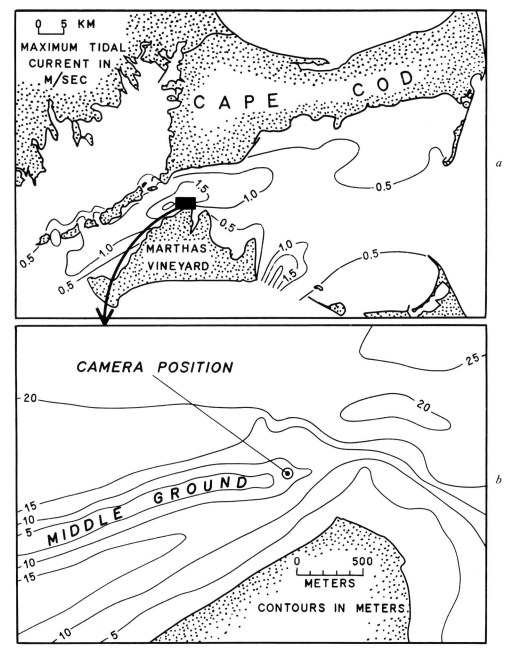

Figure 14–2. Area of time-lapse photography of the sea floor (41°29.15′N or S, 70°36.50′E or W; 8.5 m). (*a*) Vineyard Sound with contours that show the speed of maximum tidal current at maximum flood tide (three hours after low tide at Boston), according to Haight (1938). (*b*) Enlargement of part of Vineyard Sound showing bottom topography with contours in meters; based upon soundings of Coast and Geodetic Survey chart no. 260.

During the six-day period, a total of 4,857 frames of 35-mm, black-and-white, negative film was exposed at a rate of one frame each 106 sec, or 1.77 min. Each frame on the original negative (24 × 36-mm) was elongate along the length of the film for maximum detail in enlargement. In order for the material to be projected as an easily handled 16-mm motion picture, it was "converted" by commercial film-processing companies. Two basic steps were involved in the conversion. In the first treatment (involving the VistaVision process) each frame on the print was rotated 90° with respect to the original negative and slightly reduced to a 35-mm *cine* format, so that its long dimension lay across the film. A second laboratory performed a further conventional reduction in the copy from 35-mm cine to 16-mm cine. Correct spacing between the frames was maintained for successful projection.

When projected at a speed of 16 frames per sec, each second of viewing time corresponds to 28 min of actual time of photography, an increase in rate of about 1,700 times.

Each frame shows an area of bottom equal to 70 by 105 cm. Projection of the film as a movie reveals periods of high turbulence with much sand in suspension alternating with quiet periods of clear water above a stable bottom. Ripple marks were present throughout the entire six days, exhibiting a variety of forms and directions of movements. The nature of the ripple marks and of the currents that produced them were studied by running the film at slow speed through a projector equipped with a frame counter and recording various characteristics for critical sections of the film.

14–4. Results

Ripple marks

Three kinds of ripple marks were observed, each having a different size, shape, position, and time of movement. The largest of these would be better termed sand-waves, having heights of 20 to 30 cm, crest-to-crest wavelengths of about 150 cm, and extending across the entire picture. They moved chiefly during the periods of fast flood and ebb currents, and apparently are permanent features, part of the large sand shoal of Middle Ground (fig. 14–2). Probably one end of each sand wave was attached to the shoal and the other was free to swing back and forth with the changing tidal current. The lateral movement caused the lee side of each sand wave to be steep, at the angle of repose of the sand.

The smallest ripple marks were only about 1 cm high and 6 cm wide, and they lay mostly in the trough between the sand waves and close to their lee slopes. Their presence and movement were evident only during waning stages of the current, after the sand wave had stopped, but before movement of the intermediate-size ripple marks had ceased.

Ripple marks of intermediate size were most common (figs. 14–3 and 14–4). They appeared to be 1 to 2 cm high with a wavelength of about 8 cm, and they generally occurred throughout the area of the picture and at all stages of the current. These ripples were considered to be better indicators of the current direction than were the large sand waves (which were not always present in the field of view) or the small ripple marks (whose position and movement were functions of currents that were partly controlled by the position of the sand waves). Commonly, these intermediate ripple marks exhibited two forms. When the current was fast, the ripple marks were irregular and short crested. When the current slowed and before it again became fast, the ripple marks were long and straight or broadly curved. In the irregular form, sections of ripple marks moved independently and rather rapidly with respect to the rest of the bottom. The movement usually was for one full wavelength, after which the ripple marks suddenly stopped, with a previous ripple mark located down current. In other words, a short section of crest (say a 15-cm length) broke away from a longer crest and advanced one wavelength (about 8 cm) to align itself with the next crest. In the straight form, and especially after ebb currents, the ripple marks moved uniformly and slowly across the field of view in much the same manner as water ripples move across the surface of a pond.

All of the ripple marks moved in the same direction as the current that produced them; thus, they are dunes (not antidunes) in the sense described by Kennedy (1963) and others. The current that produces ripple marks of this type is slower than that which produces antidunes, none of which were observed. In fact, the existing current speeds were insufficient to form even the transitional type of bottom having neither dunes nor antidunes (Simons & Richardson, 1961).

The ripple marks of all sizes tended to remain unchanged throughout the period of slack water. No benthic animals such as mollusks or crabs were observed, probably because the environment of shifting sand was unfavorable to them. Some markings on the sand were made when loose masses of brown algae dragged across the bottom. More interesting were markings made by the tails and fins of three fishes (fig. 14–3b), probably the common, or northern, sea robin, *Prionotis carolinus* (L.). These fishes remained almost motionless on the bottom in the field of view for periods of 16, 60, and 64 min, respectively.

Two sand samples were collected from the site of the

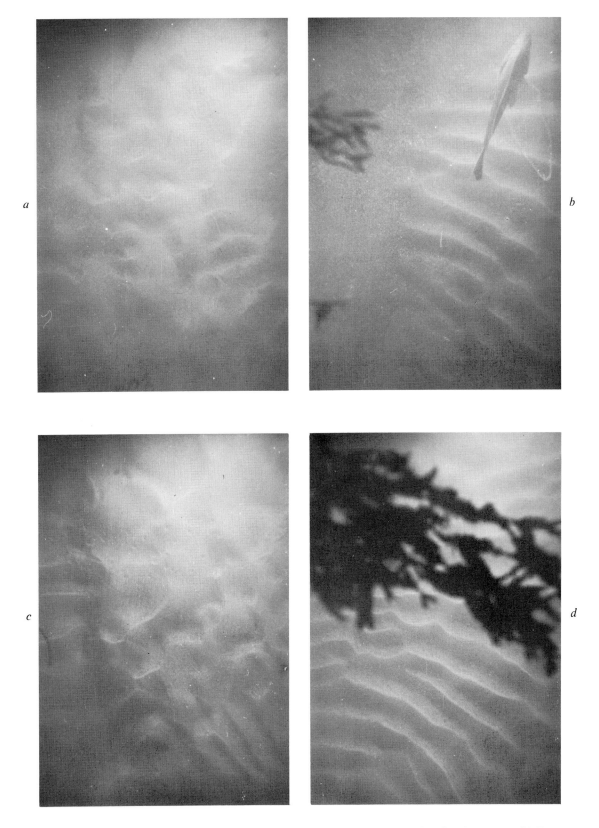

Figure 14–3. Ripple-marked bottom at slack water. (*a*) Frame no. 3380, after flood current. (*b*) Frame no. 3590, after ebb current. (*c*) Frame no. 3790, after flood current. (*d*) Frame no. 3990, after ebb current.

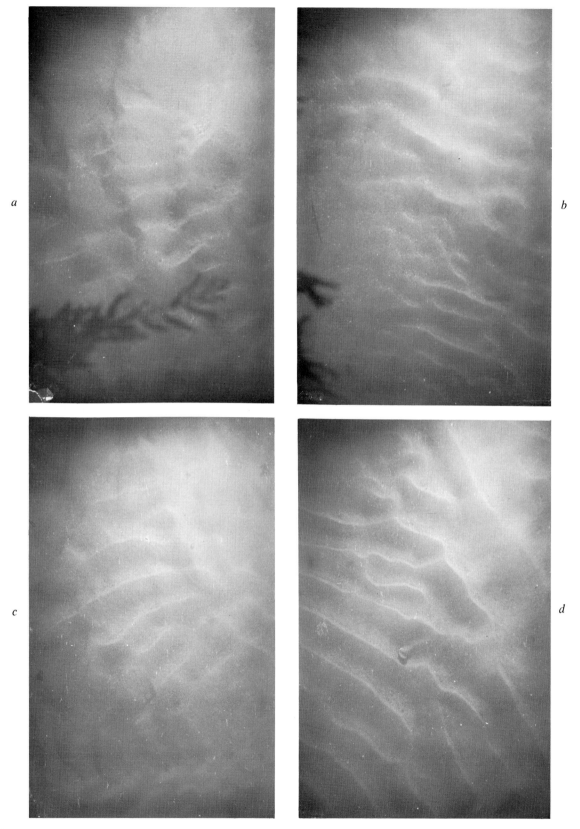

Figure 14-4. Ripple-marked bottom at slack water (continuation of fig. 14-3). (*a*) Frame no. 4210, after flood current. (*b*) Frame no. 4380, after ebb current. (*c*) Frame no. 4610, after flood current. (*d*) Frame no. 4828, after ebb current.

camera on 6 January 1965, ten years after the photography. They proved to be well sorted quartz-feldspar sand having median diameters of 0.20 and 0.50 mm and phi standard deviations of 0.42 and 0.45, respectively. According to J. D. Smith (Woods Hole Oceanographic Institution), who has made a general study of Middle Ground, these samples are typical of the general area.

Tidal currents

Vineyard Sound and the larger Nantucket Sound on its east have unusually fast and complicated tidal currents that are produced by flood and ebb through their several entrances (Redfield, 1956). The camera position is in one of the areas of fastest current, but the only precise data for the currents in the vicinity are based upon measurements made by the Coast and Geodetic Survey in 1934 (Haight, 1938, figs. 56–66; see also Coast and Geodetic Survey, no date); these measurements are republished annually as predicted currents by White (1953, for example) in a piloting handbook. Current speeds are stated for each hour relative to high and low tide at Boston. According to these data, the current reaches 3.3 knots (1.7 m/sec) at flood and 2.3 knots (1.2 m/sec) at ebb. A curve was plotted showing the predicted tidal current for the period of photography (fig. 14–5 — *top*).

Even casual viewing of the film reveals the presence of alternating periods of turbulent and of quiet water and that the current is approximately oppositely directed during alternate periods of turbulence. The ripple marks also face opposite directions during alternate periods of quiet water. Because the periods were fairly regular and equaled the number of tidal cycles that should have occurred during the six-day period, the tidal characteristics shown by the film were carefully investigated using the frame counter. The frame number and direction of movement were recorded for the following events in each cycle: near-cessation of sand movement, complete stability of ripple marks, initiation of sand movement, and greatest turbulence of the water above the bottom.

When the data were plotted according to frame number (fig. 14–5 — *middle*), a fairly uniform cyclical curve resulted; note that the speed of flood and ebb currents could not be measured from the film, so these times were plotted along arbitrary lines. The curve corresponds fairly closely with the curve of predicted current. Discrepancies that exist between the times of predicted and of observed slack water may be due to effects of winds and to variations in behavior of the tidal currents during the biweekly cycle (it was noted that a full moon, and thus spring tides, occurred on 14 August, the date of the greatest discrepancy of about two hours). Counts of the number of frames between times of stable bottom showed that eleven full

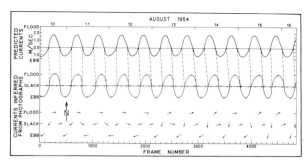

Figure 14–5. Tidal currents at the position of the time-lapse camera. *Top:* Currents predicted by Haight (1938) and White (1953) for position of the camera. *Middle:* Times of currents inferred from photographic record: maximum flood, ending of sand movement, stable ripples, start of sand movement, and maximum ebb current. *Bottom:* Direction of sand movement at maximum flood and maximum ebb currents, and directions of steep side of ripple marks at slack tides.

periods of flood current averaged 6.15 hr and eleven periods of ebb current 6.32 hr.

The direction of movement of the water at the times of greatest turbulence and the direction faced by the steeper side of the ripple marks at the times of stable bottom were noted with respect to the top of the projection screen. The orientation of the camera was not determined in the field, but the directional measurements were adjusted to a best-fit orientation, considering the general bottom topography of fig. 14–2. If this orientation is correct, the top of the projection screen corresponds to a northeasterly direction in the field. The maximum current at flood tide thus ranged between east and northeast (fig. 14–5 — *bottom*), and that at maximum ebb ranged between west southwest and south southwest. The direction of both flood and ebb currents exhibited a gradual progression to the left during the six-day period, perhaps because of the same wind or spring tidal cycle cause of the difference between predicted and observed times of slack water. The steep sides of the ripple marks at slack water commonly lay 5° to 30° to the right of the preceding maximum flood current and to the left of the preceding maximum ebb current. The reason for the difference is probably the changing influence of local bottom topography upon high and low speed currents.

The speed of the surface current at the time that the sand stopped moving and at the time that it again began to move can be estimated from the curves of fig. 14–5. These times are indicated by the small circles in the middle part of the figure on either side of the curve itself. The broken lines connect slack water of the curve inferred from the photographs and of the curve that is based upon predicted currents. If we similarly transfer the difference between the time of slack water and the time at which the sand just stopped moving or just started moving, we can obtain the speed of the surface current at the time that sand movement ended or began. Although it is only a rough approximation, the method yields an average surface current speed of 0.66 m/sec (1.3 knots) for cessation of sand movement and 1.25 m/sec (2.4 knots) for initiation of movement. Such a difference accords with the knowledge that sand can be carried by a current of much lower speed than that required to erode it (Inman, 1949). These speeds, however, are only relative, because the speed of the water that is in contact with the bottom was not measured. Inman's graphs indicate that the current that permits deposition of sand having the same grain size as that at the camera site should be about 0.02 m/sec, and that for initiation of movement is about 0.09 m/sec. These speeds are about 3 to 7 per cent of the estimated surface currents when sand movement ceased and when it began again. If this photographic method of studying the bottom were to be combined with accurate measurements of bottom currents, a precise evaluation of the effects of currents upon sediment movement *in situ* might be obtained.

References

Coast and Geodetic Survey, no date: Tidal current charts, Narragansett Bay to Nantucket Sound, 3rd ed., Serial no. 628, 12 charts.

Haight, F. J., 1938: Currents in Narragansett Bay, Buzzards Bay, and Nantucket and Vineyard Sounds. Coast and Geodetic Survey Special Publication no. 208, 1–101.

Inman, Douglas L., 1949: Sorting of sediments in the light of fluid mechanics. *J. Sedimentary Petrology*, **19**, no. 2, 51–70.

Kennedy, John F., 1963: The mechanics of dunes and antidunes in erodible-bed channels. *J. Fluid Mechanics*, **16**, 521–44.

Redfield, Alfred C., 1953: Interference phenomena in the tides of the Woods Hole region. *J. Mar. Res.*, **12**, 121–40.

Simons, D. B., & E. V. Richardson, 1961: Forms of bed roughness in alluvial channels. *Proc. Am. Soc. Civil Engineers, J. Hydraulics Div.*, **87**, HY3, 87–105.

White, R. E., 1953: *Eldridge tide and pilot book*, 1954. R. E. White, publisher, Boston, 1–200.

15. Current markings on the continental slope[*]

David M. Owen and K. O. Emery *Woods Hole Oceanographic Institution, Woods Hole, Massachusetts*

Abstract

About half of a series of nineteen stereophotographs of the continental slope south of Marthas Vineyard, Massachusetts, show scour features that may be due to the downslope passage of two separate small turbidity currents. Each current was sufficient to erase most of the normal markings made by benthic animals and to leave streamlined ridges at the lee side of small resistant masses of sediment. Long-wavelength, low-amplitude ripple marks were also locally eroded into the bottom sediment.

15–1. Introduction

Much has been learned about significant small features of the ocean floor through the use of bottom photographs. A third dimension for examination of these features is provided by stereoscopic pairs of photographs. Stereophotographs of special geological interest were obtained during a single lowering of a stereocamera on 24 May 1961 during *Atlantis* cruise 264. The stereocamera was a multishot device built by Owen in 1960 and described by him in chap. 8. One of each pair of photographs was in black-and-white and the other was in color.

During this particular lowering, nineteen pairs of photographs were made by repeated contact of the camera with the bottom during a 37-min period of drifting. These pairs are designated by numerals 1 through 19. The distance over which the camera photographed the bottom was about 1.3 km, thus the photographs average about 67 m apart. A Loran-A fix at time 2000, about one-third through the photographic sequence, yielded a position of 39°45′N, 70°44′W, but this position had to be adjusted about 4 km northward to allow the sounding of 835 fm (or 1,523 m, corrected for sound velocity in the water) to fit the known topography (fig. 15–1). The site is 175 km south of Marthas Vineyard on the conti-

nental slope. In this area the continental slope is cut by gullies; the closest submarine canyon (unnamed) is about 18 km to the east. The continental slope here begins at the shelf-break (about 140-m water depth), and it gradually steepens with greater depth so that near its contact with the continental rise (about 1,900 m) it is steepest. At the position of the photographs the continental slope has an average steepness of about 8.5° judging from the contours of fig. 15–1. Each photograph portrays an area 39 cm wide in the foreground (bottom of print), 46 cm wide in the background (top of print), and 43 cm long (front to back).

Sediments in dredge samples taken shortly before the camera lowering, and in a small corer attached to the camera frame, had the characteristics shown in table 15–1.

TABLE 15–1.

	Station		Corer on camera frame
	F	E	
Depth	1,500 m	823 m	835 m
Median diameter	10.4 μ	80.5 μ	4.5 μ
Standard deviation (φ units)	2.02	1.88	3.46
Composition (per cent)			
Sand (>62 μ)	3.0	48.1	3.4
Silt (62–4 μ)	69.6	44.5	48.2
Clay (4–1 μ)	18.4	4.9	18.7
Colloid (<1 μ)	9.0	2.5	29.7
Calcium carbonate	14.3	7.0	14.5
Organic carbon	1.03	0.52	0.86

The photographs showed the presence on the bottom of twenty-eight brittle stars (probably of two species), two small sea stars, one holothurian, and one sea urchin. Within the dredge samples the same forms were represented plus many other animals too small to be identified in the photographs (such as amphipods), or ones that lived within the sediment (such as polychaete worms).

15–2. Results

All nineteen of the pairs of photographs show a sediment-floored bottom. Two of them (exposures 11, 12) are

[*] WHOI Contribution Number 1757.

Cruise 264 of R/V *Atlantis* was funded by National Science Foundation grant 8918. The camera used was built under Office of Naval Research contract Nonr-2196(00) NR 083–004.

167

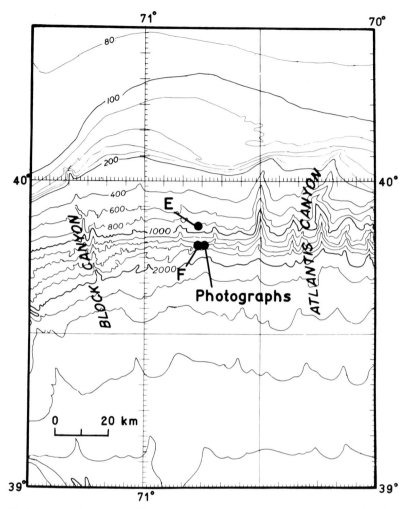

Figure 15–1. Area of photographic sequence on the continental slope south of Marthas Vineyard, Massachusetts. Contours are in meters.

obscured by dense clouds of suspended sediment, probably caused by dragging the camera frame over the bottom at these sites. The other photographs are clear, or only slightly turbid, and they serve as the basis for fig. 15–2, a plot of the results.

About half of the photographs show that the bottom is undisturbed except by the animals that live upon and in it. Evidence of their activity is exhibited by numerous tracks, trails, burrows, and burrow heaps (fig. 15–2) (fig. 15–3a) and by small irregularities possibly due to faecal matter. Among the most abundant of the markings are shallow five-rayed depressions where brittle stars have buried themselves. The remainder of the photographs present clear-cut evidence of erosion by current action. For exposures 2 through 6 (figs. 15–4a and 15–4b), the current appears to have "sand blasted" most of the bottom, leaving small masses protruding a centimeter or so above the general level; these masses possibly are remnants of resistant burrow heaps. Small parallel linear

ridges lie on the same side of each of the residual masses and taper away from them, thereby indicating that the current was unidirectional throughout the area of each photograph. Two photographs (exposures 8, 9 of fig. 15–2) (fig. 15–3b) reveal a similar kind of bottom but one that has been somewhat smoothed, as though by widespread deposition of a film of sediment following the period of erosion. Related features were described by Heezen & Hollister (1964) as crag and tail, in photographs from a flat-floored trough in the Indian Ocean, from the continental rise off Greenland, and from Drake Passage. None appear to have previously been reported from continental slopes, and examination of other photographs from continental slopes failed to disclose additional ones.

A larger but less obvious feature is exhibited by exposures 3 through 6 (figs. 15–4a, and 15–4b). This is a series of parallel wide low undulations on which the lee ridges are superimposed. The undulations lie approximately at

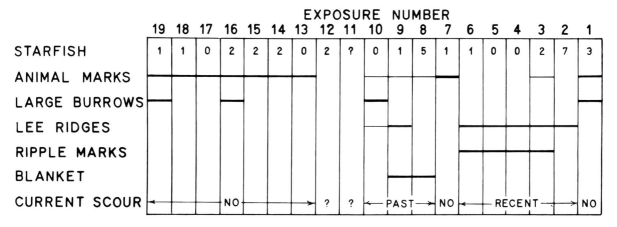

Figure 15–2. Plot of bottom features noted in the photographic sequence.

right angles to the lee ridges and their crests have the appearance of erosional ripple marks similar to sastrugi formed in snow.

15–3. Interpretation

The line of nineteen stereophotographs of the continental slope begins and ends with views of the bottom that present evidence only of the activity of animals. Within the sequence are two series of photographs that exhibit modification of the bottom by unidirectional current scour. In one of these series the current must have acted several days or weeks earlier because the bottom subsequently received a smoothing film of sediment that in turn became lightly marked by tracks and trails. The other sequence (exposures 2–6) is fresh-appearing with no cover. In both series the bottom water is so clear that erosion at the time of the photograph is unlikely; moreover, brittle stars lie atop the bottom in several of the photographs.

The general area in which the photographs were taken is one of the steepest parts of the continental slope in the region, about 8.5°. The camera frame carried no compass for orienting the photographs; nevertheless, the presence of unidirectional currents in two separate parts of the sequence is suggestive of small turbidity currents moving southward down the continental slope. A rough estimate of the speed of the current may be provided by a graph compiled by Heezen & Hollister (1964, fig. 18). According to this graph, the 10-μ median diameter of the sediment near the photographic site could have been deposited if the current carrying it decreased to below 0.01 cm/sec. This speed is so low that deposited sediment is not likely to form sharply defined lee ridges. On the other hand, if the lee ridges are erosional remnants, the graph indicates that the eroding current would have been about 80 cm/sec. It could have been slower if sand grains were

carried in suspension to serve as eroding tools. The 80 cm/sec estimate is well below the speeds that have been estimated for turbidity currents that have broken submarine cables off the Grand Banks (Heezen & Ewing, 1952; Kuenen, 1952; Shepard, 1961) and off Fiji (Houtz & Wellman, 1962).

Evidence from consolidated strata indicate that turbidity currents typically erode linear *depressions* into the sediment; these are elongate in the direction of the current (Crowell, 1955; Kuenen, 1957; Dzulynski & Sanders, 1962). They become filled with sandy sediment during waning phases of the turbidity current, and thus they should not be visible in underwater photographs. The presence of elongate *ridges* in the photographic sequence merely means that if they were formed by turbidity currents, the currents were probably small and of short duration, and that the currents had only begun to erode the bottom when they ceased.

References

Crowell, John C., 1955: Directional-current structures from the prealpine Elysch, Switzerland. *Bull. Geol. Soc. Am.*, **66**, 1351–84.

Dzulynski, Stanislaw, & John E. Sanders, 1962: Current marks on firm mud bottoms. *Trans. Conn. Acad. Arts and Sciences*, **42**, 57–96.

Heezen, Bruce C., & Charles Hollister, 1964: Deep-sea current evidence from abyssal sediments. *Marine Geology*, **1**, 141–74.

———, & Maurice Ewing, 1952: Turbidity currents and submarine slumps, and the 1929 Grand Banks earthquake. *Am. J. Sci.*, **250**, 849–73.

Houtz, R. E., & H. W. Wellman, 1962: Turbidity current at Kadavu Passage, Fiji. *Geol. Mag.*, **99**, 57–62.

Kuenen, Ph. H., 1952: Estimated size of the Grand Banks turbidity current. *Am. J. Sci.*, **250**, 874–84.

———, 1957: Sole markings of graded graywacke beds. *J. Geol.*, **65**, 231–58.

Shepard, F. P., 1961: Deep-sea sands. 21st. Int. Geol. Congress, Copenhagen, Pt. 23, 26–42.

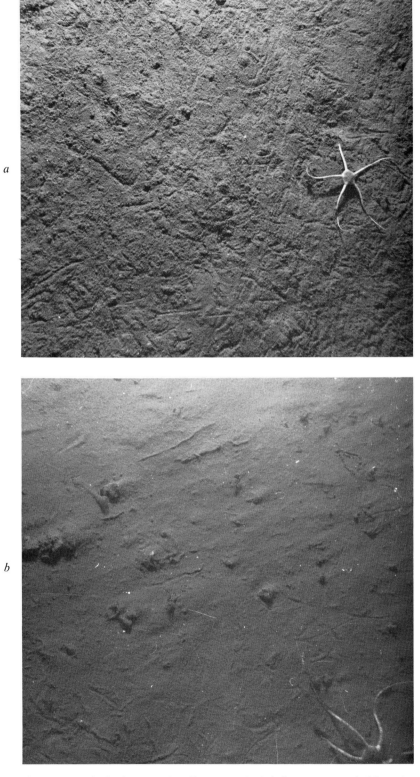

Figure 15–3. Single photographs of bottom. The brittle stars are probably *Ophio-musium lymani* Thompson. (*a*) Area having only animal tracks and trails—exposure 18 of fig. 15–2. (*b*) Area that was scoured and then smoothed by deposition of a film of sediment—exposure 9 of fig. 15–2.

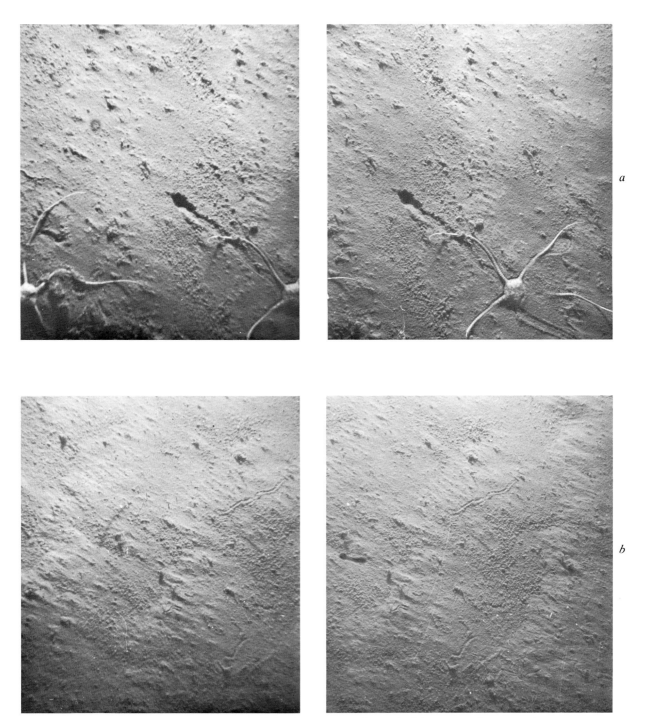

Figure 15–4. Stereophotographs of the bottom recently marked by scour of currents that moved toward the lower right corner. (*a*) Lee ridges and broad ripple marks that are being modified by the activities of two brittle stars, probably *Ophiomusium lymani*—exposure 3 of fig. 15–2. (*b*) Lee ridges and broad ripple marks on which a few trails were later made by moving animals—exposure 4 of fig. 15–2.

Figure 15–5. Photograph of recently scoured bottom—exposure 5, of fig. 15–2 reproduced from color film. The clearly defined lee ridges and ripple marks indicate movement of an eroding current toward the lower right corner. The impression of a brittle star was made after the period of erosion. The red overprint shows contours at 0.5-cm intervals drawn from the two stereophotographs at the station after they were rectified for tilt of the camera. The photograph (left-hand one of the pair) was also rectified by tilting the enlarger, so that contours and photograph could be superimposed. Contours were prepared by the U.S. Navy Oceanographic Office.

16. Estimation of relative grain size from sediment clouds[*]

William D. Athearn   *Woods Hole Oceanographic Institution, Woods Hole, Massachusetts*

Abstract

The sediment cloud raised by the tripping weight of an underwater camera may be used in estimating the grain size of bottom sediments. This is shown in studies of two areas from which sediment samples were taken simultaneously with bottom photographs. The first survey area was in Cape Cod Bay, Massachusetts, a region of glacially derived quartz and feldspar sediments, and the second was the Tongue of the Ocean, Bahamas, a region of primarily detrital carbonate sediments. The observations show that within a given geographical area it is possible to distinguish finer from coarser sediments on the basis of relative sediment-cloud densities, but that it is not practical to compare clouds from widely separated areas, because of differences in composition, color, shape, and cohesiveness of the minerals comprising the sediments.

16–1. Introduction

For years preceding the development of fully automatic equipment, underwater camera shutters were tripped by contact between a tripping weight and the sea bottom. This weight was suspended at the correct distance below the camera lens so that the bottom was in proper focus when the shutter tripped. One of the concerns of the designer was to keep the weight, and the resulting mud cloud if the weight struck unconsolidated sediment, out of the field of view in order to show as much of the bottom as possible in the photograph. Of course with the advent of automatic cameras, whose precise distance off bottom can be determined sonically and can be controlled by the winch operator, tripping weights are no longer essential and the cloud nuisance no longer exists.

There are, however, occasions when a sediment cloud is of sufficient practical value that it may be deliberately produced. This is because within a given sedimentary environment there should be a definite relationship between the average grain size of the surface sediment and the size and density of such a cloud. Shepard & Emery (1946) first suggested such a consideration of the sediment cloud in a discussion of submarine photographs from off the California coast. They stated that "the mud content of the sediment is sometimes indicated by the incipient mud cloud which is raised by the weight's hitting bottom." It may be inferred that the higher the ratio of mud (silt and clay sizes) to sand in the sediment, the larger and denser the cloud should be. Hunkins *et al.* (1960) briefly mentioned sediment clouds — stirred up when their camera struck bottom accidentally — as indicative of soft and fine-grained sediment. There seem to be no further published references specifically concerned with the relation of cloud to sediment type, although there are occasional comments as to the use of sediment clouds in estimating the velocities of bottom currents, particularly if the camera can be held stationary and the cloud viewed in successive photographs so that its rate of drift may be computed (Hunkins *et al.*, 1960, p. 160; and Pratt, 1963, p. 246). It is also conceivable that under the latter circumstances, where a cloud has been produced beneath a stationary camera taking a series of photographs, an estimate might be made as to sediment-grain-size distribution based on the rate of settling back to the bottom, but in most cases this would probably be complicated by bottom currents.

16–2. Comparative studies of sediment clouds and grain sizes

On two recent occasions attempts have been made to relate the apparent size and density of the sediment cloud shown in photographs to the actual size distribution of the sediment at the site of the photographs. These are discussed in unpublished reports of the Woods Hole Oceanographic Institution (Zeigler, Oostdam, & Owen, 1960; Athearn & Owen, 1962). In the first study the photographs were taken in Cape Cod Bay, Massachusetts,

* WHOI Contribution Number 1758.
  The studies reported above were variously supported by the U.S. Navy, under contracts N 298(122)16639, N 140(122)70867B, and Nonr-2196(00) NR 083–004.

173

over an area of 8 square miles located some 15 miles northwest of Provincetown, and at about 60 m depth. Mounted on the camera frame, designed by David Owen and described in chap. 8, was a small core tube arranged to collect a sediment sample an instant after the camera fired. As long as only one picture was taken at each lowering, a sample, uncontaminated by sediment from subsequent groundings of the core tube, could be obtained from each site for analysis.

Grain-size distribution analyses were carried out on twenty samples which had corresponding photographs showing the sediment cloud initiated by the tripping weight. Per cent by weight of silt + clay as compared to per cent by weight of sand was used as a criterion for comparison of the sediment clouds. In the Cape Cod Bay area the silt + clay content ranged between 4 per cent and 89 per cent. Mineral composition was chiefly quartz and feldspar with a little mica, a typical sediment derived from the glaciation of metasedimentary and igneous rocks of northern New England. Fig. 16–1 shows a typical silty sand from the area (Camera station 57, 28 per cent silt + clay), while fig. 16–2 shows a sandy silt (Camera station 43, 62 per cent silt + clay). In fig. 16–1 the cloud ring is clearly thinner, and apparently less dense, than that of fig. 16–2, as should be expected. Unfortunately no photographs were obtained of sample locations, having silt + clay percentages either higher than 62 per cent or lower than 28 per cent, to allow comparison of more extreme cases. Contrast between the clouds from sediments whose silt + clay contents are less different than those illustrated

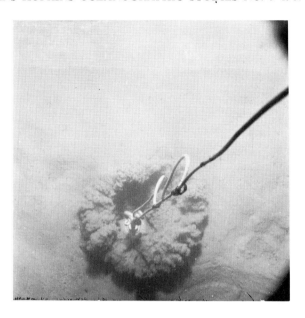

Figure 16–2. Sediment cloud produced on a sandy silt bottom (62 per cent silt + clay) in Cape Cod Bay, Massachusetts (Station 43).

depends a great deal on the quality of photographic reproduction and on the size of enlargement. A sediment with a silt + clay content of about 45 per cent could probably be distinguished from the two figured here at the present scale (1 to 7.5). With enlargements used in the laboratory (1 to 3), one can easily discriminate 10 per cent differences.

The second study concerns predominantly calcareous sediments from the Tongue of the Ocean, Bahamas. Seventy-eight bottom photographs were taken at fifteen stations throughout the area, some 2,000 square miles, at depths ranging between 1,180 and 1,680 m, using an arrangement of camera and corer similar to that used in Cape Cod Bay. Several photographs were taken at each station while the ship slowly drifted, by alternately lowering the camera and raising it to allow the film to advance and the flash to recharge. Photographs were taken at approximately 3-min intervals. Because the corer penetrated bottom at each lowering, only the top 2 cm were used for comparison with the first photograph at each sample site. The two photographs reproduced here (figs. 16–3 and 16–4) were taken at depths of 1,513 and 1,304 m, respectively. Fig. 16–3 (Station 29, Frame 1) shows a silty sand (33 per cent silt + clay) from near the base of the slope at the eastern margin of the area. The sample is composed principally of reef detritus from the shallow banks south of New Providence Island, plus shells of pelagic foraminefera. The sediment of fig. 16–4 (Station 48, Frame 1) is a silt-clay whose silt + clay content is 89 per cent. This sample is from the lower slope at the western side of the Tongue of the Ocean and is also com-

Figure 16–1. Sediment cloud produced on a silty sand bottom (28 per cent silt + clay) in Cape Cod Bay, Massachusetts (Station 57). (Field of view in each figure is 23 in. (58 cm) square.)

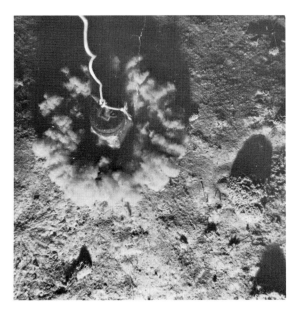

Figure 16–3. Sediment cloud produced on a silty sand bottom (33 per cent silt + clay) in the Tongue of the Ocean, Bahamas (Station 29, Frame 1).

Figure 16–4. Sediment cloud produced on a silt-clay bottom (89 per cent silt + clay) in the Tongue of the Ocean, Bahamas (Station 48, Frame 1).

posed almost entirely of calcium carbonate. Comparison of sediment clouds in these two figures shows that in the cloud in fig. 16–3, the silty sand is less dense than that in fig. 16–4, the silt-clay. Although the difference in percentage of silt + clay is greater between these latter examples (56 per cent), than between those from Cape Cod Bay (34 per cent), the contrast between the cloud sizes and densities from the Bahamian sediments is not proportionately greater. Intermediate distinctions of finer than 15 per cent probably cannot be made in these areas of carbonate sediments.

16–3. Summary

To conclude, it is feasible up to a point to distinguish finer from coarser sediments by comparison of sediment clouds. The degree of grain-size discrimination that can be made, if sampling and analysis procedures are standardized, is probably influenced mostly by cohesiveness, density, shape, and color of the constituents of the sediment. Probably no useful comparison could be made between photographs from such remote environments as Cape Cod Bay and the Tongue of the Ocean, as the sediments are very different in composition. But within a specific area, whether it be Chesapeake Bay or the Gulf of Venezuela, if field time is short and investigations are to be limited to the surface sediments of the sea bottom, a photographic survey including the sediment cloud technique would probably yield the greatest amount of information for the time available. Not only are physical

features and biological organisms observed but also information as to relative-grain-size distribution is obtained. If the ship is allowed to drift across the area, a series of photographs may be taken by a semi-automatic camera in a manner similar to that used in the Tongue of the Ocean. With appropriate spacing a very complete coverage could be made. For instance, if the vessel's rate of drift is ½ knot, approximately 50 ft (15 m) of bottom will be traversed each minute, and photographs might be taken at 1-min intervals, or perhaps spaced farther apart, depending on the desired coverage and expected bottom variability. It should be possible to determine trends in grain-size distribution with this sort of spacing even if the sediment clouds do not change much from photograph to photograph. With a little extra ingenuity it should be possible to mount a multiple sampler on the camera frame and to program it so that it would take a sediment sample at the start of, perhaps at the middle of, and at the end of a traverse, thus providing calibration points for the clouds observed.

References

Athearn, William D., & David M. Owen, 1962: Bathymetric and sediment survey of the Tongue of the Ocean, Bahamas. Part 2, Bottom photographs. Woods Hole Oceanographic Institution Ref. No. 62–27, 4 pp., 27 pls. (unpublished manuscript).

Hunkins, Kenneth L., Maurice Ewing, Bruce C. Heezen, & Robert J. Menzies, 1960: Biological and geological observations on the first photographs of the Arctic Ocean deep-sea floor. *Limnology and Oceanography*, **5,** no. 2, 154–61.

Pratt, Richard M., 1963: Bottom currents on the Blake
    Plateau. *Deep-Sea Res.*, **10,** 245–49.
Shepard, F. P., & K. O. Emery, 1946: Submarine photography
    off the California coast. *J. Geol.*, **54,** no. 5, 306–21 .
Ziegler, John M., Bernard Oostdam, & David M. Owen, 1960:

A study of the bathymetry and sediments of the U.S. Navy
torpedo range in Cape Cod Bay. Woods Hole Oceanographic Institution Ref. No. 60–35, 10 pp., 6 figs. (unpublished manuscript).

17. The floor of the Bellingshausen Sea[*]

Charles D. Hollister and Bruce C. Heezen *Department of Geology and Lamont Geological Observatory, Columbia University, Palisades, New York*

Abstract

Over 200,000 km of precision soundings and 8,000 bottom photographs (440 stations) were obtained by the USNS *Eltanin* in the Bellingshausen Sea.

The rocky crest of the Mid-Oceanic Ridge in the western Bellingshausen Sea is dextrally displaced 1,000 km by the Eltanin Fracture Zone. The Bellingshausen Abyssal Plain lies at the base of a wide continental rise.

South of the polar front, muddy bottom is seen in nearly all photographs, whereas to the north both ridge and basin photographs show either scattered rocks, outcrops (pillow lava), or nodules. Boulders occur beneath the pack ice and north of the front. Anomalously smooth bottom showing weak to strong scour lies between the pack limit and the front and demonstrates the presence of significant bottom currents. Manganese nodules north of the front become larger and more numerous from west to east, and finally give way to strong scour and ripples (greater than 4,000 m) in the constricted Drake Passage.

17–1. Introduction

In 1963 the USNS *Eltanin* commenced a systematic study of the Pacific sector of the Southern Ocean. At the present time 440 bottom-photograph stations have been obtained in the Bellingshausen Sea (fig. 17–1). No other part of the ocean floor has been photographed so systematically and extensively. This collection of photographs offers a unique possibility of delineating regional variations in microtopography. The oceanographic contrasts as well as the physiographic contrasts in this region are strong,

and one might expect that variations in microtopography would reflect oceanographic and geologic patterns.

The Bellingshausen Sea extends from the Drake Passage at about 70°W to the Mid-Oceanic Ridge at about 150°W, and from the Antarctic continent at about 75°S to the Chile Ridge at about 40°S. The Antarctic Polar Front passes through this area at approximately 60°S (Deacon, 1937). The limit of pack ice is at about 62°S, and further south, at about 70°S near the edge of the Antarctic continental shelf, lies the limit of fast, impenetrable ice (fig. 17–1). The bottom photographs of the Bellingshausen Sea are found to reflect the influences of these major oceanographic and morphologic features. All the photographs in this collection were taken with a bottom-contact, multiple-exposure, electronic flash camera (Thorndike, 1959). An area of approximately 5 m² is recorded by each photograph.

17–2. Sediment-covered bottom

The sediment cover of the Bellingshausen Sea is abundantly marked with tracks and trails (figs. 17–2, 17–3, and 17–4). Animal life is frequently seen (figs. 17–5, 17–6, and 17–7). The relatively abundant abyssal benthic population of the southern ocean stands in rather marked contrast to that of tropic and temperate seas where the lower surface productivity is inadequate to maintain a large benthic population. On the basis of the abundance of life, abyssal photographs from the floor of the Bellingshausen Sea could very easily be confused with photographs from shallower depths.

Life and its evidence are relatively abundant. Holothurians (fig. 17–5) are sometimes seen in the act of producing trails across the sea floor; fecal knots (figs. 17–2 and 17–14) apparently produced by the holothurians are often associated with characteristic trails. Echinoids are seen plowing across the sediment bottom. Tiny spiral fecal coils are occasionally seen. Large, attached, fernlike organisms draw circles in the sediment as they swing from their base (figs. 17–8, 17–9, and 17–10). Stalkless crinoids are seen attached to rock outcrops (fig. 17–28),

* The writers gratefully acknowledge the efforts of the many *Eltanin* scientists who contributed to the sea-floor photography program. The assistance and encouragement of A. P. Crary, Chief Scientist, and R. R. Hinchcliffe, Vessel Coordinating Officer, U.S. Antarctic Research Program, National Science Foundation, is greatly appreciated.

This study was supported by National Science Foundation grant NSF GA 305 and is Lamont Geological Observatory, Columbia University, Contribution Number 1001.

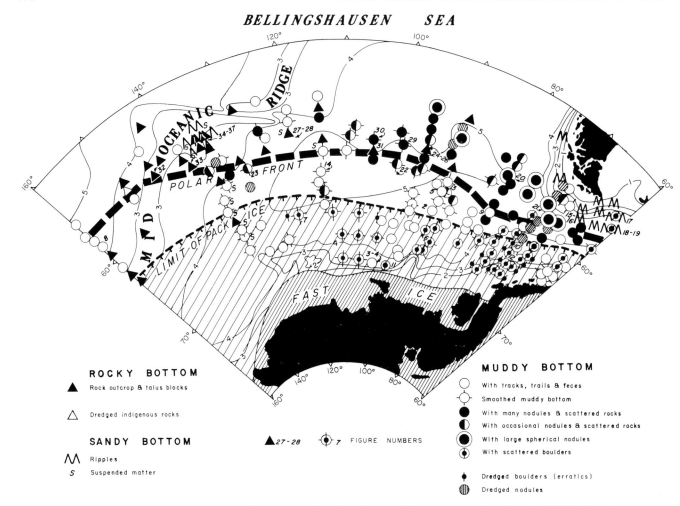

Figure 17–1. Bottom types in the Bellingshausen Sea. Based on sea-floor photographs taken by USNS *Eltanin*.

and stalked crinoids rise from the muddy and nodule-strewn bottom (fig. 17–23). Sponges occur even on sandy, rippled bottom (fig. 17–19). Coelenterates are fairly common on muddy and rock-strewn bottom (figs. 17–5, 17–7, and 17–27). Fish (fig. 17–6) swimming near the bottom cast shadows on the sea floor below. Holes with radial patterns suggesting buried shellfish are frequently seen. However, most of the tracks, particularly the grander and more bizarre multiple tracks and hiero-glyphic-like trails, remain unidentified. Fecal strings (fig. 17–31) and pellets of all sizes and descriptions (fig. 17–14) are seen on the sea floor, and one can only imagine that if cameras with greater resolution were used, much of the grainy detail of the sea floor would turn out to be small fecal pellets and minute animals.

17–3. Scattered boulders

Near the Antarctic continent, particularly in the area just west of the Antarctic peninsula, the sea floor is littered

with boulders and pebbles of all shapes and sizes: some rounded, some angular (figs. 17–11 and 17–12). These boulders appear to be glacially rafted. It seems probable that they were transported from the Antarctic peninsula to the west in response to the high-latitude easterly winds.

In many cases (fig. 17–11) the boulders lie on a tranquil, muddy bottom devoid of current evidence. However, scour is frequently dramatically illustrated by the occurrence of rather sharp and deep moats surrounding larger boulders (figs. 17–17 and 17–18). These often take the form of rock nests when a group of both pebbles and boulders occurs (fig. 17–19). Ripple marks and sediment streamers are associated with rock nests.

17–4. Smoothed, muddy bottom

Many of the photographs in the area between the polar front and the limit of pack ice have a smoothed appearance (figs. 17–13 and 17–14). Larger feces are often partially winnowed away: sediment streamers occur on the

Figure 17-2. Muddy bottom on the Antarctic continental rise. The trail associated with the cone-shaped organism (right center) is apparently composed of two crenulated, shallow furrows and a median ridge. Similar trails have been seen associated with echinoids. Note large, segmented feces in upper lefthand corner and abundant smaller and less distinct litter of feces over entire bottom. Taken at a depth of 4,747 m; 64°03'S, 89°49'W; on the lower continental rise of Antarctica northeast of Thurston Island.

lee of occasional lumps, solid objects, or larger feces. The sea floor has distinctly fewer tracks and trails, although the evidence of life in the form of partially destroyed tracks and trails is perhaps as abundant as in the areas where this scouring and smoothing is not seen. In many areas the older trails and mounds apparently have been obliterated.

17-5. Ripples

Ripple marks (figs. 17-15 and 17-16) are observed in many of the photographs taken in the northern part of the Drake Passage to depths over 4,000 m (Heezen & Hollister, 1964). This area is unique in the occurrence of

sharp, short-crested ripple marks at such great depths. The ripples are sometimes associated with rocks and rock outcrops. Many types and patterns of ripples occur. Ripples are lunate, linguoid, or straight, and the orientation of the ripples seems to change radically even within the area of one photograph (fig. 17-15 and 17-16). On the crest of the Mid-Oceanic Ridge, ripple marks have also been observed in a depth of over 3,000 m (figs 17-36 and 17-37).

17-6. Nodules and scattered rocks

Manganese nodules vary from perfectly symmetrical, large, orange-sized spheres (figs. 17-20 and 17-21) to irregularly shaped, nodular bodies which apparently are mixed with quantities of rafted rocks only thinly coated with manganese (fig. 17-24). The nodules appear to be lying on a sediment bottom. They vary widely in density of distribution. Occasionally a round nodule is seen on an otherwise soft and muddy bottom. In other areas approximately twenty nodules and scattered rocks occur in each square meter (figs. 17-22 and 17-23). In still other areas a close-packed arrangement is seen which reminds one of an apple barrel (figs. 17-20 and 17-21). The relative abundance of life in the areas of nodule accumulation is anomalous as far as the world distribution is concerned, for one normally associates nodules with an area of low bottom life.

Nodules commonly carpet the bottom (figs. 17-24, 17-27, 17-29, and 17-30). In the area there is apparently

Figure 17-3. Tracks and trails on the Antarctic continental rise. Large (~0.25 m wide), discontinuous furrows, often with lateral crenulations, are occasionally seen on soft sediment south of the polar front. Note apparent absence of current scour or smoothing. Taken at a depth of 3,840 m; 70°07'S, 102°56'W; on the Antarctic continental rise, north of Thurston Island.

little correlation between nodule size and water depth. In some cases currents have scoured moats around nodules and have left streamers of sediment in their lee. In other nodule-covered bottoms, little or no evidence of currents can be seen. However, one would, in general, associate appreciable current with the occurrence of nodules. The distribution of perfectly rounded, spherical nodules is limited to a belt which runs nearly due west from the area of well-developed ripple marks in the Drake Passage (fig. 17–1). The transition from large, perfect

nodules to distinct current ripples in a linear west-to-east pattern may reflect an easterly-flowing bottom current gradually restricted and thus accelerated by the Drake Passage.

17–7. Suspended matter

At over thirty locations, on both rocky bottom and on soft bottom (fig. 17–1), the sea water is speckled with

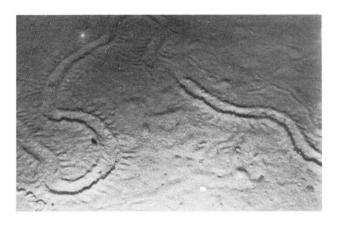

Figure 17–4. Tracks and trails on the Antarctic continental rise. Large (∼0.25 m wide), discontinuous furrows. (See fig. 17–3).

Figure 17–5. Abundance of benthic life is characteristic of the muddy deep-sea floor beneath and south of the Antarctic polar front. Often more than one large organism is seen in each photograph. This picture shows a holothurian and a coelenterate, and was taken at 61°58'S, 90°01'W, at a depth of 4,840 m.

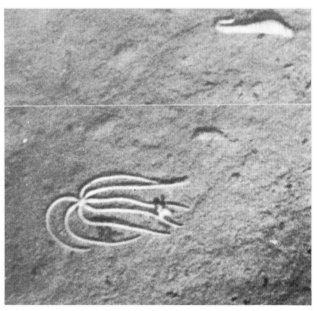

Figure 17–6. A fish and an echinoderm taken on the lower continental rise of Antarctica, 66°20'S, 90°24'W, at a depth of 4,525 m.

Figure 17–7. A coelenterate and a stalked echinoderm, taken in the west central Bellingshausen Sea, north of Cape Dart, at 65°36'S, 121°11'W, at a depth of 4,873 m.

Figure 17–8. Attached benthic life on the northwestern flank of the Mid-Oceanic Ridge. Unidentified, fernlike, attached organisms are seen in the process of forming circular impressions (see also figs. 17–9 and 17–10). The impressions are approximately 0.25 to 0.5 m in diameter. Taken at 58°03′S, 160°12′W, at a depth of 4,170 m.

white spots (fig. 17–28). One might blame this on poor photography if it were not for the fact that many of these photographs are quite clear and it is certain that the specks represent neither hypo dust nor dirt on the camera window and were not produced in the printing of the photograph. Apparently, bottom currents were traveling at such a velocity as to carry considerable material in suspension.

17–8. Outcrops and talus blocks

Rock outcrops are found both on the ocean-basin floor north of the polar front (figs. 17–25 and 17–31) and on the higher parts of the Mid-Oceanic Ridge (figs. 17–32, 17–33, 17–34 and 17–35). In some areas talus blocks 0.25 m to 1.0 m in maximum dimensions are littered on a sediment-covered bottom (fig. 17–26). In other areas, the outcrops are pillows of lava (figs. 17–32 and 17–33). A few photographs of fractured pillow lava establish beyond doubt this identification (figs. 17–34 and 17–35). Although the rock outcrops on the ocean-basin floor north of the polar front are remarkable enough, those that lie south of the polar front are even more striking. The known rates of deposition in this area would seem to require either very recent extrusion of these basalts or extensive current denudation of the Mid-Oceanic Ridge (or perhaps both).

The occurrence of the rock outcrops (fig. 17–31) in the floor of the Bellingshausen Sea in depths of approximately 5,000 m is quite surprising. Submarine geologists

have become accustomed to observing outcrops of rock, and even pillow lavas, on the crest of the Mid-Oceanic Ridge (Heezen *et al.*, 1959), but photographs of pillows and apparently freshly fractured rock in depths of over 4,000 m would not have been expected by most geologists. The deep-basin outcrops usually occur in areas of abyssal hills.

Relatively large, lateral variations can occur with rock outcrops, scattered rock, ripple marks, and soft, muddy bottom all occurring within a few hundred meters.

Figure 17–9. Attached benthic life on the continental rise of Antarctica. (See also figs. 17–8 and 17–10.) Taken at 61°57′S, 78°58′W, at a depth of 4,798 m.

Figure 17–10. Attached benthic life in the southern Drake Passage. (See also figs. 17–8 and 17–9.) Taken at 63°14′S, 71°34′W, at a depth of 3,766 m.

Figure 17–11. Scattered boulders beneath pack ice on the upper continental rise of Antarctica. This picture shows a tranquil bottom just north of the limit of fast ice, and was taken at 70°25'S, 99°36'W, at a depth of 3,680 m.

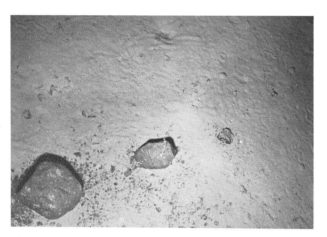

Figure 17–12. Scattered boulders beneath pack ice in the southern Drake Passage, west of Antarctic peninsula. The sea floor south of the pack ice limit is characteristically strewn with rocks of various shapes and sizes. Dredges from this area contain assorted igneous, volcanic, and metamorphic rocks. Taken at 63°58'S, 71°13'W, at a depth of 3,548 m.

17–9. Summary

Nearly all the major types of bottom are found north of the Antarctic Polar Front, including rocky outcrops, littered rocks and nodules (with or without current evidence), and soft, undisturbed muddy bottom. In the deeper waters of the Bellingshausen Sea, scattered rocks and nodules, together with rock outcrops, are seen in the vast majority of the photographs. Of the nearly 100 camera stations from the crest and upper flanks of the Mid-Oceanic Ridge, nearly half reveal rock outcrops, many of them craggy, and several are clearly pillow lava.

Immediately to the north of the polar front, rocks and rock outcrops are found on the ocean-basin floor as well

as on the Mid-Oceanic Ridge. Many of the photographs from the ocean-basin floor reveal evidences of current winnowing or scour. However, between the Antarctic polar front and the limit of pack ice, rock outcrops and, in fact, scattered rocks, are rare, and most of the photographs reveal a muddy bottom which is anomalously smooth. South of the limit of pack ice there are increasing numbers of scattered, probably ice-rafted, boulders on the muddy sea floor. Immediately off the southern tip of South America, ripple marks and strong scour marks have been observed in depths over 4,000 m (Heezen & Hollister, 1964). South of this distribution of ripple marks is a prominent distributional pattern of manganese nodules, and still further to the south the bottom is characteristically muddy with occasional ice-rafted boulders.

Figure 17–13. Smoothed, muddy bottom in the east central Bellingshausen Sea, south of the polar front. Smoothed and scoured sediment is characteristic of the region between the Antarctic polar front and the limit of pack ice. Note the leeside deposits. Taken at 61°16'S, 89°49'W, at a depth of 4,898 m.

Figure 17–14. Smoothed, fecal-strewn, muddy bottom in the west central Bellingshausen Sea, south of the polar front. Taken at 62°00'S, 115°14'W, at a depth of 5,139 m.

Ice-rafted boulders appear more frequently in the sector immediately to the west of the Antarctic peninsula. In this sector, pack ice is penetrable to the shoreline each year, whereas further to the west, fast ice surrounds the continent even during the Antarctic summer. It is conceivable that in the area west of the Antarctic peninsula the easterly Antarctic winds drift the bergs to the west before they are finally swept into the west-wind drift north of the ice front.

Current evidence determined from the 440 *Eltanin* camera stations (cruises 4 through 23) shows a pattern of erosion on the Mid-Oceanic Ridge and along the northern portion of the Bellingshausen Basin that extends through the Drake Passage and into the South Atlantic (fig. 17–38). Erosion is indicated by clean bare rock outcrops (fig. 17–25), scour marks (fig. 17–17), ripple marks and manganese nodules (fig. 17–20). South of this

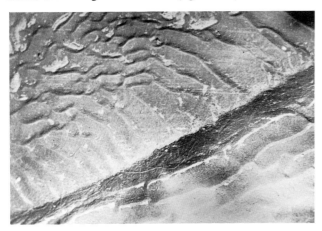

Figure 17–15. Ripples in the northern Drake Passage, southeast of Tierra del Fuego. Note the abrupt change in ripple orientation. In areas of strong currents where well-defined, short-crested ripples prevail, abrupt changes in orientation are occasionally seen (see also fig. 17–16). Taken at 57°28′S, 64°51′W, at a depth of 4,531 m.

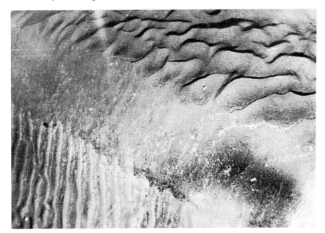

Figure 17–16. Ripples in the northern Drake Passage (see fig. 17–15).

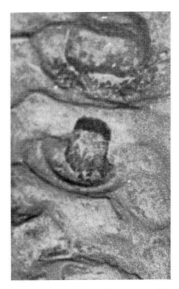

Figure 17–17. Scoured moats and scattered boulders in the northern Drake Passage, southeast of Tierra del Fuego. Evidence for strong bottom currents, in the form of ripples and scoured moats, is characteristic of camera stations in the northern portion of the Drake Passage to depths exceeding 4,000 m. In this photograph strong scour has developed moats surrounding the rocks. The rocks are probably ice-rafted. Taken at 55°59′S, 61°43′W, at a depth of 4,220 m.

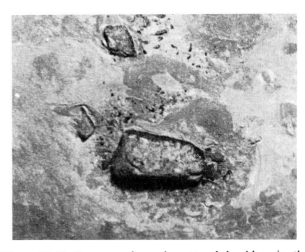

Figure 17–18. Scour marks and scattered boulders in the northern Drake Passage, southeast of Tierra del Fuego. The rocks are probably ice-rafted (see also fig. 17–17). Taken at 56°03′S, 60°48′W, at a depth of 4,190 m.

zone photographs show current lineation and suspended sediment (fig. 17–13) that indicate current scour and sediment transport. These zones correlate with the axis of eastward spreading bottom water (fig. 17–39).

This cold water which originates in the Ross Sea is deflected to the left by the Coriolis force and flows along the southern margin of the Mid-Oceanic Ridge to the bottom of the Bellingshausen Basin. Branches of this flow pass through the fracture zones to the southwest

Figure 17–19. Rock nests and sponge in the Drake Passage. The rocks are probably ice-rafted. Note the sponge swept to the left. (See fig. 17–18.)

Figure 17–20. Large, spherical, closely packed nodules in the Drake Passage west southwest of Tierra del Fuego. These orange-sized nodules occur at eight stations along an elongate belt, north of the polar front 90°W. East of this belt stations in the northern portion of the Drake Passage reveal ripples in depths over 4,000 m. Sediment apparently thrown in suspension by the camera trigger weight is seen moving to the left in response to an appreciable bottom current. Taken at 58°11′S, 79°11′W, at a depth of 4,716 m.

Pacific, but the coldest water continues to flow to the east along the northern slope of the basin. We infer a weaker westerly flow along the upper continental rise in the southern portion of the basin. Here weaker current evidence, in the form of smoothing and lineations, is frequently observed, and the region appears to be a depositional environment.

Suspended sediment, current lineations, scour marks, and bare rock outcrops along the northern portion of the

Weddell Basin demonstrate the high velocity and extreme competence of the Antarctic bottom current which sweeps around the Scotia arc into the South Atlantic and flows along the western margin of the Atlantic Ocean. This strong cold current is considered responsible for the shaping of the continental rise and other major sediment accumulations in the western Atlantic (Heezen & Hollister, 1964; Heezen, Hollister, & Ruddiman, 1966).

At only 6 of 440 camera stations was a compass included in the photographs. In view of the unusual abundance of directional features observed in the area this lack of a means to orient the photographs is particularly regrettable.

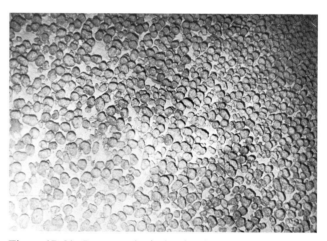

Figure 17–21. Large, spherical, closely packed nodules in the northwestern Drake Passage, south of Tierra del Fuego. Note scour marks around the nodules in the upper left. Taken at 57°59′S, 70°44′W, at a depth of 3,924 m.

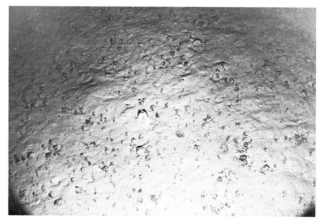

Figure 17–22. Nodules and scattered rocks on the ocean-basin floor. Photograph station taken between the polar front and the pack-ice limit characteristically reveals a current-scoured mud bottom. In this area of high surface productivity organisms and fecal pellets are abundant. Note the scoured moats. Taken in the central Bellingshausen Sea at 61°01′S, 99°59′W, at a depth of 4,966 m.

In the future a compass must be included in each sea-floor photograph so that directional features can be oriented, for the orientation of current lineations, scour marks, ripple marks, and other directional features may provide the key to many problems related to movements of sediment and water at the abyssal benthic boundary.

References

Deacon, G. E. R., 1937: The hydrology of the Southern Ocean. *Discovery Reports,* **15,** 1–124. Cambridge University Press, London.

Gordon, A., 1967: Potential temperature, oxygen and circulation of bottom water in the Southern Ocean (in press).

Heezen, Bruce C., Marie Tharp, & Maurice Ewing, 1959: The floors of the ocean, I. Special Paper 65, The Geological Society of America, New York, 122 p.

——, & Charles Hollister, 1964: Deep-sea current evidence from abyssal sediments. *Marine Geology,* **1,** no. 2, 141–74.

——, ——, & William Ruddiman, 1966: Shaping of the continental rise by deep geostrophic contour currents. *Science,* **152,** 502–8.

Jacobs, Stanley S., 1965: Physical and chemical oceanographic observations in the southern oceans. Lamont Geological Observatory of Columbia University. Technical Report No. 1–CU–1–65.

——, 1966: Physical and chemical oceanographic observations in the southern oceans. Lamont Geological Observatory of Columbia University. Technical Report No. 1–CU–1–66.

Thorndike, E. M., 1959: Deep-sea cameras of the Lamont Observatory. *Deep-Sea Res.,* **5,** 234–37.

Figure 17–23. Scattered rocks and nodules on the south-eastern flank of the Mid-Oceanic Ridge. Note the stalked crinoid in lower right. Sediment streamers occur as lee deposits. This photograph was taken near the polar front in the western Bellingshausen Sea, at 60°07′S, 128°54′W, and a depth of 4,336 m.

Figure 17–24. Nodules and scattered rocks on the ocean-basin floor north of the polar front. Manganese nodules, scattered rocks, and rock outcrops are characteristic of the Bellingshausen Sea north of the polar front. Dredges from this area contain manganese nodules, manganese encrusted rocks, and pebbles of shale, quartzite, schist, granite and pumice. (See also figs. 17–25, 17–26, and 17–27.) Taken in the northern Bellingshausen Sea, west of Tierra del Fuego, at 58°54′S, 95°08′W, and a depth of 3,981 m.

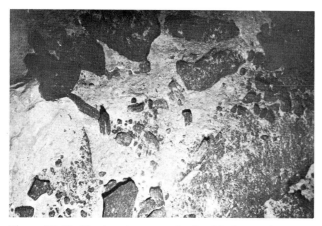

Figure 17–25. Rock outcrops and talus blocks on the ocean-basin floor. The vast majority of photographs on the lower flanks of the Mid-Oceanic Ridge or in the ocean basin floor reveal a mud bottom. These unusual examples of rock outcrops are probably the result of continued non-deposition resulting from bottom-current scour. (See figs. 17–24 and 17–26.)

Figure 17–26. Talus blocks on the ocean-basin floor (see figs. 17–24 and 17–25).

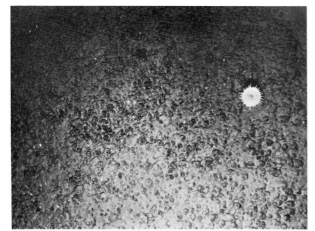

Figure 17–27. Nodules and scattered-rock carpet on the eastern flank of the Mid-Oceanic Ridge. Note scour marks around the nodules and coelenterate. Angular, ice-rafted rocks are also apparent. Light specks may represent suspended matter. (See fig. 17–28.) Taken in the western Bellingshausen Sea at 57°59′S, 120°03′W, and a depth of 4,825 m.

Figure 17–28. Rock outcrops, suspended matter, and crinoids on the ocean-basin floor. Note the suspended sediment cloud (See fig. 17–27.)

Figure 17–29. Nodules and scattered-rock carpet on the ocean-basin floor beneath the easterly bottom current. Note the light color fecal knot and fecal strings. Taken at 59°01′S, 99°54′W and a depth of 5,072 m.

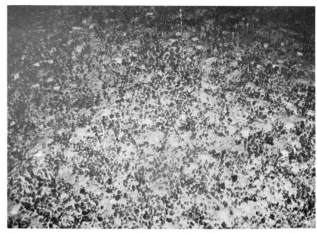

Figure 17–30. Nodules and scattered rocks on the ocean-basin floor beneath the easterly bottom current. Light colored sediment appears between nodules. Note the stalked organism in top center. Taken at 59°07′S, 105°03′W, at a depth of 3,904 m.

Figure 17–31. Rock outcrops on the ocean-basin floor. Note the fecal string or worm tube draped on the outcrop. Taken in the northwestern Bellingshausen Sea, at 60°01′S, 104°52′W, and a depth of 4,802 m.

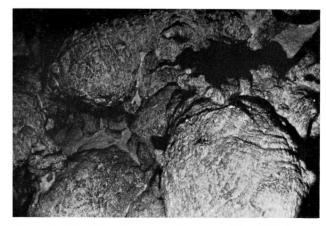

Figure 17–32. Pillow lava near the crest of the Mid-Oceanic Ridge. Note the characteristic radial jointing patterns of pillow lava. Taken at 56°07′S, 144°59′W, at a depth of 2,629 m.

Figure 17–33. Pillow lava near the crest of the Mid-Oceanic Ridge, western Bellingshausen Sea. Taken at 57°35'S, 138° 51'W, at a depth of 2,921 m.

Figure 17–34. Fractured pillow lava on the crest of the Mid-Oceanic Ridge, western Bellingshausen Sea. Note the striking radial jointing characteristic of pillow lavas. (See also figs. 17–35, 17–36, and 17–37.) Taken at 55°59'S, 134°27'W, at a depth of 3,157 m.

Figure 17–35. Pillow lavas on the crest of the Mid-Oceanic Ridge. This photograph, taken a few hundred meters from fig. 17–34 shows a smooth, slightly jointed pillow. Without the photograph shown in fig. 17–34 this identification would have been difficult. (See figs. 17–34, 17–36, and 17–37.)

Figure 17–36. Ripples on the Mid-Oceanic Ridge. Photographs in sediment-covered regions near the crest of the Mid-Oceanic Ridge often show well developed, straight-crested ripples. Other photographs from this station reveal rock outcrops. (See figs. 17–34, 17–35, and 17–37.)

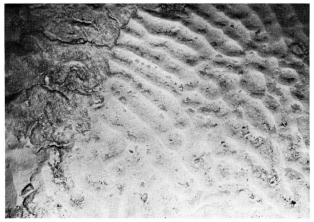

Figure 17–37. Ripples and outcrop on the Mid Oceanic Ridge. (See figs. 17–34, 17–35, and 17–36.)

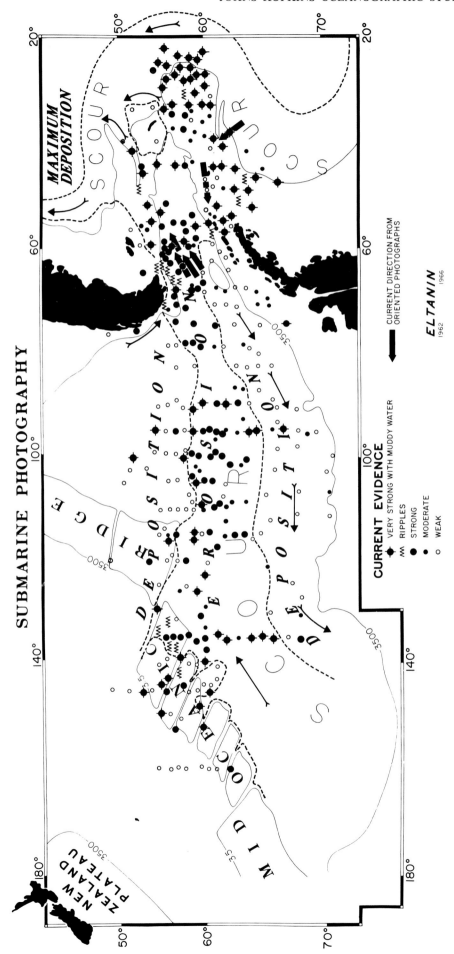

Figure 17–38. Current evidence observed in sea-floor photographs. Based on 440 USNS *Eltanin* camera stations.

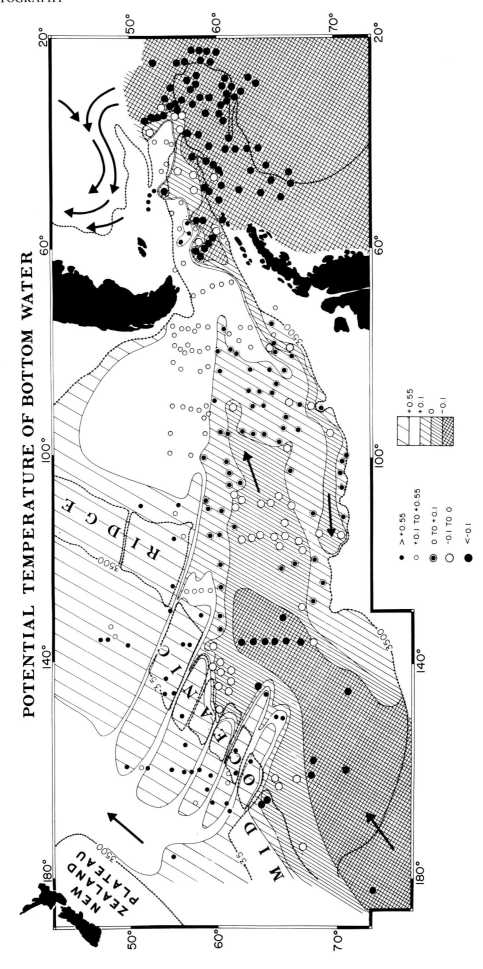

Figure 17–39. Bottom water circulation inferred from bottom temperature distribution. Based principally on observations by *Eltanin* (Jacobs, 1965, 1966; Gordon [in press]).

18. Underwater photography of the Carlsberg Ridge[*]

A. S. Laughton *National Institute of Oceanography, Wormley, Godalming, Surrey, England*

Abstract

Surveying, dredging, and underwater photography were carried out in three contrasting areas on the Carlsberg Ridge in the northwest Indian Ocean: (a) an area on the crest of the ridge, including the median valley, (b) an area in the southern foothills of the ridge, and (c) an area on the axis of the ridge but northwest of the Owen fracture zone, where the mountains are continental in character.

The photographs are described relative to the bathymetric profiles beneath the stations and with reference to the rocks dredged in the vicinity. On the crest of the ridge the rocks are fresh, brecciated, and metamorphosed lavas with a thin manganese crust; in the foothills the lavas are more altered and the manganese crust is thick; and in the third area the rocks are not volcanic at all, but are more similar to continental limestones.

18–1. Introduction

Three areas of the Carlsberg Ridge in the northwest Indian Ocean were studied in detail during the cruise of RRS *Discovery* in 1963, as part of the International Indian Ocean Expedition. The aim of this part of the cruise was to study the Carlsberg Ridge and its relationship with the Owen fracture zone (Matthews, 1966), and the Gulf of Aden (Laughton, 1966) in order to understand the way in which a mid-ocean ridge meets a continental mass.

Surveys by HMS *Owen* in 1961 and 1962 (Hydrographer, 1963) showed that the Carlsberg Ridge is in many ways similar to the Mid-Atlantic Ridge, especially in having a median valley characterized by a large magnetic anomaly. However, only a few rock samples had previously been obtained from the ridge and therefore two areas (4a and 4c of fig 18–1) were chosen for detailed geological and geophysical study. Area 4a (280 square

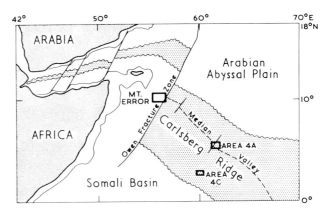

Figure 18–1. Location of areas photographed and sampled on the Carlsberg Ridge. (Light stippling indicates Carlsberg Ridge and its extension into the Gulf of Aden.)

miles) was chosen as representative of the crest of the ridge, including the median valley. Area 4c (100 square miles) was chosen as representative of the foothills of the ridge, and it was hoped to obtain here rocks which could be dated in order to test the Hess-Dietz theory of ridge formation. These two areas were surveyed by HMS *Owen* in 1962 using anchored radar beacons and charts of bathymetry; magnetic anomaly field and gravity field measurements were made (Hydrographer, 1966). RRS *Discovery* used these charts as a basis for rock dredging, bottom photography, coring, and heat-flow measurements in 1963.

A third area, Mount Error, lies on the axis of the Carlsberg Ridge and was visited twice by RRS *Discovery* in 1963, when it was both surveyed and sampled. Mount Error differs in nearly all respects from the mountains in area 4a. It lies northwest of the Owen fracture zone and is similar to a continental fragment.

In each of these three areas rock samples were obtained by dredging. Underwater photographs were taken in order to indicate where to dredge, the way in which the rocks lay on the bottom, and to assist in the interpretation. The rocks of area 4a have been described by Cann (1965) and those from the other areas are still being examined.

The photographs were taken with a bottoming camera (Laughton, 1957) at a height of 10 ft (3.1 m) above the

* Acknowledgment must be made of the tireless efforts of J. M. Jopling who maintained and operated the cameras with which these pictures were taken, to M. N. Hill who organized and led the scientific party, and to the Master and crew of RRS *Discovery* who brought the ship to the exact point required and kept it there.

191

bottom (station 5222 at 5 ft [1.5 m] above bottom). The area was illuminated from below the camera, which pointed 50° from the vertical. Shadows behind upstanding objects are therefore seen behind them. At each station about fifty photographs were taken at intervals of one or two minutes while the ship drifted. The drift tracks were fixed by radar from anchored buoys. As far as was possible, the station was arranged so that the drift of the ship enabled a section of photographs to be taken over a region (e.g., a cliff) near to one which had been sampled by dredging. However, in the strong surface currents which prevail over the Carlsberg Ridge, especially at the northwestern end, the differences in ship control during dredging and photographing resulted in a rather poor correlation between the two activities.

This chapter presents the photographic data obtained in these three areas and discusses the differences between them and the usefulness of underwater photographs as an aid to submarine geology.

18-2. Median valley area (4a)

The survey by HMS *Owen* covered 280 square miles, with tracks at one-mile spacings controlled by radar from anchored buoys. The topography (fig. 18-2) and the magnetics showed that the median valley in the area was displaced 10 miles (16 km) by a right lateral fault (Matthews, Vine, & Cann, 1965). The transverse valley

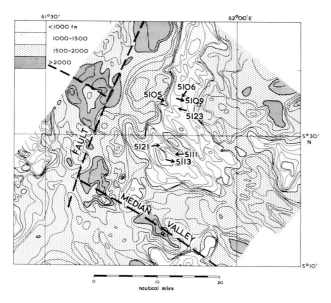

Figure 18-2. Bathymetric chart of area 4a on crest of Carlsberg Ridge. Depths in fathoms (assuming sounding velocity 800 fm/sec). Contour interval 100 fm. — trend of median valley and transverse fault. Camera stations: 5105, 5109, 5113, 5121. Dredge stations: 5106, 5111, 5123.

resulting from this displacement is flanked on the southeast by a ridge. Four camera stations were made, one near the foot of the west-facing slope of this ridge, one on the crest of the ridge, and two near the crest of the associated ridge to the south which has a west-northwest-east-southeast lineation. Profiles of these stations are shown in fig. 18-3, and the descriptions of the photographs are given in table 18-1.

These pictures present two interesting points. Much brecciation has taken place, and the fragments have in many places been rounded by the strong bottom currents which are indicated not only by the ripple marks but also by the movement of relatively large pebbles into the troughs of ripples and into bands at right angles to them.

Secondly, in stations 5113 and 5121, the massive outcropping rocks are rounded and frequently show surface lineations, sometimes with two sets at right angles. These were at first thought to be surface expressions of the schistosity revealed in some dredged rocks but are now believed to be reticulate fractures cutting across the glassy coating of lava pillows. Similar markings have been found on Icelandic pillow lava flows (Thorarinsson, personal communication).

The dredged samples show that all rocks are covered with a thin (1-5 mm) skin of manganese oxide which therefore partially obscures the true texture and color of the rocks photographed.

In area 4a, rocks were dredged at stations 5106, 5123, and 5111. They are described by Cann & Vine (1966) and comprise fresh olivine basalts; brecciated and cemented basalts; altered basaltic lava and tuff; and some foliated, dynamically metamorphosed basalts.

18-3. Southern foothills area (4c)

The survey of area 4c was conducted by HMS *Owen* in 1962, following the same plan as in area 4a, and again a good bathymetric chart was available before RRS *Discovery* started her program of dredging and photography. However, this program was beset with difficulties in maintaining an anchored buoy in position, due to the strong surface currents. In spite of this, four camera stations and three successful dredge hauls were made, but not in the positions planned. The topography consisted of elongated mountains and valleys running parallel to the axis of the Carlsberg Ridge (fig. 18-7), and the magnetic anomaly pattern showing similar trends suggested that the mountains are related to volcanic features. The elongation was more pronounced than in the area 4a. Two camera stations were made on a ridge on the east side of the area but no successful dredge hauls were obtained there. The other two camera stations were made

TABLE 18–1. Camera stations on crest of Carlsberg Ridge — area 4a (for profiles see fig. 18–3).

Station 5105	Forty pictures at 14-ft (4.3 m) range.
05°35.5′N 61°48.1′E	Near base of fault scarp.
Depth range	1,750–1,680 fm (uncorrected) (3,200–3,070 m).
Description	All but one picture show flat, mottled, sediment disturbed by burrowing benthos and surface tracks (fig. 18–4a). Fig. 18–4d shows fractured edge of lava flow.
Station 5109	Sixty-two pictures at 14-ft (4.3 m) range.
05°35.7′N 61°51.2′E	On ridge east of fault scarp.
Depth range	1,270–1,360 fm (uncorrected) (2,320–2,490 m).
Description	The first sixteen pictures show bedrock outcrops (fig. 18–6a), rounded boulders (fig. 18–6c), and gravel lying on a sand that is sometimes rippled. Two other groups of photographs (26–28 and 35–37) show outcropping bedrock (fig. 18–5a) and ripple-marked sand.
	All of the remaining pictures (except the last four) show well developed ripple markings in sand (fig. 18–4d). Sometimes these are long-crested and sometimes short, with a wavelength of about 15 cm. In many cases the troughs are filled with dark coarse material (fig. 18–4c); in others the coarse material almost obscures the ripple marks and is drawn out in streaks parallel to the current.
	The bedrock outcrops show no evidence of the cooling cracks found in stations 5113 and 5121.
	The last four pictures show pelagic ooze.
Station 5113	Fifty pictures at 14-ft (4.3 m) range.
05°26.2′N 61°49.1′E	Near crest of ridge at south end of fault scarp.
Depth range	960–1,180 fm (uncorrected) (1,760–2,160 m).
Description	The first fifteen pictures show outcrops of bedrock, with surface markings (fig. 18–5c), large angular boulders (fig. 18–6b) and ripple-marked sediments.
	The center group of pictures (16–46) are of sloping mottled sediment.
	The last four pictures show more surface-marked bedrock and ripple-marked sand.
Station 5121	Fifty-two pictures at 14-ft (4.3 m) range.
05°28.3′N 61°47.1′E	Near crest of ridge at south end of fault scarp.
Depth range	1,150–1,220 fm (uncorrected) (2,100–2,230 m).
Description	The first thirty-three pictures show dense accumulations of boulders, mostly rounded, and surface-marked bedrock (fig. 18–5b and d) with very little sediment.
	The remainder show well-developed short-crested ripple marks in sand.

on a less elongated mountain, and the three dredge hauls were in the same general area.

Profiles of the camera stations are shown in fig. 18–8, and table 18–2 summarizes the results of the photography.

In all stations manganese nodules or heavy manganese encrustation was found. The nodules were in some cases extremely uniform in size, although varying considerably

TABLE 18–2. Camera stations on foothills of Carlsberg Ridge — area 4c (for profiles see fig. 18–8).

Station 5127	Thirty-five pictures at 14-ft (4.3 m) range.
02°47.5′N 60°20.7′E	South cliff of eastern ridge.
Depth range	2,210–2,250 fm (uncorrected) (4,040–4,110 m).
Description	Dense concentration of rounded, uniform size (5 cm diameter), manganese nodules (fig. 18–9a and b) in the first half of the station. On a cliff and a rise there is massive, rounded bedrock (fig. 18–10a). Subsequent dredging showed this to be a manganese crust (10 cm thick) covering a brecciated and altered lava flow.
	The last thirteen pictures show reworked ooze.
Station 5128	Twenty pictures at 14-ft (4.3 m) range.
02°47.8′N 60°21.4′E	On eastern ridge.
Depth range	2,050–2,070 fm (uncorrected) (3,750–3,790 m).
Description	All pictures show massive, rounded bedrock together with partly cemented conglomerate of boulders encrusted by manganese (fig. 18–10c and d).
Station 5132	Fifty-one pictures at 14-ft (4.3 m) range.
02°46.2′N 60°02.5′E	Near crest of southwest ridge.
Depth range	2,060–2,160 fm (uncorrected) (3,770–3,950 m).
Description	Twenty-one pictures of small (3 cm diameter), somewhat angular, manganese nodules (fig. 18–9d), followed by seven of reworked ooze. A steep rise to the crest shows large, rounded boulders (0.5 m in diameter) sometimes joined to form a continuous rock (fig. 18–11a, b, and c).
	A terrace of ooze follows, the last few pictures of which show more manganese nodules, larger and more rounded than the first group (fig. 18–9c).
	The last eight pictures on a cliff show massive, rounded bedrock, sometimes very steep, and the surface made continuous by a thick manganese layer (fig. 18–10b).
Station 5137	Fifty-one pictures at 14-ft (4.3 m) range.
02°46.0′N 59°53.0′E	South cliff of southwest ridge.
Depth range	1,980–2,150 fm (uncorrected) (3,620–3,930 m).
Description	Steady slope of reworked ooze in all pictures, except across two cliffs where there are massive rounded boulders and bedrock (fig. 18–11d).

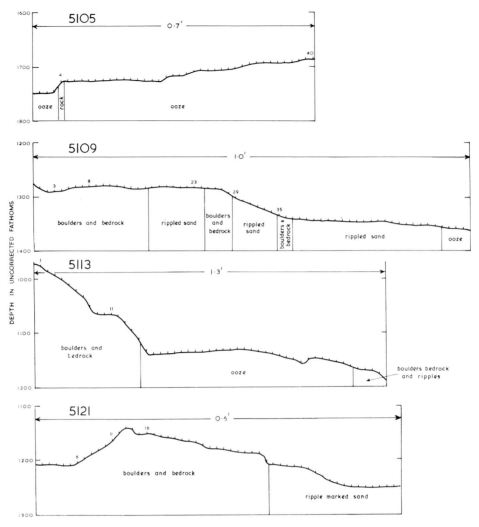

Figure 18–3. Profiles of camera stations in area 4a, deduced from length of wire payed out, wire angle and echo-sounding depth. Ticks on profile indicate position of successive pictures (numbers show pictures illustrated in figs. 18–8 to 18–10). Lengths of profiles determined by radar fixes during stations. Vertical exaggeration between 1:1 and 2:1.

in concentration, and in other cases they varied in size and were more angular. In the fields of uniform round nodules they appear to be all lying on the surface, whereas in the more varied fields many are almost completely buried. The areas in which the nodules are found are well defined, adjacent sediment areas being little different from the usual reworked abyssal ooze. The nodule areas are more commonly found on the crest or flanks of the ridges, whereas the lower slopes are devoid of them. This may result from variations in the rate of sedimentation and the strength of bottom currents, although no ripple marks were found in the area. Many nodules were recovered in the dredge hauls.

Where massive bedrock was photographed, it was always well rounded and in some cases clearly represented a thick crust of manganese over a highly weathered massive lava. This was confirmed by 10-cm thick manganese encrustations brought up in the dredge. The core material beneath the crust was in all cases a friable white clay, probably the end product of advanced hydrothermal alteration of the original basaltic lava. No fresh basalts were obtained from which a potassium-argon date could be measured.

18–4. Mount Error

Mount Error lies immediately to the northwest of the Owen fracture zone on the center line of the extension of the Carlsberg Ridge. It was first discovered, although not mapped in detail, by HMS *Owen* in 1962 and was visited twice during *Discovery*'s cruise in 1963. On the second visit the flat top was surveyed in detail, a seismic refraction station was made, and two successful dredge hauls and many pictures on two camera stations were obtained.

Topographically, the seamount differed from those of the Carlsberg Ridge to the southeast (fig. 18–12). It was much larger, being nearly 100 miles (160 km) long on its base, the vertical relief being 2,300 fm (4,200 m). The flat top, some 15 by 5 miles (24 by 8 km), lies at 230 fm (420 m) and is swept by strong currents. Evidence for these came partly from the strong surface currents (up to 2 knots) and partly from a patch of sand waves observed on the echo sounder. The edge of the flat top fell steeply away and at the southwest end was incised by what appeared to be canyons. The north-facing side was exceptionally steep, averaging 1 in 2 in places.

Both camera stations started on the flat top, and photographs were taken as the ship was carried down slope by the current. A profile of station 5222 is shown in fig. 18–13 and descriptions of the photographs in table 18–3. Photographs of station 5220 are not illustrated as they were taken at long range (40 ft, 12 m), and detail is difficult to see.

The rocks seen on Mount Error are quite different from those of areas 4a and 4c. On the edge of the flat top, fractured slabs of flat-bedded rocks were seen on which

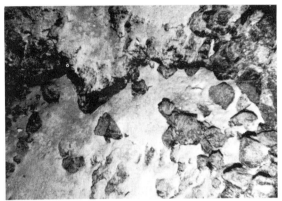

TABLE 18–3. Camera station on Mount Error (for profile see fig. 18–13).

Station 5222	Thirty-four pictures at 14-ft (4.3 m) range.
10°15.8′N 56°00.3′E	Southwest edge of plateau and on cliff.
Depth range	210–300 fm (uncorrected) (380–550 m).
Description	The first two pictures on the flat top show linear fractured, flat, rock slabs (fig. 18–14a). On the cliff there is a variety of bottom types which include small manganese-covered rock outcrops (fig. 18–14c), live coral and coral debris (fig. 18–14d and a), semiconsolidated sediments (fig. 18–15a), and hard rock cliffs, fractures, and dykes (fig. 18–15d, b, and c).
	All pictures show a considerable benthic community and an absence of loose sediment cover, doubtless resulting from the strong currents in the area.

Figure 18–4. On the crest of the Carlsberg Ridge (area 4a). (a) 5105.40 (1,690 fm, 3,090 m). Pelagic sediment, mottled and reworked by benthic fauna. Track of unidentified surface-living animal. (b) 5105.4 (1,730 fm, 3,160 m). Fractured edge of lava flow, outcropping on only cliff in this profile. (c) 5109.23 (1,280 fm, 2,340 m). Ripple-marked sand (wavelength 15 cm). Coarse material in troughs is probably manganese-coated fragments of altered basalt. (d) 5109.29 (1,300 fm, 2,380 m). Exceptionally long-crested ripple marks in sand (wavelength 15 cm). (Area of pictures 3 by 4 m: for station details see table 18–1.)

lived a variety of benthic fauna, but there was remarkably little sediment. On the cliff face the rock had the appearance of an eroded sedimentary rock partly encrusted with corals and usually darkened by a thin film of manganese. Where the sediment pockets adjoined rock outcrops, this coating was absent, revealing the true lighter color of the rock beneath. This absence of manganese coating may be due to abrasion by current-moved sediment. Elsewhere, narrow linear fractures and upstanding dykes are visible, suggesting the type of fracturing and subsequent infilling by harder deposits that is found in limestone on land. At the base of slopes, and in small pockets, the sediment contains large quantities of coral debris and other unidentifiable fragments, often appearing to be semiconsolidated.

Dredged samples from this cliff, which were particularly hard to get, were hard, pink, porcellanous limestones coated with a manganese film and encrusted with many living and dead organisms. From the dredged material, the underwater photographs, and from topographic, magnetic, and seismic studies, Mount Error is more akin to a piece of foundered continent than to a volcanic seamount. The huge displacements of the oceanic crust which resulted in the Gulf of Aden (Laughton, 1966) and the Owen fracture zone (Matthews, 1966) may have shifted Mount Error, together with the Socotra shelf, from an original position joining the Arabian continent near the Kuria Maria Islands.

18–5. Discussion

The three areas of the Carlsberg Ridge studied have each shown a characteristic set of both rocks and photographs. On the crest, fresh, brecciated, and metamorphosed lavas are common, covered with only a thin skin of manganese seldom more than 1 cm thick. The photographs therefore reflect something of the true texture of the rock, its angularity or roundness, and its surface markings. By contrast, on the foothills these features are almost totally obscured by the thick manganese crust smoothing and rounding the bedrock surface. Here the bedrock has been reduced to clay by hydrothermal alteration. On Mount Error, the photographs are strikingly different from areas 4a and 4c. There are no lava flows or rounded boulders. Fractures are linear and sometimes show as hard veins that have resisted erosion. Rock surfaces are often eroded or abraded to remove a thin manganese film and to reveal the true rock color and texture. The increased density of benthic fauna can be attributed to the shallowness of the water.

In correlating the picture sequences with the depth profiles, most of the pictures of exposed rocks are either on the crest of small hills or on local steep cliffs, as one would expect. The manganese nodules appear where the slopes are high and presumably where the sedimentation rate is low due to bottom winnowing by currents.

At first sight it is surprising to find such a small percentage of rocks on what often appear to be steep slopes. The vertical exaggeration associated with the usual echo-sounding record is sometimes misleading, and the multiple hyperbolae often indicate only undulating sediment accumulations. Rock exposures are commonest near to the top of seamounts, and the slopes beneath the cliffs where one might expect to find screes are usually sediment covered. In rock collection, whether by dredging or shallow drilling, this should be remembered.

Bottom photography is only a tool in the interpretation of submarine geology and cannot do more than give a small sample of the area being studied. Controlled dredging can give the geologist rocks to work on in the laboratory and drilling can give the geological picture in greater depth, but visual observation by remote photography, or, in the near future, by personal inspection from a bathyscaph, is necessary to examine regional differences, small structures, textures, and the environmental setting from which the rocks are taken.

References

Cann, J. R., & F. J. Vine, 1965: An area on the crest of the Carlsberg Ridge: petrology and magnetic survey. *Phil. Trans. Roy. Soc.*, ser. A, **259**, 198–217.

Hydrographer, 1963; Bathymetric, magnetic and gravity investigations, H.M.S. *Owen*, 1961–1962. *Adm. Mar. Sci. Pub.* no. 4, parts 1 and 2.

———, 1966: Bathymetric, magnetic and gravity investigations, H.M.S. *Owen*, 1962–1963. *Adm. Mar. Sci. Pub.* no. 9, parts 1 and 2.

Laughton, A. S., 1957; A new deep-sea underwater camera. *Deep-Sea Res.*, 4, 120–25.

———, 1966: The Gulf of Aden. *Phil. Trans. Roy. Soc.*, ser. A, **259**, 150–71.

Matthews, D. H., 1966: The Owen fracture zone and the northern end of the Carlsberg Ridge. *Phil. Trans. Roy. Soc.*, ser. A, **259**, 172–86.

———, F. J. Vine, & J. R. Cann, 1965: Geology of an area of the Carlsberg Ridge, Indian Ocean. *Bull. Geol. Soc. Am.*, **76**, 675–82.

Figure 18-5. On the crest of the Carlsberg Ridge (area 4*a*). (*a*) 5109.35 (1,330 fm, 2,430 m). Outcrops of volcanic basalt surrounded by small broken fragments. Ripple marks may result from funneling of current between rocks (note crinoid on the rock). (*b*) 5121.6 (1,210 fm, 2,210 m). Volcanic basalt thinly encrusted with manganese. Note lack of sediment in between boulders. (*c*) 5113.1 (970 fm, 1,770 m). Massive pillow lavas with surface markings due to extrusion and cooling fractures. (*d*) 5121.16 (1,150 fm, 2,100 m). Massive pillow lavas with extrusion markings. (Area of pictures 3 by 4 m: for station details see table 18–1).

a

b

c

d

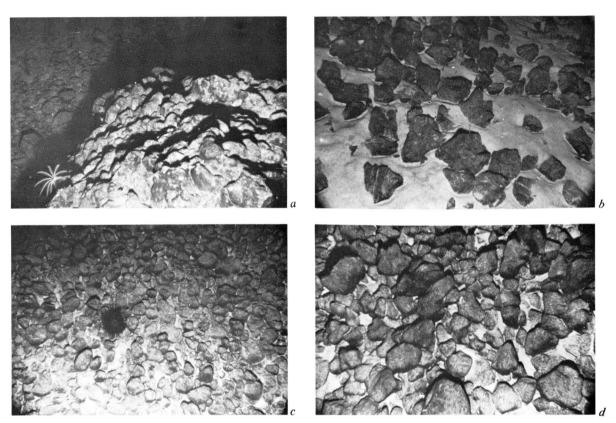

Figure 18–6. On the crest of the Carlsberg Ridge (area 4*a*). (*a*) 5109.3 (1,280 fm, 2,340 m). Outcrop of brecciated and meta-morphosed volcanic lava with loose boulders in the background (note crinoid). (*b*) 5113.11 (1,070 fm, 1,960 m). Angular boulders of fractured volcanic lava. Note the scour pits around the boulders due to bottom currents. (*c*) 5109.8 (1,270 fm, 2,320 m). Accumulation of boulders rounded either by manganese encrustation or by bottom current abrasion (note echinoid). (*d*) 5121.11 (1,160 fm, 2,120 m). Conglomerate of rounded boulders typical of this station showing signs of cementation. (Area of pictures 3 by 4 m: for station details see table 18–1.)

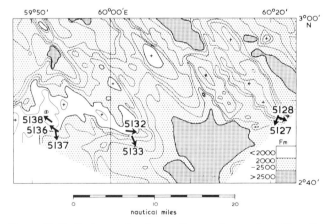

Figure 18–7. Bathymetric chart of area 4*c* on foothills of Carlsberg Ridge. Depths in fathoms (assuming sounding velocity 800 fm/sec). Contour interval 100 fm. Camera stations: 5127, 5128, 5132, 5137.   Dredge stations: 5133, 5136, 5138.

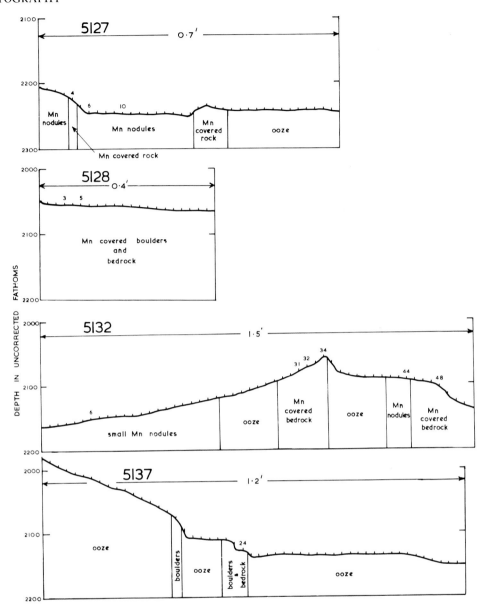

Figure 18–8. Profiles of camera stations in area 4c, deduced from length of wire payed out, wire angle, and echo-sounding depth. Ticks on profile indicate position of successive pictures (numbers show pictures illustrated in figs. 18–11 to 18–13). Lengths of profiles determined by radar fixes during stations. Vertical exaggeration between 1:1 and 2:1.

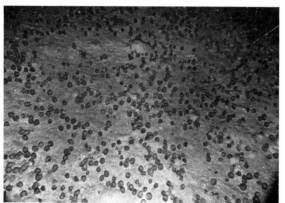

a

Figure 18–9. On the foothills of the Carlsberg Ridge (area 4c). (a) 5127.6 (2,240 fm, 4,100 m). High concentration of uniform, rounded manganese nodules of 5-cm diameter. Most are on the sediment surface. (b) 5127.10 (2,240 fm, 4,100 m). Low concentration of uniform, rounded manganese nodules of 5-cm diameter. Most are on the sediment surface. (c) 5132.44 (2,090 fm, 3,820 m). High concentration of rounded manganese nodules of two predominant diameters (5 cm and 3 cm). Many nodules are partially or totally covered by sediment. (d) 5132.6 (2,160 fm, 3,950 m). Small angular manganese nodules probably in a more youthful stage of development than the rounded ones in the other pictures. (Area of pictures 3 by 4 m: for station details see table 18–2.)

b

c

d

Figure 18–10. On the foothills of the Carlsberg Ridge (area 4c). (a) 5127.4 (2,230 fm, 4,080 m). Highly altered volcanic lava thickly encrusted with manganese to form smooth rounded surface. (b) 5132.48 (2,110 fm, 3,860 m). Lava flow encrusted with manganese 10 cm thick to form smooth rounded surface obscuring brecciated rock surface beneath (a sample of this crust was dredged at station 5136). (c) 5128.3 (2,280 fm, 4,170 m). Manganese-encrusted lava flow. (d) 5128.5 (2,290 fm, 4,190 m). Manganese-encrusted lava flow with an upstanding angular rock in foreground. (Area of pictures 3 by 4 m: for station details see table 18–2.)

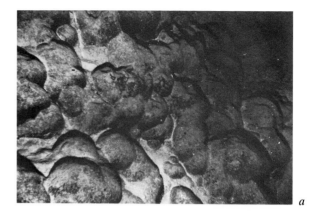

a

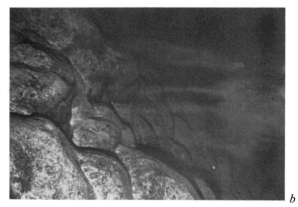

b

c

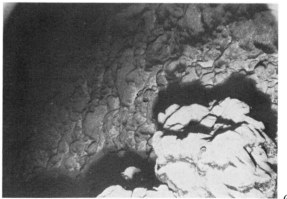

d

Figure 18–11. On the foothills of the Carlsberg Ridge (area 4c). (a) 5132.31 (2,080 fm, 3,800 m). (b) 5132.32 (2,070 fm, 3,790 m). (c) 5132.34 (2,060 fm, 3,770 m). All these pictures show large rounded boulders or bedrock smoothed by thick manganese encrustation and covered by varying amounts of sediment. All three pictures are near the crest of a ridge. (d) 5137.24 (2,120 fm, 3,880 m). Rounded boulders similar to those above, on a cliff outcrop on the south side of the ridge. (Area of pictures 3 by 4 m: for station details see table 18–2.)

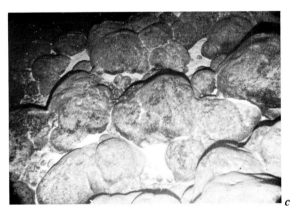

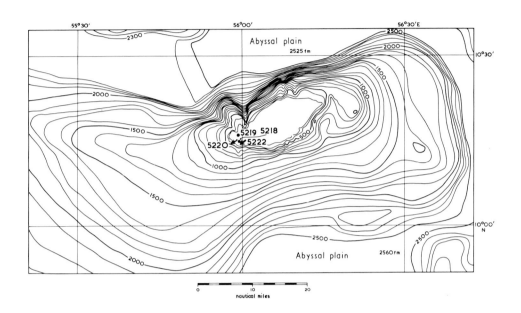

Figure 18–12. Bathymetric chart of Mount Error. Depths in fathoms (assuming sounding velocity 800 fm/sec). Contour interval 100 fm. Camera stations: 5220, 5222. Dredge stations: 5218, 5219.

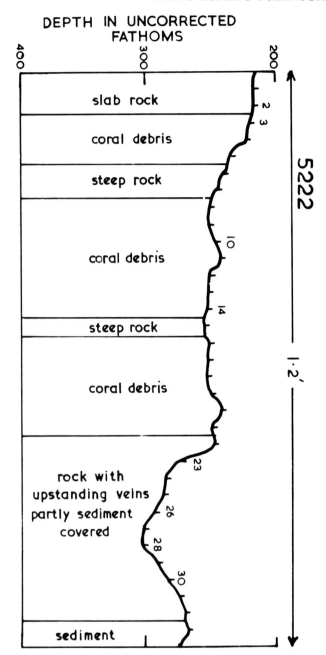

Figure 18–13. Profile of camera station 5222 on Mount Error, deduced from length of wire paid out, wire angle, and echo-sounding depth. Ticks on profile indicate position of successive pictures (numbers show pictures illustrated in figs. 18–14 and 18–15). Length of profile determined by radar fixes during station. Vertical exaggeration 3:1.

Figure 18–14. Mount Error. (*a*) 5222.2 (220 fm, 402 m).
Slabs of flat rock with parallel fractures on plateau of seamount. Rock face is devoid of sediment cover but carries a
dense fauna. (*b*) 5222.3 (225 fm, 411 m). Slope off the edge of
the plateau strewn with coral debris. (*c*) 5222.10 (246 fm,
450 m). Outcrops of eroded bedrock of pale color. These
are darkened by manganese encrustation where abrasion by
sediment is not active. There is much coral debris, some live
coral, and a holothurian. (*d*) 5222.14 (258 fm, 472 m). Slope
encrusted with some live coral and much coral debris. The
fish has not been identified. (Area of pictures 1.5 by 2 m: for
station details see table 18–3.)

a

b

c

d

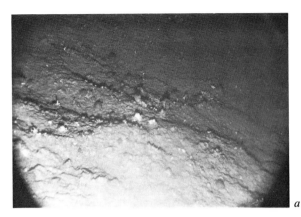

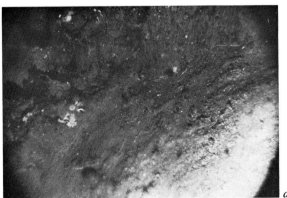

Figure 18–15. Mount Error. (*a*) 5222.23 (284 fm, 519 m). Slope of semiconsolidated (or eroded) sedimentary rock supporting dense fauna. (*b*) 5222.26 (304 fm, 556 m). Thin sediment cover over vertically fractured rock. Width of fracture is about 5 cm. (*c*) 5222.28 (313 fm, 572 m). Vertical dyke some 2 cm wide and 15 cm high protruding through sediment slope. (*d*) 5222.30 (293 fm, 536 m). Vertical rock face possibly exposed by slumping or faulting. Note the vertical striae on the vertical face. (Area of pictures 1.5 by 2 m: for station details see table 18–3.)

19. Biological applications of sea-floor photography*

H. Barraclough Fell   *Museum of Comparative Zoology, Harvard University, Cambridge, Massachusetts*

Abstract

Underwater photography from either ships or submersibles is an indispensable technique in both qualitative and quantitative studies of sea-floor life. Besides showing this life in its natural state, photography enables more random sampling of epifauna than do methods involving the collection of samples by dredge, grab or trawl. To obtain maximum information, photography (in monochrome and color) and collecting methods must be used together. In this way much can be learned about such aspects as the ecology, feeding habits, methods of locomotion, reaction to light and temperature, community formation, and population densities of underwater life.

19–1. Introduction

Some notable photographs of populations of sea-floor animals have been obtained by Laughton (1959) in deep water, and by a number of others working in shallower seas; yet surprisingly little use has been made of the material, and the method, by biologists. Photographs of the Antarctic sea floor in color and in monochrome were published by Bullivant (1959) before the first Arctic sea-floor photographs. Some of Bullivant's pictures showed massive sponge formations on the Ross Sea floor; when Kuznetsov (1960) first reported "colossal accumulations of sponges" from the floor of the North Atlantic, he could compare them with similar formations already known from the Antarctic.

Yet it is true that many thousands of excellent sea-floor photographs have been taken in the northern hemisphere since the technique of remotely controlled underwater cameras was developed in 1940. That few biologists have studied them is probably due to the fact that many of the pictures were taken without obtaining corresponding samples of animals from the same stations by means of dredge or grab or trawl; hence the identification of the indistinct images in the pictures has necessarily been a difficult task which few systematists would care to undertake, save only in the most general terms. To give a confident determination of species from photographs requires very close familiarity. However, where biologists have been directly associated with a photographic program relating to a group in which they were experienced, or a concomitant program of biological and photographic sampling has been carried out, some detailed work has emerged (Vevers, 1951, 1952; Fell, 1961; Clark, 1963; Marshall & Bourne, 1964). There can be little doubt that photographic techniques will occupy a significant place in future biological oceanography. Collateral photographic programs have also been initiated using underwater television, a method not discussed in this chapter, and photography from the port of bathyscaphs (see chap. 25). The latter method is, of course, restricted to very few users, but it has resulted in some pictures of outstanding interest, as for example some published by LaFond (1962; see also chap. 5). Photographs taken from the US bathyscaph *Trieste* have been studied by the writer, with the kind cooperation of U.S. Navy scientists, and some are reproduced here. Utilization of drifting ice floes by members of the Lamont Geological Observatory enabled some notable photographs to be taken of the deep Arctic floor (Hunkins, Ewing, Heezen, & Menzies, 1960). Though the pictures are primarily of non-biological subjects, some biological interest attaches to the traces observed in sediments. The elaborate equipment required for some of these techniques may have discouraged smaller institutions from undertaking sea-floor photography; indeed Longhurst (1964) implies that the majority of laboratories should concentrate their efforts on improving grabs, since costs will exclude the other methods for many years. This viewpoint overlooks the important fact that seaworthy underwater cameras can be constructed at a relatively low cost, and that some of the authors cited have used photographs obtained by very

* This research was supported in part by the National Science Foundation grant GB-3532. I am grateful to the following for generous cooperation in providing material for study: Dr. J. B. Hersey, Woods Hole Oceanographic Institution (figs. 19–1, 19–4a, 19–10); Dr. J. S. Bullivant, and the New Zealand Oceanographic Institute (figs. 19–2, 19–5, 19–6b, 19–7 to 19–9, 19–11); Drs. Eugene LaFond and Eric Barham, U.S. Navy Electronics Laboratory, San Diego (figs. 19–3, 19–4b, 19–6a). Dr. John H. Dearborn kindly assisted me in some photographic processing. The identification of *Scotoplanes* in fig. 19–3 is due to Dr. David L. Pawson.

modest equipment. A further development has been the utilization of stereo cameras, notably by Hersey using methods developed by Edgerton. Stereo pictures greatly enhance the perception of fine detail of importance in determining species, as stressed by Marshall & Bourne (1964; see also chap. 23). Examples of such pictures are shown in fig. 19–1, and also in fig. 19–4a, where the stereo presentation aids interpretation of the biology, and in

fig. 19–10, where it aids the systematic recognition of species seen.

Some of the current activity in marine biology is directed toward determining the feeding habits of deep-water animals, and their bearing on other problems of oceanography (Sokolova, 1958), and toward an elucidation of vertical zonation (Vinogradova, 1961). Quantitative and qualitative surveys of the sea-floor faunas are

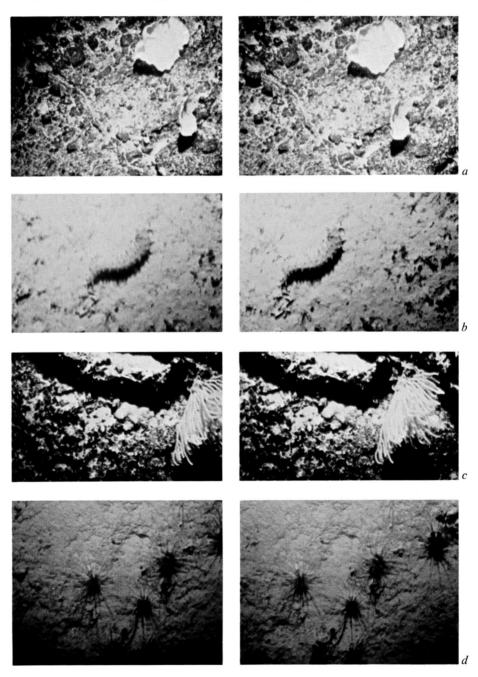

Figure 19–1. Archibenthic invertebrates, Plantagenet Bank, near Bermuda: (*a*) fan-sponge *Phakellia;* (*b*) holothurian; (*c*) gorgonian, ? *Calyptrophora;* (*d*) echinoid *Aspidodiadema*. Depths (*a*), (*c*), 750 m; (*b*), 800 m; (*d*) 450 m (J. B. Hersey *et al.*).

Figure 19–2. Ross Sea shelf fauna; sponges, coelenterates, polyzoans and crinoid *Promachocrinus;* depth, 110 m (J. S. Bullivant, New Zealand Oceanographic Institute).

in progress in many areas. In all these phases, photography can play a significant part. To be effective, sea-floor photographs should be taken at frequent intervals along the track of any ship which is trawling or dredging for biological specimens. Photographs should not be taken without biological samples, for the identification is diffi-

cult without comparative material. Neither should biological samples ideally be taken without photographs, because photographs throw much light on the environment and ecology of the animals collected. Photographs sometimes also reveal unsuspected circumstances, such as animals in an environment not previously thought appro-

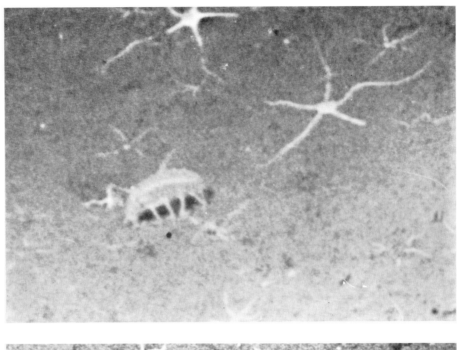

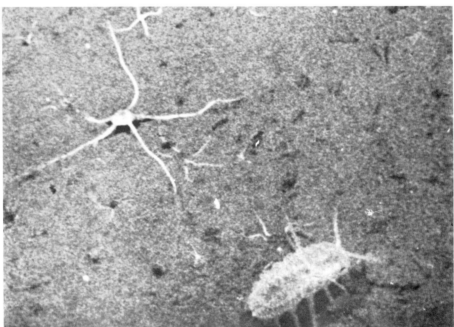

Figure 19–3. Soft-bottom association of ophiuroids (mainly *Ophiomusium*) and holothurians (*Scotoplanes*) on the floor of the San Diego Trench, as seen from the port of the bathyscaph *Trieste*, at 1,280 m (U.S. Navy Photo, Eric Barham).

priate for them, an example being Hurley's (1959) discovery of *Pyrosoma* on the sea floor at 160 m.

19–2. Suggested desiderata

Difficulties in interpreting sea-floor photographs of animal populations prompts a list of possible improve-

ments. It is not always easy to determine from the position of a shadow whether or not the animal is touching the bottom, or is located just above the bottom; stereo photographs solve this problem. To determine a species or interpret the activity of an animal may require gross enlargement or intensification of the original image; hence a fine grain is desirable, and fine-grain developers should be used. The photographs of the elasipod holo-

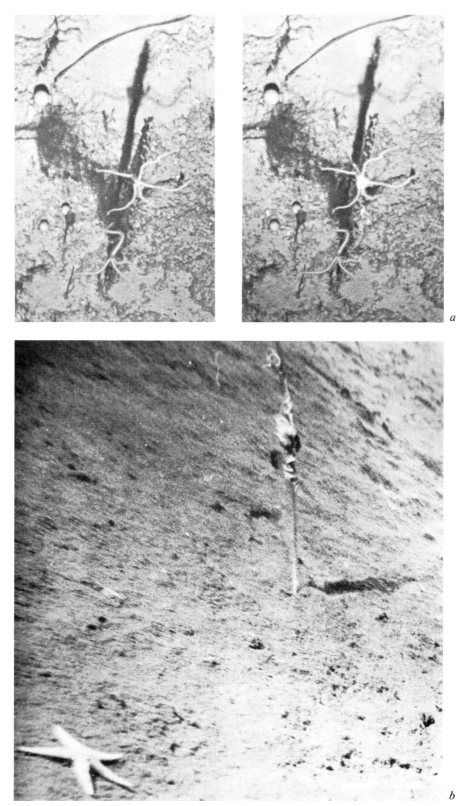

Figure 19–4. Epizoic ophiuroids in characteristic fishing postures. (*a*) probably *Ophiacantha*, on either a gorgonian or fan-sponge, Plantagenet Bank, at 300 m (J. B. Hersey *et al.*); (*b*) probably *Asteronyx*, on a gorgonian, San Diego Trench, at 600 m (U.S. Navy Photo, Martin and Dill); unidentified asteroid to left.

Figure 19–5. Ophiuroid *Ophiurolepis gelida* parasitized by sponge *Iophon radiatus;* gorgonian and asteroid (Odontaster) to left; Ross Sea, 75 m (J. S, Bullivant, New Zealand Oceanographic Institute).

thurian in figs. 19–3 and 19–6 are enlarged from very small images on the original negative, but they serve to reveal many unexpected and surprising features, such as the evident metachronal movement of the tube-feet and the antenna-like tube-feet held above the mouth and over the body, features reminiscent of annelids or arthropods, rather than echinoderms. Photographs are sometimes out of focus at critical levels: a battery of cameras could perhaps be designed to take pictures focused in different planes. Critical areas are sometimes overexposed, or underexposed: this could be overcome by use of a wide-latitude film or perhaps by using cameras with different apertures, or films of different speeds, all mounted in one unit. On ships of sufficient size a photographic laboratory could be included, and films developed as a transect proceeds. If the photographs do not appear to match the trawl contents, the transect could then be repeated. A unit of length, or area, is desirable in the field. Some cameras operate at a predetermined distance above the sea floor, yielding a side of known length to each frame, but some are erratic in operation. In some bathyscaph photographs a chance beer can, sent to the

sea bed from a passing ship, has served as the best unit of measure. If stereoscopic photographs are taken, the scale is available from the simple measurement technique described in chap. 4.

The relative value of color and monochrome photography has received limited attention in published work prior to this book. The writer regards color photography as a valuable means of identification in marine biological investigations. I have found (Fell [1962]), for example, that in detailing the successive steps by which exact determinations were made of the species of sea stars visible in photographs of the Antarctic sea bed, color patterns revealed by color photography were important features. The dredged samples, preserved as usual in fluids which bleach the colors of the animals, were white by the time the specimens reached a laboratory and could be identified. However, the specimens were photographed on the deck of the research vessel, thus providing data on the natural colors before preservation. Armed with this information, it was much easier to identify the rather indistinct images of the same species as seen in sea-floor photos taken at the photographic stations. Had black-

and-white film been used, the resultant pictures would have rendered the color patterns of the animals incorrectly (blue appearing as light grey, orange as light grey, hence these two colors being indistinguishable). It would not have been possible, therefore, to distinguish species of genera such as *Ophioceres* and *Ophiacantha*, where differential patterns of blue and orange occur. With color photos, on the other hand, such species are clearly identified. Similarly, dark-red animals, such as crustaceans and the echinoid *Sterechinus*, would not appear different from dark-green members of genera of the same groups when photographed on black-and-white film, but they are immediately distinguishable on color transparencies. These examples, based on actual experience, are probably typical and show that it would be a retrograde step not to use color film.

The case of *Sterechinus* (in Antarctic seas) has not yet been reported, though publication is in progress. All existing literature gives a totally incorrect account of the colors of these animals in life. The source of the error is traceable to the fact that, prior to the International Geophysical Year, all expeditions to the Antarctic had inevitably operated under severely adverse conditions; all biological samples were hurriedly obtained, and preserved immediately, for study in later years in the laboratories in the countries mounting the expeditions. As bleaching had occurred, the brilliant red echinoids were transformed to pale green, pale violet, or white. Such colors are carefully detailed in the standard monograph by Mortensen (1928–1952). However, this error was corrected when the first color photos were received from the Antarctic, revealing that all observed species of the genus are red. This color at once distinguishes them from green and purple species of other genera known to enter circumpolar waters in the American sector of the sub-Antarctic. Hence color photos of bottom populations are easy to interpret, since no confusion of the genera will now occur. Under conditions imposed by black-and-white photography, the various genera of regular echinoids would be hopelessly confused. It is obvious that similar arguments must apply to numerous other faunas where detailed work has not yet been undertaken. Color photography is an important aspect of sea-floor studies and should be carried out at all biological stations.

19–3. Interpretation of bottom photographs

Relationship of organism to substrate

While some benthic organisms tolerate a wide range of sea-floor conditions, many are restricted to particular types of bottom. This is not always evident from the

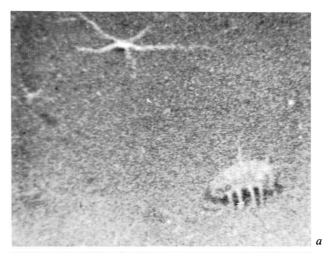

a

b

Figure 19-6. Soft-bottom ecology: (*a*) illustrating false sediment grain produced by excessive enlargement, but revealing systematic characters of the organisms; (*b*) characteristic castings of holothurians on soft substrate—features which, when found as fossils, have been commonly misinterpreted as "worm traces." (U.S. Navy photos, Eric Barham.)

collections taken from trawls and dredges, since these instruments may traverse a variety of bottom during one transect. Here a useful adjunct to the collection procedure is provided by taking periodic bottom photographs during the progress of the transect. Bottom may vary from naked, or thinly silted, rock (fig. 19-1*c*) to coarse or fine gravel (fig. 19-1*a*), or soft silt or ooze (fig. 19-1*b* and *d*.) On the harder substrates many of the organisms will be anchored to the hard objects, either permanently, as in the cases of the sponge and gorgonian coral illustrated in fig. 19-1, or they may be facultatively mobile, attaching themselves by limbs or by clasping organs such as cirri in the case of some crinoids (fig. 19-2). The epifauna on a soft substrate may itself provide a hard substrate for such mobile elements as crinoids or ophiuroids (fig. 19-4).

Soft substrate may form flat submarine plains, or the

Figure 19–7. Evidence of apparent supply of organic particles from above, or minute organisms descending from above; the holothurians (*Cucumaria*) have adopted anemone-like postures, and the sabellid polychaete also has tentacles directed upward. Ross Sea, at 75 m (J. S. Bullivant, New Zealand Oceanographic Institute).

floor of a submarine canyon, carrying a characteristic population of organisms so constructed that they either do not sink into the soft material, or else have means of extricating themselves from it. A typical population of this type, in deep water, will comprise in large part echinoderms, especially ophiuroids and holothurians. Fig. 19–3 shows two samples of a large population of ophiuroids (the larger members *Ophiomusium*, the smaller members being partly young stages, and partly certain other small genera) inhabiting the San Diego Trench, together with the curious elasipod holothurian *Scoto-planes*. The ophiuroids do not sink into the sediment because their weight is distributed by their long horizontal arms. The holothurians are very delicately constructed, consisting mainly of watery fluids and, as shown by these remarkable photographs from the U.S. bathyscaph *Trieste*, apparently they can actually walk over the sediment on the tips of the lateral rows of tube-feet.

Some bottom animals such as crinoids and echinoids develop very elongated cirri or spines. The purpose of these is to provide a stable support by sinking just so far into the sediment as is necessary to reach a state of equilibrium, without submerging the vital organs in the substrate. The echinoid *Aspidodiadema* possesses exceptionally long, curved spines for this purpose (fig. 19–1*d*). The spines are very lightly constructed, being hollow, so they do not add appreciably to the weight of the animal, yet are very effective in supporting it in its favored environment. These echinoids, restricted to deep water, are almost invariably shattered in the process of bringing them to the surface, and the population photographed in fig. 19–1 is doubtless the first ever seen in its natural state — another example of the special advantage of sea-floor photography, in this case stereo photography, in the study of the animal *per se*.

Although Belyayev & Ushakov (1957) disclosed the astonishingly high incidence of sponge and sponge-coelenterate-polyzoan complexes in the shelf fauna of Antarctica, it was not until Bullivant (1959) obtained his first remote-controlled photographs of this fauna that one could fully realize its richness and actual appearance in life; one of Bullivant's photos is shown in fig. 19–2. A fauna of this type must reflect the rich phytoplankton from which the majority of the bottom animals draw their food, either directly or indirectly.

A soft substrate is not always a flat substrate. Fig. 19–4*b*, taken in the same submarine canyon as fig. 19–3 shows a small part of a steep hillside. In parts of the original photograph not included here some crevasse-like openings are visible, suggesting that periodic slumping must occur. Doubtless many fossil populations, in which many animals of the same species and age-group are found preserved on the same bedding plane, may have been entombed through sudden slumping of soft substrate of this type.

Feeding habits are related to substrate, but these aspects are covered under a separate heading. One consequence of the feeding of holothurian populations on soft substrate may, however, be noted, since it is relevant to many bottom photographs: this is the production of copious castings of egested material. Probably the main source of nutritive material in the actual substrate in the bathyal environment is the bacterial content, and its

Figure 19–8. Evidence of gentle bottom current. McMurdo Sound, Antarctica, at 260 m. Note that while all the polychaetes face the current (from upper left), the sponge on the right is not dependent on current, and its growth pattern is centrifugal (J. S. Bullivant, New Zealand Oceanographic Institute).

by-products. To secure enough nutritives, the holothurians swallow relatively immense quantities of mud, thus continuously moving the overlying layers and preventing any precise stratification of matter falling from above. Hence considerable care is needed in distinguishing biocoenoses from thanatocoenoses in grab or core samples, and it is highly improbable that annual increments of sediment can survive in horizontal layers on any part of the sea floor, except under unusual abiotic conditions. The castings of holothurians are among the commonest objects seen in bottom photographs from soft substrates (fig. 19–6*b*). The interpretation of bottom photographs often requires considerable enlargement, or intensification by chemical means, of the original images on the film. It is perhaps not always evident to another observer when such enlargement has been carried to the limit; and thus, in order to forestall any risk of erroneous

Figure 19–9. Contrasted feeding responses in the presence of bottom current (from upper right); polychaetes are current-oriented, ophiuroids are not. Five different feeding patterns are seen, for details refer to pages 217–19 (J. S. Bullivant, New Zealand Oceanographic Institute).

interpretation of particle size in the substrate, when the metallic grain of the photographic emulsion is exhibited this fact should be stated. Careful examination of any photograph where this has been necessary (fig. 19–6a) will, however, show the nature of the granular effect.

Detection of bottom current

Examination of the orientation of organisms seen in bottom photographs will often disclose a nonrandom pattern. Sometimes the organisms which are dependent upon food materials delivered to them by the environment itself (fixed animals, or errant forms which prefer a sedentary life when conditions permit) are seen to be directing the feeding organs upward (fig. 19–7). Presumably in such cases there is relatively little movement in the water mass, and fragments of organic materials, or organisms of minute size, are falling from above. Under these conditions normally mobile holothurians may adopt the posture of a sea-anemone, and stand upright, with the oral tentacles held out in a ring. Random orientation of such erect organisms suggests the presence of micro-nekton capable of being captured from various directions, the water remaining still.

Fig. 19–8 illustrates a rich Antarctic sea-floor community, in which the dominant member is a tubicolous polychaete. The general orientation is toward the upper left corner of the photograph, from which direction presumably the food material must be arriving, whence we infer a gentle bottom current. Sponges, on the other hand, create their own feeding currents by the vigorous action of the flagellate cells lining the gastral chambers; hence we can account for the fact that the large hexactinellid sponge on the right of the field has its various oscula directed in different directions, essentially radially disposed, while the minute ostia inhale water from all directions.

Main feeding behavior patterns

Both substrate and bottom current affect the feeding of bottom organisms. The effect varies widely, however, according to the nature of the food, and the mobility of the organism. Fig. 19–9 shows part of a photograph of

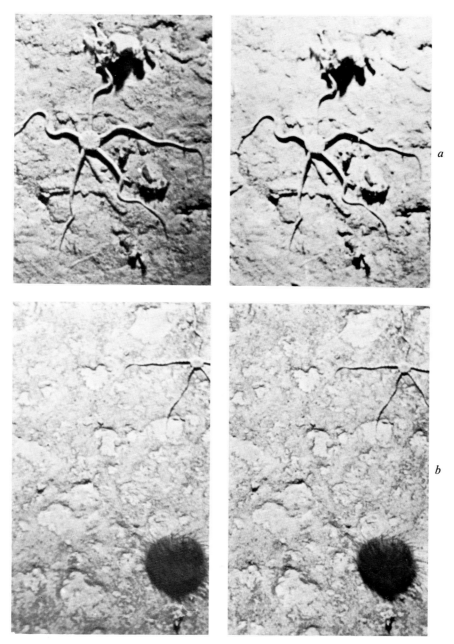

Figure 19–10. Stereo pairs illustrating value of this technique in facilitating identification and interpretation; in (a), the tips of the arms of the ophiuroid are seen to be held above the substrate; in (b), the curved, hoof-spines of the echinoid permit its determination as an echinothuroid, probably *Sperosoma;* Plantagenet Bank, 500 m (J. B. Hersey *et al.*).

the Antarctic sea floor, where there is soft substrate. It follows from the inference made in the preceding paragraph that there must be a bottom current passing diagonally across the field from the upper right-hand corner; this accounts for the orientation of the tube-worms and polyzoans, and also for the posture of the long sea-anemone near the lower right corner. The mobile animals evidently disregard the current. At the lower center is seen an ophiuroid (*Ophiurolepsis gelida*) creeping across the sea bed toward the viewer, two of the arms being bent into the characteristic shape adopted when they are being used as "oars" or "pushers." This animal is seeking out food by its own activity, as may be deduced by examining the second specimen of the same species, located near the top left corner. The latter has thrust one of its arms into the substrate, where it has raised a small

Figure 19–11. Characteristic postures of antedoid crinoids on the Antarctic sea floor;   (*a*) the poly-brachiate species *Promachocrinus kerguelensis*;   (*b*) ten-armed species, probably *Anthometra adriani*; both species habitually remain attached by the dorsal cirri, the mouth directed upward or into the current if current is present; Ross Sea, 360 m (J. S. Bullivant, New Zealand Oceanographic Institute).

mound, and it is actively seeking for detrital or living material in the silt, the mouth directed downward to receive whatever is found. Immediately to the right of this specimen, at the upper edge, lies a specimen of *Astrotoma agassizii*, one of the euryaline ophiuroids which possesses the power of coiling its arms ventrally.

In this case we see the coils are directed upward, whence it follows that the mouth must be directed upward, so that the animal is lying on its dorsum. Study of this photo-graph, and of biological samples from the same area, led to the inference (Fell, 1961) that *Astrotoma agassizii* must "fish" the overlying water, presumably collecting small particles or micronekton on the sticky tube-feet, and then drawing them across the mouth, to transfer the food. A little further to the right, and a little way in from the upper margin, stands an erect sponge, on which an ophiuroid is seen to have climbed, apparently holding on by one arm, while the others are held out in the water. Other sea-floor photographs from the same region showed that the ophiuroid is *Ophioceres incipiens*, and that it habitually climbs erect elements of the epifauna, such as tubicolous annelids and sponges. It, too, must fish the surrounding water. Near the center of the field, a little to the left, is a nudibranch mollusk, creeping toward the lower edge of the field; this carnivorous animal is in no way dependent upon currents for its food and disregards the current.

Turning back, now, to some of the photographs already examined, we find in fig. 19-2 an example of a crinoid in the typical feeding posture; it is anchored (on a sponge) by its cirri, while the arms are undulating rhythmically, alternate arms raised and lowered. The crinoid (*Promachocrinus kerguelensis*) illustrates the essentially sedentary habit of feather stars, which will only swim if conditions become unfavorable; in fig. 19-11 are seen other examples, similarly anchored. Although it is widely believed that feather stars are free-swimming crinoids, we have never seen any swimming example in photographs. Fig. 19-4a shows what is evidently the fishing habit already seen in Antarctic waters, in this case displayed by an ophiacanthid ophiuroid on the Plantagenet Bank, near Bermuda. The animals here are perched horizontally on the upper edge of the fan-sponge or gorgonian, their arms held out in the water as if to catch material falling from above. Had they been taken in a trawl or dredge, they would (if their hold were maintained on the host organism) doubtless have coiled the arms about it and might have been interpreted as feeding upon the host. The lower photo, fig. 19-4b, is ambiguous, in that it is not clear what the ophiuroid (*Asteronyx*) is actually doing. Specimens are commonly trawled, coiled in this same manner about gorgonians and similar animals. Possibly *Asteronyx* feeds upon the zooids (or fishes from a convenient perch) but coils itself tightly around the stem at other times.

Turbidity

The transparency of bottom water differs from place to place but is less than that at mid-depth (see chap. 10). Transient changes in transparency may be caused by animals living near or in the sea floor, or by the camera apparatus touching the bottom. Extreme clarity is rarely obtained. Fig. 19-5 is exceptional, the definition being so good that it is possible to recognize the parasitic sponge *Iophon* present on the ophiuroid, *Ophiurolepis*

gelida. Vassière & Fredj (1964) using a Troika to convey their camera, mention that they seldom obtained photographs of animals on the sea floor, though they frequently obtained photographs of the cloud of sediment caused by the sudden departure of animals, disturbed by their apparatus. However, hundreds of uniformly clear and excellent photographs, have been taken by Cousteau (personal communication) using the Troika. Also, Marshall & Bourne (1964), studying photographs obtained by Graham, have reported up to nine successive shots of the same animal (an eel) taken without disturbing it, and they note that the fishes they studied seemed little affected by the camera flash, and that they seldom saw the puffs of sediment raised by fishes darting off the bottom. Photographs published by Hartmann & Barnard (1958) appear to be excessively fogged by turbidity; they were working over an impoverished fauna located in a basin covered by fine silt. The animals photographed include feather stars, pointing to an environment free from suspended matter, whence it must be inferred that the turbidity was artificially produced, or due to some defect in the emulsion or developer. At shallow depths one method employed to overcome turbidity is to construct a cone in front of the camera, containing sealed clear water. The method, necessarily cumbrous, might be difficult to employ at depth on account of engineering problems.

Locomotor habits

On hard substrate, animals with locomotor appendages are able to move about freely. On soft substrate, the same animals may become temporarily established, provided a hard bottom is available in the shape of epifauna projecting above the substrate or a shell bank formed from mollusks or brachiopods which inhabit soft substrate. This accounts for the occasional presence of hard-bottom sea urchins as fossils in soft elastic matrices, with sudden extinctions of small communities whenever the shell bed is overwhelmed by mud. Suitable modification of locomotor appendages may lead to the development of species inhabiting soft substrate, though derived from forms adapted to hard substrate. The elongate, slender spines of *Aspidodiadema* (fig. 19-1d) are such modified structures. Some cidaroids are adapted to move about on mud, using their long spines as stilts, and leaving characteristic streak-like markings on the silt; *Cidaris cidaris* has been photographed in this environment at a depth of 270 m in the Mediterranean by Vassière & Fredj (1964). The peculiar feeding patterns of certain ophiuroids reported above are directly related to their powers of movement, and the manner in which the muscles of the arms operate. Utilizing photography, Marshall & Bourne (1964) have ascertained certain characteristic postures adopted by halosaurs and macrourids when hovering or swimming, and have related their observations to the dynamics of the fin pattern. These, and similar examples, show the wide range of application of remote-controlled photography to problems of marine biology intractable to other techniques.

Community formation and recognition

Photographs of sea-floor communities, such as those illustrating the present chapter, speak for themselves. Obviously such photographs, combined with biological samples permitting precise identification of a large part of the photographed fauna, offer a means of describing sea-floor communities and the nature of the various populations present in far greater detail and with much greater clarity. Trawl and dredge samples blend mixtures of soft-substrate faunules with others from hard substrate traversed on the same transect, and a series of photographs at intervals along the transect permits the jumbled samples to be disentangled, and overcomes many of the disadvantages inherent in the traditional collection procedures.

Population densities

Transects utilizing photographic techniques can provide good data on epifaunal elements. In most areas there are likely also to be infaunal elements, such as burrowing sea urchins, and these may be almost completely overlooked by a technique relying mainly on photography (see chap. 21). Consequently the photographic method can under no circumstances replace dredging; the biological samples, as already stressed, are required to permit reliable determination of photographic images, and they are clearly needed to provide data on the infauna. Conversely, without photography, small but numerous, even dominant, members of the epifauna can be selectively lost or reduced in proportion in dredge samples, if the size of the mesh happens to pass the small but abundant species. Similarly, highly mobile species escape from the trawl. The existence of such elements in a population can be determined by photographic methods when others fail. Notable use of the photographic technique in establishing the nature of sea-floor populations, including quantitative estimates, has been made by Vevers (1951, 1952) and by Hurley (1959), both authors establishing the existence of extremely rich populations of ophiuroids of various genera on sea floors supplied with incoming food particles carried on bottom currents. Fell (1961) has also utilized photographs to determine local maxima in Antarctic ophiuroid populations. Laughton (in Marr, 1963) has published a remarkable photograph of a dense popu-

lation of antedonid crinoids on the sea floor off Portugal. On the basis of comatulids taken by *Discovery*, and earlier collections from Antarctica, Marr (1963) has named a crinoid complex in terms of its supposed dominant genera, *Promachocrinus* and *Anthometra*. This inference seems to have been made on the basis of Laughton's photograph, rather than on Antarctic seafloor photographs. Data obtained photographically and by conventional sampling procedures in the Antarctic by the New Zealand Oceanographic Institute indicates that crinoids are only a minor element in the shelf faunas, and we have seen no evidence suggesting that communities of crinoids exist (although they may exist). The dubious feature of Marr's inference would appear to be that it was based on an Atlantic photograph, unique in character, and taken out of context by being applied to Antarctica. With ordinary controls, there is every reason to trust the photographic evidence. The high incidence of crinoids in Antarctica trawls is probably due to the fact that they are disturbed by the trawl and begin to carry out slow swimming undulations, thus becoming ready victims to the trawl or dredge. The much vaster population of polyzoans, ophiuroids, and other animals is either attached to the bottom as in the case of the former, or small and easily lost through the mesh in the case of the latter. Hence the trawl sample is not a true or random sample, whereas sea-floor photographs are random samples. As discussed further in chap. 21 it must be concluded that although photography is an important, indeed indispensable, technique in estimating benthic populations it must not be relied upon alone, since it cannot sample the infauna.

Food chains

By virtue of the fact that photography yields a random quantitative and qualitative sample of the epifauna, which a trawl or dredge does not do, any theoretical study on food chains of benthic populations ought to include photographic data as primary factual material. For example, the faunule illustrated in fig. 19–2 would certainly not have been collected in the same proportions had the sampler been a trawl. A great part of the sedentary attached epifauna would merely be pressed down and passed over, while the more easily detachable or bulkier elements would have been collected.

Further, the kind of information yielded by figs. 19–7, 19–8, and 19–9 is clearly presented photographically, whereas it is hard to imagine how it could have been obtained at all by traditional sampling procedures. Sufficient has already been said about these photographs to establish the prime need for photography in biological oceanography.

19–4. Summary

Photography of sea-floor organisms in life is a valuable, and sometimes essential, adjunct to conventional biological sampling. Used on its own, without collateral biological samples, photography loses much of its special value, and interpretation may be difficult or ambiguous. Conventional biological sampling, by dredge, trawl, or grab, is no longer adequate to provide sufficient data for the marine biologist, and faulty sampling may result from differential loss of faunal elements through meshes of the nets, or mobility of the species favoring the escape of some elements. A combination of the techniques provides a system of checks not otherwise available. Ideally, every biological station should also be a photographic station (preferably including color photography) and pictures should be obtained at regular intervals along transects where the trawl or dredge is employed.

References

Belyayev, G. M., & Ushakov, 1957: Nekotorie zakonomernosti kolychestvennovo raspedeleniya donnoi fauni v vodach Antarktiki. *Doklady Acad. Sci. U.S.S.R.*, **122**, no. 1, 137, (in Russian).

Bullivant, J. S., 1959: An oceanographic survey of the Ross Sea. *Nature*, **184**, 422–23.

Clark, H. E. S., 1963: Fauna of the Ross Sea, Part 3: Asteroidea. *Memoirs of the New Zealand Oceanographic Inst.*, **21**, 1–84, pl. 1–15.

Fell, H. B., 1962: Fauna of the Ross Sea, Part 1: Ophiuroidea. *Memoirs of the New Zealand Oceanographic Inst.*, **18**, 1–79, pl. 1–19.

Hartman, Olga, & J. Laurens Barnard, 1958: The benthic fauna of the deep basins off Southern California. *Allan Hancock Pacific Expeditions*, **22**, no. 1, 1–67, pl. 1–2. University of Southern California Press, Los Angeles, California.

Hunkins, Kenneth L., Maurice Ewing, Bruce C. Heezen, & Robert J. Menzies, 1960: Biological and geological observations on the first photographs of the Arctic Ocean deep-sea floor. *Limnology and Oceanography*, **5**, no. 2, 154–61.

Hurley, D. E., 1959: Some features of the benthic environment of Cook Strait. *New Zealand J. Sci.*, **2**, no. 1, 137–47.

———, & D. G. McKnight, Occurrence of *Pyrosoma* on the sea floor 160 meters deep. *Nature*, **183**, 554–55.

Kuznetsov, A., 1960: Data on the quantitative distribution of bottom fauna on the floor of the Atlantic Ocean. *Doklady Acad. Sci. U.S.S.R.*, **130**, no. 6 (in Russian); U.S. Dept. Commerce, OTS, JPRS 5948, 1–11 (English translation).

LaFond, Eugene, 1962: Dive eighty-four. *Sea Frontiers*, **8**, no. 2, 94–102.

Laughton, A. S., 1959: Photography of the ocean floor. *Endeavour*, **18**, no. 72, 178–85.

Longhurst, Alan R., 1964: A review of the present situation in benthic synecology. *Bull. de l'Institute Océanographique de Monaco*, **63**, no. 1317, 1–54.

Marr, J. S. W., 1963: Unstalked crinoids of the Antarctic continental shelf. Notes on their natural history and distribution. *Phil. Trans. Roy. Soc.*, London, **246**, series B, 734, 327–79.

Marshall, N. B., & D. W. Bourne, 1964: A photographic survey of benthic fishes in the Red Sea and Gulf of Aden, with observations on their population density, diversity, and habits. *Bull. Mus. Comp. Zool.*, Harvard University, **132**, no. 2, 225–44, pl. 1–4.

Mortensen, Th., 1928–52: *A monograph of the Echinoidea, I–V*. Reitzel, Copenhagen.

Sokolova, M. N., 1959: On the distribution of bottom animals in relation to their feeding habits and the character of sedimentation. *Deep-Sea Res.*, **6**, 1–4.

Vassière, Raymond, & Gaston Fredj, 1964: Etude photographique préliminaire de l'étage bathyal dans la région de Saint-Tropez (ensemble A). *Bull. de l'Institut Océanographique de Monaco*, **64**, no. 1323, 1–70.

Vevers, H. G., 1951: Photography of the sea floor. *J. Mar. Biol. Ass. U.K.*, **30**, no. 1, 101–11.

———, 1952: A photographic survey of certain areas of sea floor near Plymouth, *J. Mar. Biol. Ass. U.K.*, **31**, no. 2, 215–21, pl. 1.

Vinogradova, N., 1961: Vertical zonation in the distribution of the deep-sea benthic fauna in the ocean. *Deep-Sea Res.*, **8**, 245–50.

20. Optically triggered underwater cameras for marine biology[*]

L. R. Breslau *Saclant A.S.W. Research Center, La Spezia, Italy*

G. L. Clarke *The Biological Laboratories, Harvard University, Cambridge, Massachusetts*

H. E. Edgerton *Massachusetts Institute of Technology, Cambridge, Massachusetts*

Abstract

A submarine camera — the luminescence camera — was designed and constructed for the purpose of investigating marine bioluminescence. The camera and strobe-light unit contained an electronic control system which utilized a photomultiplier tube as the primary detector. Upon "seeing" a bioluminescent flash, the camera is triggered and a picture of the organism causing the bioluminescence is obtained. This camera has been used extensively at sea and has produced many pictures.

Another submarine camera — the interruption camera — was designed and constructed for the purpose of investigating the abundance of macroscopic life in the ocean. This camera contained an electronic control system which utilized a light beam and crystal-photocell primary detector. When a macroscopic organism interrupts the light beam, the camera is triggered and a picture of the organism is obtained. This camera is more recent than the bioluminescence camera and has not yet been used extensively at sea.

20–1. Introduction

Photography of deep-sea animals in mid water by any means is a difficult and challenging project, which even today is not done with conspicuous success except from deep submersibles. The challenge is to place the camera in an advantageous position for taking a good photograph, and to know when to take the picture. Johnson,

* WHOI Contribution Number 1759.
   The cameras described here were designed and constructed in the Stroboscope Laboratory at MIT. The field work was performed aboard ships of the Woods Hole Oceanographic Institution and the Institut Océanographique de Monaco. This work was supported by funds from National Science Foundation Grant No. 3838, the Office of Naval Research under contract Nonr-1720(00) NR 083–116, and National Geographic Society contribution No. 1001 to the Woods Hole Oceanographic Institution.

Backus, Hersey, & Owen (1956) described a camera system which depended on sonar echo location of subjects, and in this chapter two cameras are described, each of which is automatically actuated by an optical signal from the subject and may be suspended from a ship. In all likelihood these three methods will be largely discarded in favor of photography from the several manned submersibles now in use or under development. Also, recent improvements in underwater television devices considerably ease the problem of finding subjects in mid water. Nevertheless, when certain physical properties, either optical or acoustical, are to be selected for study, it seems quite probable that one or another of the ideas on which these cameras are based will prove useful in the future.

20–2. The luminescence camera

This instrument was designed and constructed to trigger an underwater camera device by flashes of luminescence in the sea, thus taking photographs of the animals producing the flashes. A photomultiplier detector is stimulated by the light from a luminescent animal and triggers an electronic flash and camera combination. Simultaneously, a sonar pulse is emitted from an attached transducer, announcing that a picture has been taken. After this the camera automatically advances its film to complete the cycle. The instrument is entirely self-contained, can be suspended by an ordinary steel cable, and has been designed to operate at any depth down to 6,000 m. The film capacity is sufficient for 800 photographs.

The instrument was designed particularly to investigate the light flashes previously observed by Clarke & Wertheim (1956), Clarke & Backus (1956), and Clarke & Hubbard (1959), while using the bathyphotometer (Hubbard, 1956). Information from their work (fig. 20-1) showed that the luminescence detector should be triggered by flashes with an irradiance of $10^{-5} \mu$ watt/cm^2 or

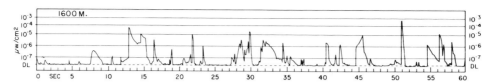

Figure 20–1. Section of bathyphotometer record showing bioluminescence.

greater in order to obtain a satisfactory number of photographs. A photomultiplier trigger was required to obtain this high sensitivity. Photomultiplier rube R.C.A. 6199 was chosen because of its sensitivity, convenient size, and a spectral response peaked in the region of maximum transmission of sea water, which is probably also the region of most bioluminescent emission. The photosensitive surface is the same as that employed in the bathyphotometer. Were this not so, it would be impossible to form any correlation between the results of the two instruments. The device can be triggered by a pulse of light as small as $5 \times 10^{-6}\,\mu$ watt/cm² as measured with the bathyphotometer. Provision is made for manual preselection of the threshold level of the photomultiplier trigger.

The camera and film-advancing circuitry are contained in one cylindrical housing, and the electronic flash and photomultiplier trigger in another (fig. 20–2). These fit into a rigid cradle which fixes the over-all geometry of the instrument. A small watch in a water-tight case was attached to the cradle so as to appear in each photograph. By the coordination of records on a time basis, the depth, location, and other data for each picture are determined.

The field of view of the photomultiplier is a right circular cone of about 25°, with its apex located at the face plate of the flash unit. A baffle attached to the cradle truncates this cone at a distance of 10 in (25 cm) from the face plate, thus delineating a sensitivity zone into which a luminescent animal must enter in order that its flash will trigger the instrument. The entire volume of water observed by the photomultiplier lies in the field of view of the camera and is illuminated by the electronic flash. Since many luminescent animals have translucent bodies, side lighting was deemed to be most desirable. The instrument incorporates an "in-line" camera which has been specially designed by Edgerton (1955) for a cylindrical housing, and which uses a standard 100-ft (30 m), 35-mm motion picture reel. Since the camera is shutterless, the electronic light flash takes the picture. The lens used is a Leitz Elmar f/3.5 with a 50-mm focal length.

Since the instrument is entirely self-contained, no electrical conductors are required in the cable. The circuitry is such that the device can operate continuously for at least ten hours before the batteries need recharging. Mercury switches are employed so that the instrument can be loaded and stored in a prepared state.

The luminescence camera was operated from the R/V *Crawford* of Woods Hole Oceanographic Institution and

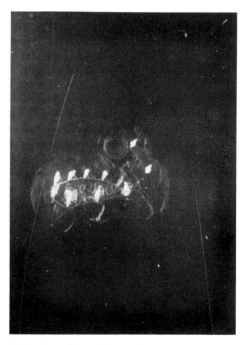

Figure 20–2. Luminescence camera. The cylindrical housing on the left is the camera, upper center the sonar pinger used to measure the depth of the instrument, and on the right the electronic flash.

Figure 20–3. A sophonophore (about 6 cm in diameter) at a depth of about 100 m. The photograph was taken using the luminescence camera, aboard the R/V *Crawford* about 170 miles east of Cape Hatteras (August, 1957).

Figure 20-4. A euphausid (about 2 cm long) at a depth of 300 m. The photograph was taken using the luminescence camera, aboard the Coast Guard cutter *Yamacraw*, about 100 miles east of Cape Hatteras (February, 1958).

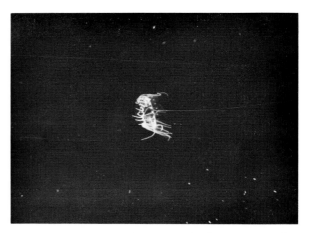

Figure 20-6. A medusa (about 1 cm in diameter) at a depth of 100 m. The photograph was taken using the luminescence camera, aboard the Coast Guard cutter *Yamacraw*, in the eastern Gulf of Mexico (February, 1958; 24°58′N 85°01′W).

Coast Guard cutter *Yamacraw* in the waters south of Woods Hole during the summer of 1957, and from the Research Vessels *Calypso* and *Winnaretta-Singer* of the Institut Océanographique de Monaco in the Mediterranean Sea during the summer of 1958 (Clarke & Breslau, 1959). The instrument was used independently on certain occasions, but improved results were obtained when it was employed in conjunction with the bathyphotometer, since the light values received by the bathyphotometer at various depths are immediately displayed on a ship-

board recorder, thus enabling the camera to be lowered to regions of maximum bioluminescent activity.

Sample photographs of animals that took their own pictures by means of the luminescence camera are reproduced in figs. 20-3 to 20-10. A photograph of a bioluminescent jellyfish taken at a depth of about 1,000 m appeared on the front cover of *Science* (Dec. 29, 1961). A photograph of a siphonophore taken 170 miles east of Cape Hatteras at a depth of about 100 m was reproduced in *Newsweek* (Oct. 7, 1957, p. 98).

20-3. The interruption camera

This instrument was designed and constructed for the general purpose of photographing macroscopic, but small, life at medium depths in the ocean. It uses a trigger

Figure 20-5. A ctenophore (about 3 cm in diameter) at a depth of 10 m. The photograph was taken using the luminescence camera, aboard the Coast Guard cutter *Yamacraw*, 100 miles southeast of Jacksonville, Florida (February, 1958; 28°16′N 79°41′W).

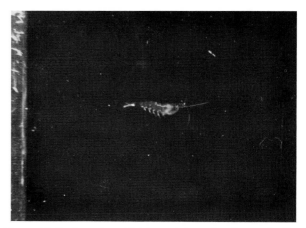

Figure 20-7. A euphausid (about 1.5 cm long) at a depth of 50 m. The photograph was taken using the luminescence camera, aboard the Coast Guard cutter *Yamacraw*, in the Straits of Florida (February, 1958; 23°52′N 82°05′W).

Figure 20–8. A medusa (about 1 cm in diameter) at a depth of 360 m. The photograph was taken using the luminescence camera, aboard the Coast Guard cutter *Yamacraw*, in the eastern Gulf of Mexico (February, 1958; 24°58′N 85°01′W).

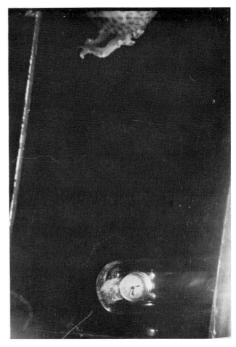

Figure 20–10. A cephalopod (arms about 2 cm long) at a depth of 10 m. The photograph was taken using the luminescence camera, aboard the R/V *Calypso* in the Mediterranean Sea south of France (June, 1958; 43°31′N 7°23′E).

system which consists of a light beam impinging upon a semiconductor photocell. When a macroscopic organism interrupts the light beam, it triggers the electronic-flash and camera combination. A sonar pulse announces that a picture has been taken, and the film is automatically advanced as in the luminescence camera. The instrument is self-contained and has been designed to operate at any depth down to 2,000 m. The film capacity is sufficient for 800 photographs.

A crystal photocell is used as the photodetector because

of its adequate sensitivity and small physical size. A cadmium sulphide crystal (Clairex Type CL-2) is used for white light and a cadmium selenide (Clairex Type CL-3) is used for deep-red and infra-red light. A standard two-cell flashlight bulb (G.E. PR-6) and reflector are used in the light source to produce a white-light beam which is 3 cm in diameter. To obtain a deep-red light beam a Kodak Wratten No. 29 filter, which transmits wavelengths of 6,000 Å and over, is placed over the light source. To obtain an infra-red light beam a Kodak

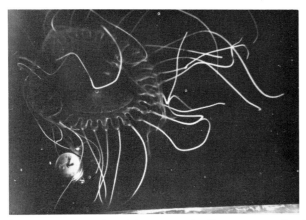

Figure 20–9. A medusa (about 7 cm in diameter) at a depth of 1,000 m. The photograph was taken using the luminescence camera, aboard the R/V *Crawford* in the Gulf of Mexico (August, 1958).

Figure 20–11. Interruption camera.

Wratten No. 898 filter, which transmits wavelengths of 6,800 Å and over, is placed over the light source.

Provision is made for manual pre-selection of the size of the object required to trigger the camera. At maximum sensitivity the camera will trigger, for all color light beams and appropriate crystals, when 1 per cent of the light beam is interrupted. This can be caused by a totally opaque subject obstructing 1 per cent of the area of the light beam, or a subject of 99 per cent transmittance obstructing the entire area of the light beam.

The camera with film-advance circuitry, and the electronic flash with trigger-control circuitry, are contained in separate large cylinders which constitute respectively the camera and flash units. These fit into a rigid cradle so designed that each cylinder can be easily detached for servicing and transportation. The semi-conductor photocell and the incandescent light are contained in separate small cylinders which constitute the underwater light receiver and are attached to the cradle by rigid arms. The instrument is shown in fig. 20–11.

The light beam is 1 m long and 3 cm in diameter. It is located in the same plane as the camera and flash units, which are sub-parallel, and at a distance of about 2 m from the camera window. When the instrument is suspended in its normal operating position the light beam is vertical. Since many of the macroscopic organisms were expected to be opaque, pesudo-front lighting was deemed to be most desirable. The camera is pointed toward the center of the light beam and has a field of 1.3 m in the vertical direction and 2 m in the horizontal direction. Consequently the entire photosensitive trigger volume is photographed. The instrument incorporates the same "in-line" Edgerton camera that was mentioned previously. The lens used in this application was a Summaron f/3.5 with a 35 mm focal length.

The instrument is entirely self-contained, can operate continuously for at least eight hours before the batteries need recharging, and employs switches as in the luminescence camera.

The instrument was operated with limited success aboard the R/V Crawford during the summer of 1958 and fig. 20–12 shows a photograph of a siphonophore taken using the interruption camera. At this time a white-light beam was used in the trigger system. Subsequently, laboratory tank tests were run at Woods Hole with local species of fishes in order to determine their response to various light-beam colors. These limited tests indicated that these species avoided passing through a beam of white light, or of red light, but would pass through a beam of infra-red. Using the infra-red beam arrangement, many pictures of fish in the tank have been obtained.

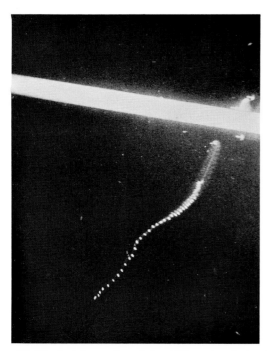

Figure 20–12. A siphonophore (about 25 cm long) at a depth of about 200 m. The photograph was taken using the interruption camera, aboard the R/V Crawford in the Gulf of Maine (August, 1958).

References

Breslau, L. R., & H. E. Edgerton, 1958: The luminescence camera. *J. Biol. Phot. Assoc.*, **26**, no. 2, 49–58.
——, & ——, 1959: The interruption camera. Woods Hole Oceanographic Institution Ref. No. 59–27 (unpublished manuscript).
Clarke, George L., & Richard H. Backus, 1956: Measurements of light penetration in relation to vertical migration and records of luminescence of deep-sea animals. *Deep-Sea Res.*, **4**, 1–14.
——, & Gunther K. Wertheim, 1956: Measurements of illumination at great depths and at night in the Atlantic Ocean by means of a new bathyphotometer. *Deep-Sea Res.*, **3**, 189–205.
——, & Lloyd R. Breslau, 1959: Measurements of bioluminescence off Monaco and Northern Corsica. *Bull. de l'Institut Océanographique de Monaco*, no. 1147, 1–13.
——, & Charles J. Hubbard, 1959: Quantitative records of the luminescent flashing of oceanic animals at great depths. *Limnology and Oceanography*, **4**, 163–80.
——, R. J. Conover, C. N. David, & J. A. C. Nicol, 1962: Comparative studies of luminescence in copepods and other pelagic marine animals. *J. Mar. Biol. Ass. U.K.*, **42**, 541–64.
——, & E. J. Denton, 1962: Light and animal life. In *The sea; ideas and observations on progress in the study of the seas*. Vol. 1, *Physical Oceanography*, M. N. Hill (ed.), Interscience Publ., New York, 456–68.
——, & Richard H. Backus, 1964: Interrelations between the vertical migration of deep scattering layers, biolumi-

nescence, and changes in daylight in the sea. *Bull. de l'Institut Océanographique de Monaco*, **64,** no. 1318, 1–36.

———, & Mahlon G. Kelly, 1964: Variation in transparency and in bioluminescence on longitudinal transects in the western Indian Ocean. *Bull. de l'Institut Océanographique de Monaco*, **64,** no. 1319, 1–20.

———, 1965: Elements of ecology, Chap. 6: Light. John Wiley and Sons, New York. (Revised printing), 534 pp.

———, & Mahlon G. Kelly, 1967: Measurements of diurnal changes in bioluminescence from the sea surface to 2,000 meters using a new photometric device. *Limnology and Oceanography*, **10** (in press).

Edgerton, H. E., & L. D. Hoadley, 1955: Cameras and lights for underwater use. *J. Soc. Motion Picture and Television Engrs.*, **64,** 345–50.

Hubbard, Charles J., 1956: A wide-range, high speed bathy-photometer. Woods Hole Oceanographic Institution Ref. No. 56–68 (unpublished manuscript).

Johnson, Henry R., Richard H. Backus, J. B. Hersey, & David M. Owen, 1956: Suspended echo-sounder and camera studies of midwater sound scatterers. *Deep-Sea Res.*, **3,** 266–72.

21. Bottom photography as a tool for estimating benthic populations[*]

David M. Owen, Howard L. Sanders, and Robert R. Hessler   *Woods Hole Oceanographic Institution, Woods Hole, Massachusetts*

Abstract

Comparative counts of bottom-dwelling animals from photographs and grab samples reveal poor numerical correlation unless account is taken only of animals which are actually visible on the bottom. Tracks and trails seen on photographs are considered unsatisfactory in giving a quantitative estimation of benthic life. It appears that the camera is a satisfactory tool only for estimating epifaunal life since it gives no indication of the abundant life below the bottom included in the grab samples.

Literally thousands of photographs of the sea floor have been made for the purpose of estimating the benthic animal life in different regions of the ocean. To determine the validity of this tool, and the range of its applicability, a series of bottom photographs was taken, concurrently with the collecting of bottom faunal samples, at a number of stations along a transect south of New England on the outer continental shelf, continental slope, continental rise, and the abyss in the region of the Sargasso Sea on cruises 264, 273, 277, and 283 of the R/V *Atlantis*, during May 20–25, 1961; Sept. 26–Oct. 4, 1961; May 22–28, 1962; and Aug. 27–31, 1962, respectively (fig. 21–1). For details concerning the camera used see chap. 8. The camera was calibrated to photograph $\frac{1}{7}$ m² of the ocean floor.

Counts of animals on the photographs and of those present in the bottom samples revealed very little numerical correlation (table 21–1). At station S2, on the upper continental slope (200 m depth), only three animals are discernible from photographs on a square meter basis. Yet the faunal sample contained an animal density of almost 13,000/m². Station C (fig. 21–2), on the outer continental shelf (97 m), gave values of 24.5 and 5,314 animals/m² for the photographs and the bottom sample, respectively.

Similarly, the other stations showed the same gross numerical discrepancies. At the three other stations on the upper slope, stations S3(300 m, figs. 21–3 and 21–4), S4(400 m), and D#1(487 m), the values were 85, 32, and 10.5 animals/m² for the photographs and 21,263; 6,081; and 8,669/m² for the bottom samples. At the three lower continental slope stations, E#3(823 m), F#1(1,500 m), and G#1(2,086 m), the values obtained from the photographs were 2.5 (fig. 21–5), less than 12 (see chap. 15 and figs. 21–6, 21–7, 21–8 and 21–9), and 3.7 (figs. 21–10 and 21–11). The bottom samples from the same three stations

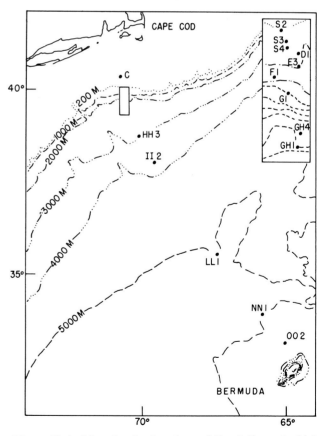

Figure 21–1. Map showing locations of the stations at which both bottom photographs and bottom samples were taken.

* WHOI Contribution Number 1760.

This research was supported by the National Science Foundation grants G-15638, G-8339, GB-563, and the Office of Naval Research, under contract Nonr-2196(00) NR 083–004.

TABLE 21-1.

Station	Depth	Latitude	Longitude	No. frames	No. visible animals	No. visible animals/m²	No. animals/m² in sample
C	97	40°20.5′	70°47′	6	21	24.5	5,314
S2	200	40°01.8′	70°42′	14	6	3	12,910
S3	300	39°58.4′	70°40.3′	16	194	85	21,263
S4	400	39°56.5′	70°39.9′	9	41	32	6,081
D#1	487	39°54.5′	70°35′	14	21	10.5	8,669
E#3	823	39°50.5′	70°35′	11	4	2.5	2,979
F#1	1,500	39°47′	70°45′	19	31–32	<12	1,719
G#1	2,086	39°42′	70°39′	13	7	3.7	2,154
GH#1	2,500	39°25.5′	70°35′	12	15	8.75	521
GH#4	2,469	39°29′	70°34′	14	12	6	467
HH#3	2,870	38°47′	70°08′	2	1	3.5	748
II#2	3,752	38°05′	69°36′	6	3	3.5	1,003
LL#1	4,977	35°35′	67°25′	11	2?	1.6	55
NN#1	4,950	33°56.5′	65°50.7′	9	1	0.75	38
OO#2	4,667	33°07′	65°02.2′	17	2	0.8	126

yielded 2,979; 1,719; and 2,154 animals respectively. The four abyssal rise stations, GH#1(2,500 m), GH#4(2,469 m), HH#3(2,870 m), and II#2(3,752 m), gave values of 8.75 (fig. 21–12), 6.0 (fig. 21–13 and 21–14), 3.5, and 3.5 for the photographs and 521, 467, 748, and 1,003 for the bottom samples. Abyssal stations LL#1(4,977 m), NN#1(4,950 m), and OO#2(4,667 m), from the faunally impoverished Sargasso Sea, gave values of 1.6 (fig. 21–15), 0.75 (fig. 21–16 and 21–17), and 0.8 for the photographs, and 55, 38, and 126/m² for the bottom samples.

Organisms less than 1 cm in diameter are difficult to discern in the photographs, but animals as small as 2 mm can be picked from the sediment samples. However, it is unlikely that this size difference is the most significant reason for the large discrepancies in the results obtained by the two methods.

If we confine our considerations to those groups that appear commonly in the photographs, a much better correlation can be achieved. At station S3 (figs. 21–3 and 21–4) 85 brittle stars were counted from the photographs on a square-meter basis, 118/m² were found in the sample. Station F#1 (figs. 21–6, 21–7, 21–8, and 21–9), revealed 10.5 brittle stars from the photographs, 8.6 from the bottom sample.

Certain photographs (figs. 21–2, 21–5, and 21–12) revealed the presence of numerous tracks, mounds, holes, burrows, and tubes, and it is tempting to use the abundance of these signs of animal activity as an approximate measure of the amount of infauna (i.e., organisms that live in rather than on the bottom) present. Unfortunately this is an unjustifiable assumption, for several reasons. In regions of low sedimentary deposition such signs may represent the sum of animal activity over a very long period of time, but in highly depositional regions these signs may be quickly buried. Numerous tracks may be made by the same individual crisscrossing the bottom in

the area photographed; a single organism may construct many mounds, tubes, or burrows, or a number of animals may dwell in a single tube or burrow. Further, many, most, or all tubes and burrows may be unoccupied, many members of the infauna leave no discernible indication of their presence at the sediment surface, and bottom currents and small-scale turbidity currents can erase all signs of animal activity (figs. 21–6 and 21–7. and photographs in chap. 15).

In summary, the number of animals observed in the bottom photographs represent, at most, the merest fraction of the animals present. Furthermore, there seems to be no correlation between the size of the benthic populations and the number of animals counted in the photographs. These conclusions are in essential agreement with the findings of Menzies, Smith, & Emery (1963) in their study off the coast of California. The basic reason for the discrepancies is that the characteristic sediments of the deep sea (and many other marine environments) are relatively soft oozes. Wherever soft sediments are found, the animals are predominantly, or even exclusively, members of the infauna, i.e., organisms that live in rather than on the bottom. On the other hand, only the epifauna are available to the camera, and in these photographs are represented by various groups of echinoderms, brittle stars, starfish, sea urchins, sea cucumbers, and the occasional sea anemone or bryozoan.

Thus it is only in those regions where the epifauna is dominant and the infauna an insignificant element that the camera has validity as a tool for quantitative studies of the benthos. Such a situation is only applicable to the hard or rocky substrates which form a negligible fraction of the ocean floor.

References

Menzies, Robert James, Logan Smith, & K. O. Emery, 1963: A combined underwater camera and bottom grab: a new tool for estimation of deep-sea benthos. *Int. Revue ges Hydrobiol.*, **48**, no. 4, 529–45.

Figure 21–2. Frame from a series of photographs taken at station C (depth 97 m), showing a starfish, tubes and, hardly discernible, the arms, but not the discs, of small brittle stars.

Figure 21–3. Frame from a series of photographs taken at station S3 (depth 300 m), showing numerous small brittle stars, holes, mounds, and some tracks.

Figure 21–4. Frame from the series taken at station S3, showing small brittle stars, tubes, and mounds.

Figure 21–5. Frame from a series of photographs taken at station E ⚹ 3 (depth 823 m), showing mounds formed by the subsurface activities of infaunal animals, and holes, tubes, and a few minute brittle stars.

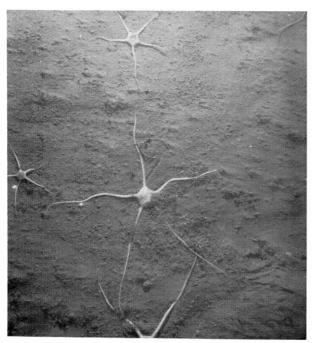

Figure 21–6. Frame from a series of photographs taken at station F ✳ 1 (depth 1,500 m), showing large and small brittle stars, and the effects of bottom currents that are in the process of erasing the signs of animal activity.

Figure 21–7. Frame from the series taken at station F ✳ 1, clearly showing bottom-current effects, and a large brittle star.

Figure 21–8. Frame from the series taken at station F ✳ 1, with bottom current absent, and showing a brittle star, numerous brittle-star tracks, and mounds.

Figure 21–9. Frame from the series taken at station F ✳ 1, showing brittle stars and brittle-star marks.

Figure 21–10. Frame from a series of photographs taken at station G ⚹ 1 (depth 2,086 m), showing a sea urchin, a brittle-star arm, brittle-star marks, animal trails, a hole, and a mound.

Figure 21–11. Frame from the series taken at station G ⚹ 1, showing bristles of a sea urchin (in lower left corner), a brittle star, brittle-star marks, a snail shell, and rocks.

Figure 21–12. Frame from a series of photographs taken at station GH ⚹ 1 (depth 2,500 m), showing small brittle stars, holes, mounds, and small tubes.

Figure 21–13. Frame from a series of photographs taken at station GH ⚹ 4 (depth 2,469 m) showing animal tracks, small tubes, and a large mound.

Figure 21–14. Frame from the series taken at station GH ⚹ 4, showing a large starfish, small brittle stars, and small tubes.

Figure 21–15. Frame from a series of photographs taken at station LL ⚹ 1 (depth 4,977 m), showing the approximately parallel orientation of many of the bottom markings.

Figure 21–16. Frame from a series of photographs taken at station NN ⚹ 1 (depth 4,950 m) showing a few small animal tracks.

Figure 21–17. Frame from the series taken at station NN ⚹ 1 showing a large animal track, and a manganese nodule (in the lower left corner).

22. Benthic animals, particularly *Hyalinoecia* (Annelida) and *Ophiomusium* (Echinodermata), in sea-bottom photographs from the continental slope[*]

Roland L. Wigley   *Bureau of Commercial Fisheries Biological Laboratory, Woods Hole, Massachusetts*

K. O. Emery   *Woods Hole Oceanographic Institution, Woods Hole, Massachusetts*

Abstract

Photographs of the sea floor generally have received only partial or superficial use by biologists, largely because of difficulty in identifying the animals. A combined underwater camera and bottom-grab sampler permits fuller use of the photographs.

An analysis of forty photographs of the sea bottom, together with their accompanying quantitative samples of the benthos, revealed a sparse to unexpectedly rich benthic fauna inhabiting the continental slope between Georges Bank and Cape Hatteras.

The polychaete worm *Hyalinoecia tubicola* and the brittle star *Ophiomusium lymani* were the two species shown most commonly and clearly in the photographs; they were also the dominant epibenthic forms in the grab samples. Notes on their size, distribution pattern, body position, tracks, and related ecological information are included in this report.

Large surface-dwelling species comprised a major portion of the total benthic biomass. In both the surface fauna and total benthos, weights and numbers of individuals were largest in the shallowest zone (200 to 500 m) and least in the deepest zone (2,000 to 3,310 m).

A comparison of total benthic biomass (based on grab-sampler data and photographs) clearly indicates that the surface-dwelling component was underestimated using the grab-sampler technique. It is concluded that sea-bottom photographs are a desirable supplement to grab samples in studies of benthic biomass. Furthermore, the photographic method is judged to be superior to the conventional method for determining the biomass of large surface-dwelling organisms.

22–1. Introduction

The number of sea-bottom photographs is increasing rapidly as a result of major technical developments in submarine photography during the past decade (Edgerton, 1963; Shipek, 1960) and substantially increased oceanographic research. Most photographs are not analyzed for their biological content, or are studied only superficially, chiefly because of the inability of biologists to identify the organisms that are disclosed. The rather small size (less than 1 or 2 cm) or thin bodies of many surface-dwelling organisms contribute to the difficulty. Moreover, many animals are partially buried in the sediments or inhabit tubes or cases that frequently leave little, if any, of the body or appendages exposed to view. These factors, the drab coloration that blends with bottom debris, and superficial similarity in appearance of many diverse kinds of organisms (even from different phyla) add to the difficulty of identification of all but a small proportion of the animals.

The identification of animals in photographs is facilitated if a combined camera and bottom sampler are used to collect benthic organisms from approximately the same area of sea floor that is photographed. Such a device was used off southern California by Menzies, Smith, & Emery (1963) and on the continental shelf off New England by Emery, Merrill, & Trumbull (1965). Ideally the sampler recovers many of the same specimens that are photographed. This permits the detailed microscopic examination of the various organisms that generally is necessary for accurate identification.

This report describes the distribution, body positions, tracks, and related ecological aspects of the two principal species of benthic animals in a series of forty photo-

* WHOI Contribution Number 1761.

Our associates at the Bureau of Commercial Fisheries and the Woods Hole Oceanographic Institution, particularly B. Burns, E. B. Haynes, J. Hulsemann, R. M. Pratt, J. Schlee, R. B. Theroux, and J. V. A. Trumbull assisted in collecting and processing materials; Lowell P. Thomas identified specimens of brittle stars; H. L. Sanders read the manuscript critically; D. M. Owen designed and constructed the photographic unit; and F. A. Bailey prepared the drawing in fig. 22–5. The R/V *Gosnold* was under command of Captain Harry Seibert.

graphs collected with a camera-grab device from the continental slope off the Atlantic coast of the United States. The numerical density and biomass values derived from an analysis of the photographs are compared with those from the grab samples.

22–2. Methods

Equipment

The sampling instrument consisted of a large clam-shell grab with a camera and light attached (fig. 22–1). The

basic component was a Campbell bottom grab that covered a bottom area of 0.56 m², had a capacity of 0.20 m³ of sediment, and weighed 250 kg. A Robot Star camera loaded with 35-mm, black-and-white film was mounted in a watertight housing that had an acrylic plastic window in the side. This unit was attached inside one bucket of the grab. In order to compensate for the camera's position offset from the midline it was aimed inward toward the midline at a slight angle to the vertical. The area that was photographed was 0.48 m²; about 72 per cent of this area was encompassed by the jaws of the grab. An electronic flashtube with batteries, condensers, and switching

Figure 22–1. Bottom view of Campbell grab sampler and attached camera and high-intensity light. Camera housing is adaptable for stereoscopic equipment. Width of jaws (vertical dimension in photograph) is 57 cm.

solenoid was arranged in a watertight housing attached inside the other bucket of the sampler. The triggering mechanism consisted of a spring-loaded mercury switch and a lead weight suspended about 1 m below the grab.

When the instrument was lowered from the ship, the tripping weight struck bottom first, activating the switch that resulted in film exposure. The sampler continued its descent until it struck bottom, tripped, and enclosed a portion of the sediment and fauna, which was retrieved.

Materials

General. The photographs and bottom samples that form the basis of this report are part of a large series collected by the R/V *Gosnold* in 1963 and 1964 during a survey of the geology and benthic invertebrates of the Atlantic continental shelf and slope (Emery & Schlee, 1963). This survey was sponsored jointly by the U.S. Geological Survey, the Woods Hole Oceanographic Institution, and the U.S. Bureau of Commercial Fisheries. Quantitative samples of the benthos and their accompanying photographs were available from forty stations on the continental slope between Georges Bank and Cape Hatteras (fig. 22–2). Water depths range from 214 to 3,310 m. Bottom sediments are predominantly mixtures of silt and clay, except for those from the upper part of the slope which consist mostly of sand (table 22–1). Geophysical measurements show that these sediments are as thick as 100 m, and that they rest atop Tertiary and Mesozoic shales, marls, and sandstones which are exposed only locally where slopes are especially steep. The continental slope in this region has a general declivity of 2° to 8° (average about 4°).

Processing of samples. At sea the samples were hoisted to the surface and dumped into a large watertight metal

TABLE 22–1. Particle size and composition of sediments in twenty-six bottom samples from the continental slope of the western North Atlantic.[1] (The number of samples upon which each computation is based is shown in parentheses.)

Water depth (m)	Median diameter of particles (μ)[2]	Percentage by dry weight					
		Clay	Silt	Sand	Gravel	Calcium carbonate	Organic carbon
0–500	115	7	21	70	2	6.5	0.51
	(12)	(12)	(12)	(12)	(12)	(12)	(11)
500–1,000	7.5	43	35	22	0	8.6	0.77
	(2)	(2)	(2)	(2)	(2)	(3)	(2)
1,000–2,000	9.1	35	54	11	0	13.9	—
	(6)	(6)	(6)	(6)	(6)	(6)	
>2,000	4.6	48	45	7	0	28.1	—
	(6)	(6)	(6)	(6)	(6)	(10)	

[1] Data on particle size and composition are not available for fourteen of the forty stations from which photographs are described.
[2] Calcium carbonate-free basis.

TABLE 22–2. Number of *Hyalinoecia tubicola* per square meter in photographs or bottom samples on the continental slope of the western North Atlantic.

Station[1]	Water depth (m)	Number per square meter	
		In photo	In grab
1071	346	12	2
2078	365	272	2
1078	376	6	2
2160	400	2	0
1371	446	6	45
1358	450	29	23
1367	455	29	5
1357	564	10	0

[1] Stations are plotted in fig. 22–6. Tracks of *H. tubicola* occurred in photographs from the following stations: 1071, 1078, 1116, 1357, 1358, 1367, 1371, 2074, 2078, 2088, 2099, 2107, 2108, 2152, 2158.

box. Several small subsamples of sediment were removed for geological study, after which the remaining material was washed on a 1-mm-mesh screen. All material remaining on the screen (living animals, worm tubes, shell fragments, gravel, etc.) was preserved. The laboratory analysis consisted of sorting the animals into major groups, counting the number of individuals, and weighing each group after excess formalin had been removed by blotting paper. Nonliving components such as gastropod shells inhabited by pagurid crabs, or annelid tubes (whether inhabited or empty) are omitted from the counts and weights. The shells of living mollusks are included in the weight measurements. Portions of individual specimens appearing in the sea-bottom photographs are prorated according to the percentage of the animal showing within the area of the picture. For comparison, all specimen counts are expressed as numbers of individuals per square meter, or grams per square meter (damp weight).

22–3. Organisms in photographs and bottom samples

Hyalinoecia tubicola

One of the largest and commonest organisms appearing in the sea-bottom photographs was the tube-dwelling polychaete worm *Hyalinoecia tubicola* (Müller) — figs. 22–3, 22–4, 22–5. This species appeared in photographs at eight stations distributed from the vicinity of Georges Bank to near Cape Hatteras (table 22–2, fig. 22–6), a wide geographic range that is in keeping with the known cosmopolitan distribution of the species (Pettibone, 1963). Water depths at these stations ranged from 346 to 564 m. This restricted bathymetric range is well within the depth range of this species (274 to 1,170 m) as reported by Verrill (1885) for the western North Atlantic Ocean.

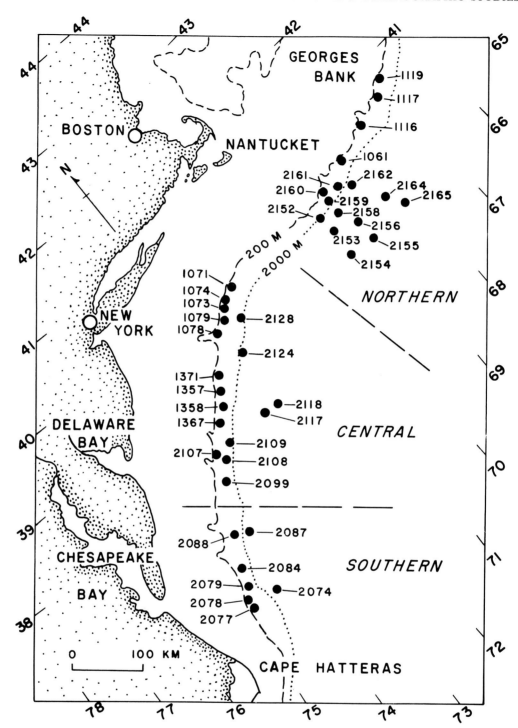

Figure 22–2. Stations occupied by R/V *Gosnold* on the western North Atlantic continental slope, where quantitative samples of the benthos and clear sea-bottom photographs were obtained during 1963 and 1964. Depth contours are shown for 200 m (broken line) and 2,000 m (dotted line). The region was divided arbitrarily into three geographic regions—northern, central, and southern.

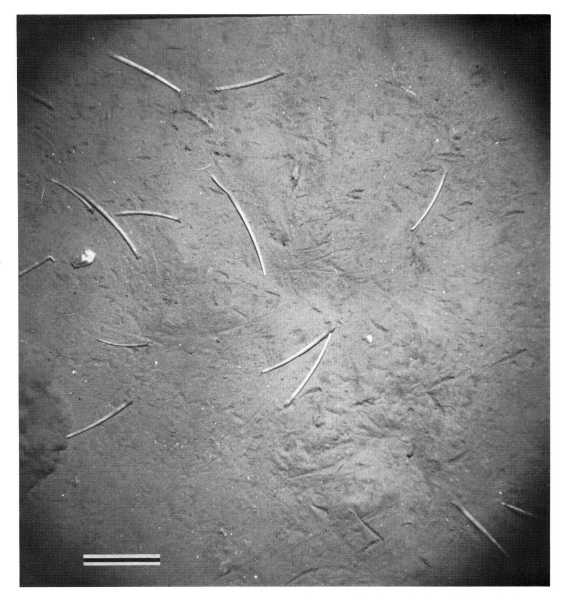

Figure 22–3. Bottom photograph showing *Hyalinoecia tubicola* at Station 1358, east of Delaware Bay (38°42.4′N, 73°01.1′W); water depth 450 m; sediment silty sand. The cloud of suspended sediment at the left-hand side of this and subsequent photographs resulted from the impact of the camera-tripping weight. Scale bar (lower left) is 10 cm.

The depth range for all oceans combined is 13 to 4,380 m (Pettibone, 1963).

The trend of increasing density of worms from the northern part of the study area southward was pronounced (fig. 22–6). The enormous concentration of living specimens at the southernmost station (Station 2078) in this series, illustrated in fig. 22–4, was totally unexpected. Although large numbers of specimens are known to have been taken in trawls dragged along the bottom (Verrill, 1885), the observed density of 272/m² is approximately two orders of magnitude larger than expected. Water

depth at this location is 365 m, which is not an unusually deep or shallow occurrence for this species.

Bottom photographs showing *Hyalinoecia* have been taken at a depth of 1,935 m on the continental slope off Baja California (Parker, 1963). Their density there was only 1.5/m², as compared with the average density of 46/m² on the Atlantic continental slope, based on photographs at only those stations where one or more specimens occurred.

Collections and photographs of *H. tubicola* disclosed a high percentage of occupied tubes. An examination of

Figure 22–4. Bottom photograph showing *Hyalinoecia tubicola* at Station 2078, southeast of Chesapeake Bay 36°21.7′N, 74°45.3′W); water depth 365 m; sediment foraminiferal silt. Scale bar (lower left) is 10 cm.

the preserved *H. tubicola* from grab-sample collections revealed that 80 per cent of the tubes were inhabited at the time collected. Unoccupied tubes were generally discolored and covered with a detrital film. A considerable number were broken or otherwise incomplete. The tube cavities of a few were completely packed with sediment, suggesting that they may have been buried. Eighty-six per cent of the tubes photographed appeared to be occupied. This value is in general agreement with the ratio observed in the grab samples.

Tube lengths of specimens in the grab samples as well as those in the photographs ranged from 6 to 15 cm, but most were 8 to 12 cm long. The tube gradually tapers

from the larger, anterior end (left end in fig. 22–5) to a rather narrow posterior end. Large tubes are nearly straight whereas small ones have a pronounced taper. Diameters at the anterior end are 3 to 8 mm (average, about 5 mm).

Elongate grooves in the sediment that were commonly less than 0.5 cm deep, about 0.5 cm wide, and of lengths reaching to a maximum of roughly 25 cm are present in seven of the eight photographs that depict *Hyalinoecia tubicola*. These grooves appear to be tracks made by the worms as they crawled along the bottom, although this is not a certainty. One reason for doubt is that the tracks rarely appear to be aligned exactly with the worm tubes or to

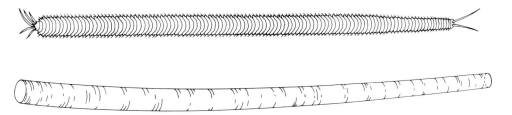

Figure 22–5. Outline drawing of a 10-cm *Hyalinoecia tubicola* and its tube. Drawing by F. A. Bailey.

extend from either the posterior or anterior ends of them. Texture of the sediment and size of the worm would be expected to affect greatly the presence or absence of tracks, and their size. Tracks were present in all photographs showing *H. tubicola*, except at Station 2160. The photograph at this station revealed no tracks, but the sediment was relatively coarse and consequently unlikely to be imprinted. *Hyalinoecia* tracks in the soft sediments of the lower continental slope off Baja California are well illustrated in a report by Parker (1963; plates 14 and 15*b*). In addition to the seven stations where photographs showed both *H. tubicola* worms and tracks, tracks alone are evident in photographs at eight stations. These stations are included in both table 22–2 and fig. 22–6. Their locations are in accord with other information concerning the distribution of this species. General and specific information on distribution, plus the rare and inconsistent occurrence in the grab samples, or photographs, of other species that would be likely to make tracks of this kind, support the conclusion that the sediment markings are tracks of *H. tubicola*.

The photographs record a low level of activity for this species of worm. Only one of the 177 specimens pictured appears to be crawling. Moreover, only about 10 per cent of the worms are partially emerged from their tube. These few are extended one-third or less of the body length from the tube's larger, anterior, end. All emerged worms seem to be exploring the sediment surface, possibly searching for food, but not burrowing or digging deeply in the sediment. A peculiar habit of stacking, in which one worm tube (most often the anterior end) is propped atop another tube like a jackstraw, is clearly shown in fig. 22–4. Absence of detritus along the midline on the upper side of otherwise detritus-covered tubes indicated a scraping action, possibly by lateral movement of one tube along the top of another.

Few large surface-dwelling animals appeared to inhabit the same bottom area with *H. tubicola*. A large-tentacled sea anemone similar to *Bolocera teudiae* was present with the polychaete in two pictures. A 14-cm fish, family Macruridae, is shown swimming in the vicinity of *H. tubicola* in another. It is the only animal photographed or taken in the grab samples that would be a likely predator on this polychaete. A large brittle star, *Ophiomusium*, occurred in four photographs that also included tracks of *H. tubicola*.

A total of 177 specimens are visible in the photographs, whereas only 46 were obtained in the grab samples at these same stations. This ratio suggests that the photographic method is superior to grab-type sampling for determining the occurrence and density of this species. The large size, light color, distinctive shape, and epibenthic habit of *H. tubicola* enhance its detection by photographic means. Without the accompanying grab samples, however, it could not have been determined whether a substantial number of specimens were buried.

Ophiomusium lymani

One of the most handsome and conspicuous creatures in the series of photographs is the brittle star *Ophiomusium lymani* Thompson — figs. 22–7, 22–8, and 22–9. Its large symmetrical form and white coloration contrast sharply with the dark olive-green sediments which it inhabits. Disc diameters of the larger specimens are 1.5 to 2.5 cm, and their arms are 10 to 15 cm long. Because of their large size they were invariably the dominant epibenthic invertebrate species at each station where they occurred. The disc of this brittle star is not circular as it is in many of its relatives, but rather is somewhat pentagonal, with nearly straight margins extending from the insertion of one arm to the next. The long slender arms are commonly straight, except for the distal one-third, which usually is curved upward or sideways.

This species of brittle star appeared in bottom photographs at eight stations (table 22–3). It was distributed from Nantucket southward to near Cape Hatteras, at stations near the base of the continental slope (fig. 22–10). The observed bathymetric range was 1,480 to 2,150 m, an unusually restricted depth interval for deep-water inhabitants of this kind. The depth range reported by Verrill (1885) for collections of this species from the western Atlantic by the *Albatross* was 443 to 3,781 m. The same or similar species occurred in bottom photographs from depths of 1,935 m on the continental slope off Baja California (Parker, 1963).

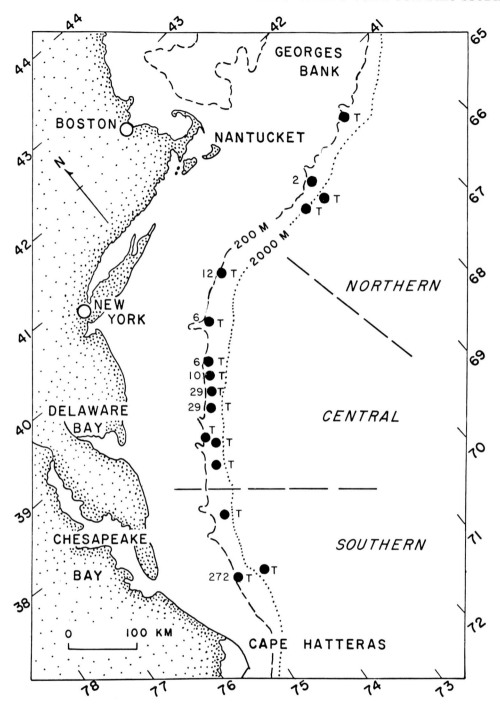

Figure 22–6. Distribution and density of *Hyalinoecia tubicola*. Figures to the left of the 200-m contour line give number per square meter, based on analyses of photographs. Stations at which tracks in the sediment made by this species could be seen in the photographs are indicated by the letter *T*.

The average *Ophiomusium* density of 3.9/m² at the eight stations where they occurred (and 0.8/m² if all forty samples are considered) is a substantial number for such large animals, particularly at these great depths. They were somewhat more numerous in samples from the central and northern areas, but this situation may be a vagary of sampling. The unusually high density of 11/m² at Station 2161 (see fig. 22–8) southeast of Nantucket is noteworthy because it represents a large benthic biomass for slope waters. The calculated weight of

TABLE 22–3. Number of large *Ophiomusium* in bottom photographs and in grab samples at eight stations on the continental slope of the western Atlantic.

Station number	Water depth (m)	Number of *Ophiomusium* per square meter	
		In photo	In grab
2159	1,480	4	—
2079	1,545	2	—
2128	1,605	3	—
2161	1,625	11	2
2108	1,660	1	—
2124	1,780	6	2
2152	1,865	2	—
2087	2,150	2	—

Ophiomusium at this station was 31 gm/m². The total benthic biomass for all stations in this region at depths between 1,000 and 2,000 m was 5.9 gm/m², a value determined from grab samples. Wet-weight values expressed here include the skeletal components, which in brittle stars constitute a major share of the total body weight. Even the dry organic weight of *Ophiomusium* at Station 2161 (calculated to be 1 or 2 gm/m²) is a substantial quantity for these depths.

The spacing of individuals varies greatly with their density. Where aggregations were most dense, however, the spacing between individuals is quite regular, seemingly having reached a minimum distance of a few centi-

Figure 22–7. Bottom photograph showing the brittle star, *Ophiomusium lymani*, at Station 2124, east of Cape May, New Jersey (30°05.4′N, 72°08.2′W); water depth 1,780 m; sediment clayey silt. Scale bar (lower left) is 10 cm.

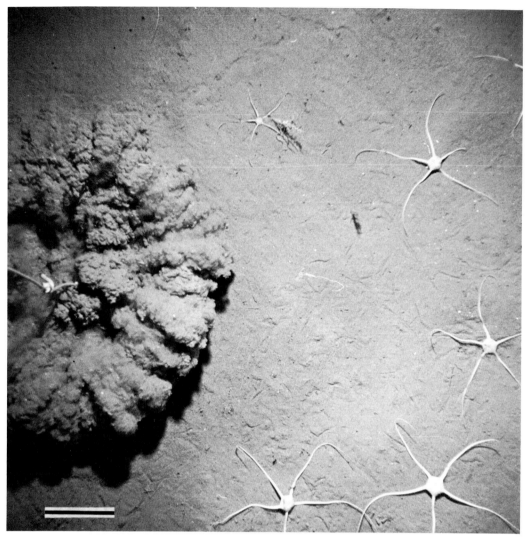

Figure 22–8. Bottom photograph showing the brittle star, *Ophiomusium lymani*, at Station 2161, southeast of Nantucket (39°56.0′N, 68°50.8′W); water depth 1,625 m; sediment clayey silt. Scale bar (lower left) is 10 cm.

meters between arm tips. The overlapping of arms, as seen at the bottom of fig. 22–8, may be a temporary result of the rapid movement of the two specimens toward the right, to avoid the disturbance caused by the camera tripping weight.

When crawling along the bottom, *Ophiomusium lymani* appears to propel itself by a sweeping motion of four arms and trails the fifth arm behind. The direction in which the arms of the specimens are curved in figs. 22–7 and 22–8 suggests that these individuals were moving diagonally from the lower right toward the upper left. Also, the photographs showing more than one individual disclose that the movement of all individuals was in the same direction.

During rapid movements pressure on the sediment is greatest at the middle of the arm or a portion between the

middle and the distal quarter (fig. 22–8). On the other hand when movement is slow, the basal three-quarters of the arm leaves an imprint and the arm is pivoted on the sediment about one-half to three-quarters of the distance out from the disc.

Tracks in the sediment surface made by *Ophiomusium lymani* are clearly evident in all photographs that show individuals of this species. Some typical examples are included in figs. 22–7 and 22–8. Tracks in the form of narrow grooves about 0.2 to 0.5 cm wide and commonly less than 5 cm long are representative for this species. The arm tips apparently are not used for body propulsion; presumably their primary function is for tactile and feeding purposes.

Photographic evidence indicates that 80 per cent of the large *Ophiomusium* lie exposed on the sediment surface;

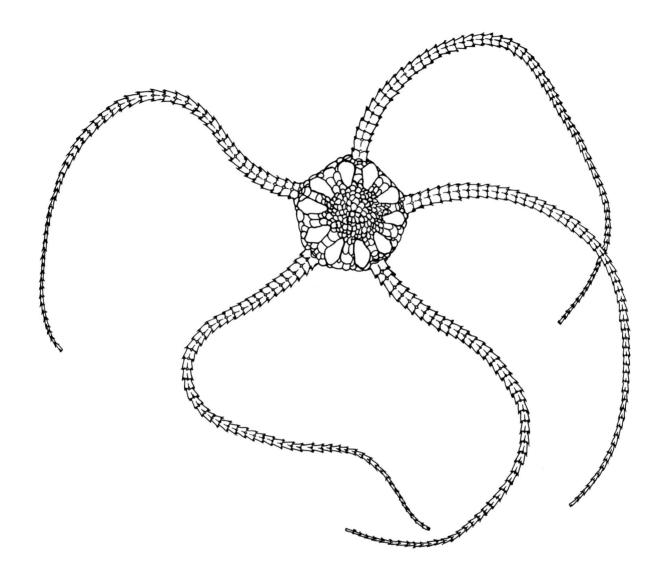

Figure 22–9. Sketch of the dorsal view of *Ophiomusium lymani*. Arm length of this specimen was 10 cm, disc diameter was 1.5 cm.

10 per cent are partially buried in the sediment; and 5 per cent are more deeply buried, with only the arm tips exposed. The remaining 5 per cent have the disc raised several cm above bottom in a peculiar stance. The arms are curved downward from the disc; the midsection of the arms is supported on the sediment and the distal half of the arms is curved upward above the sediment (fig. 22–7).

The proportionately large number of *Ophiomusium* revealed by the photographs compared to the few taken in the grab samples again demonstrated the superiority of the photographic method for determining the occurrence and density of large sedentary epibenthic animals, as compared with the bottom-grab method. Without the supplementary evidence provided by the grab samples, however, the number of specimens buried in the sediment would not be known.

Small brittle stars

In addition to the large *Ophiomusium*, just discussed, small brittle stars were moderately common in the grab samples, and a few were detected in the bottom photographs (table 22–4). The largest included in this category are those with a disc diameter less than 0.5 cm, and arms shorter than 3 cm. Several species were represented in this group, including juvenile *Ophiomusium lymani*. The commonest species appears to be *Amphilepis ingolfiana* Mortensen. All of these small brittle stars have been grouped and treated as a unit because the different species appearing in the photographs cannot be identified.

A total of seventeen small brittle stars appeared in photographs from five stations. Geographically, these stations are distributed from the northern part of the study area to the southern part. Bathymetrically, depths ranged from 214 to 2,910 m. Small brittle stars were not limited to particular depth zones or geographic areas, probably because the several species in this group have somewhat different habitat requirements.

A total of 109 individual specimens of small brittle stars were collected in the forty grab samples (estimated density, $4.9/m^2$). The density based solely on the photographic evidence was $0.9/m^2$. If only those stations are considered where small brittle stars were detected (in photographs at five stations; in grab samples at twenty-one stations) the average density from the two methods are in close agreement, $7/m^2$ in photographs and $9/m^2$ in grab samples. Individual small brittle stars were not aggregated in clusters, like the large *Ophiomusium*, but were rather widely separated. For example, the photograph taken at Station 2117 (not reproduced here) showed eight individuals. The average distance between them was 12 cm (six arm lengths).

It seems likely that many other small brittle stars were in the bottom areas photographed, but they cannot be seen. This belief is supported by the considerably greater number recovered by the grab sampler (109) than were photographed (17). Some of the grab-caught brittle stars may have been buried in the sediment, but their habit of living on or near the surface suggests that a significant number may have been camouflaged on the sea floor.

It must be concluded that the photographic technique is inadequate for quantitative studies of this group of animals.

22–4. Inferences about biomass

How accurate are quantitative estimates of benthic fauna based exclusively upon visual study of sea-bottom photographs? This question is asked more and more frequently as photographic equipment and techniques improve and as the number of bottom photographs increases. It also stems from plans to use underwater manned vehicles, such as *Trieste*, *Alvin*, *Sea-Pup*, and others, for macrobenthic research. In addition, the tedious and expensive process of collecting and processing quantitative samples of benthic fauna is much in need of simplification wherever feasible. Photographic and direct visual methods have been considered as two promising supplements or alternatives to the conventional practice of collecting and studying living or preserved specimens from deep water. The series of photographs and biological samples assembled for this study provides information on some aspects of this subject.

A quantitative comparison was made between contents of grab samples and estimates of the surface-dwelling benthic animals from nearly simultaneous photographs. Without referring to the data from grab samples, the number of living animals in the photographs was counted and the size of each was measured. The weight for individual specimens was derived from measurements of preserved specimens. Mean values for both numbers and weights were determined for northern, central, and southern areas — see fig. 22–2 — and for four water depth zones (200–500 m, 500–1,000 m, 1,000–2,000 m, and greater than 2,000 m). These data, together with comparable values from grab samples, are listed in table 22–5. Each entry is an average of all individual samples in each category.

TABLE 22–4.   Number of small brittle stars[1] in bottom photographs and in grab samples at forty stations on the continental slope of the western Atlantic.

| Water depth (m) | Number of stations | In photographs | | | In grab samples | | |
| | | Number of stations | Number/m^2 [2] | | Number of stations | Number/m^2 [2] | |
			Average	Range		Average	Range
200– 500	13	2	4	2–6	8	16	2–72
500–1,000	4	0	—	—	3	6	2–11
1,000–2,000	12	1	2	2–2	6	4	2–5
2,000–3,310	11	2	12	8–17	4	8	4–18
200–3,310	40	5	7	2–17	21	9	2–72

[1] *Amphilepsis*, juvenile *Ophiomusium*, and others.
[2] Numbers/m^2 include only the stations at which small brittle stars were seen (in photographs) or collected (in grab samples).

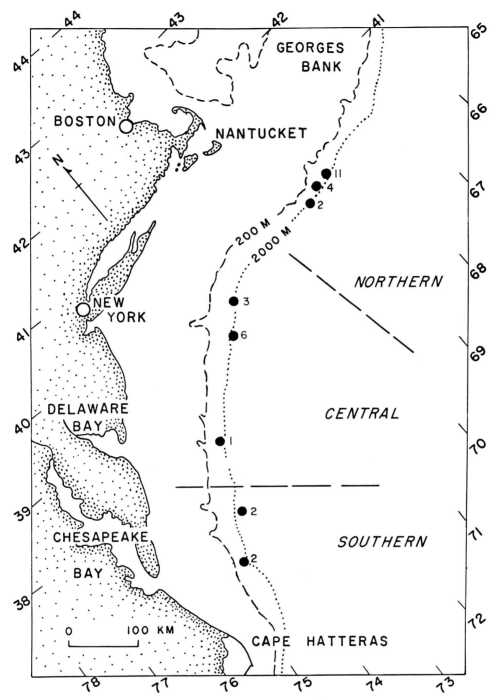

Figure 22–10. Distribution and density of adult *Ophiomusium lymani* in the study area. Density is expressed as the number per square meter, based on the analysis of bottom photographs.

Grab samples disclosed a marked decrease in the total number of individuals per unit area from shallow to deep water and from north to south. The density of specimens along the upper part of the continental slope averaged 302/m², whereas at the base of the slope and beyond, the average density was only 46/m². The decrease in numerical density from north to south was substantial, but considerably less than the depth-related change. Numerical density was roughly two to three times as great in the northern area as in the southern area. Density in the central area was intermediate.

Density of surface-dwelling individuals estimated from

TABLE 22–5. Comparison of total benthos (based on grab-sampler contents) with surface-dwelling benthos (as deduced from photographs). [Values in parentheses are the number of samples on which the computation is based.]

Geographic area	Epibenthos (photographs)				Total benthos (grab sampler)			
	Water depth (m)				Water depth (m)			
	200–500	500–1,000	1,000–2,000	>2,000	200–500	500–1,000	1,000–2,000	>2,000
Number of individuals per square meter								
Northern	59 (4)	15 (1)	8 (4)	2 (7)	422 (4)	66 (1)	92 (4)	58 (7)
Central	14 (8)	8 (2)	4 (5)	10 (2)	261 (8)	413 (2)	94 (5)	30 (2)
Southern	272 (1)	2 (1)	213 (3)	1 (2)	156 (1)	41 (1)	67 (3)	24 (2)
Average	48	8	58	3	302	233	86	46
Biomass (gm/m²)								
Northern	34.2 (4)	69.7 (1)	10.8 (4)	0.3 (7)	8.8 (4)	5.8 (1)	5.8 (4)	[1]3.7 (7)
Central	9.0 (8)	17.9 (2)	4.9 (5)	1.6 (2)	17.4 (8)	8.8 (2)	[2]7.5 (5)	3.7 (2)
Southern	187.2 (1)	1.3 (1)	46.5 (3)	6.2 (2)	10.1 (1)	5.8 (1)	[3]3.3 (3)	3.8 (2)
Average	30.5	26.7	17.3	1.7	14.2	7.3	5.9	3.7

[1] Weight does not include one unusually large specimen; uncorrected weight is 9.3 gm/m².
[2] Weight does not include one unusually large specimen; uncorrected weight is 12.5 gm/m².
[3] Weight does not include two unusually large specimens; uncorrected weight is 19.6 gm/m².

the photographs also revealed a general decrease in number as water depth increased. The densities estimated from photograph analyses were decidedly lower, however, and several were inconsistent with the general trend. The gradient in density from north to south was not evident.

Values of the benthic biomass based on the grab-sampler data are rather consistent in exhibiting a decline in weight with increased water depth. Along the upper part of the continental slope the biomass averaged approximately 14 gm/m²; at the slope base and deeper it was 3.7 gm/m²; and at intermediate depths the average generally was between 4 and 9 gm/m². There was no apparent trend of decreasing biomass from north to south, as there was for density of individuals. Actually, the central area had a slightly greater weight of benthos than the northern and southern areas.

Biomass of the large surface-dwelling species, based on photographic interpretations, agrees with the grab-sampler results in revealing a marked decrease in benthic

biomass at increased water depth. Also, the absence of a latitudinal gradient in weight of epifauna in the photographic analysis corresponds with the grab-sampler results. Principal differences between the biomass values calculated from the two sources are: estimates based on photographs resulted in substantially higher values at all three depth zones shallower than 2,000 m; greater variation in biomass estimates resulted from photograph analyses.

Both the grab and the camera sampled essentially the same fauna, except that infaunal components and small epibenthic animals taken in the sampler were not visible in the photographs. Consequently, the resulting estimates of total benthos weight should equal, or exceed, the weight of the epibenthic component alone. Inasmuch as the estimated epibenthos weight (photographic data) in seventeen of forty samples and in seven of twelve environmental areas (table 22–5) was larger than the total benthos (grab-sampler data), inaccuracies in the sampling techniques are indicated. Gross differences in biomass estimates occurred in only four samples, however, and in each one the grab failed to recover certain specimens that appeared in the photographs. We conclude that sea-bottom photographs are a valuable supplement to grab samples. They are particularly useful in biomass studies in areas where large, distinctly shaped epibenthic species constitute a substantial portion of the macrobenthos. The relative rapidity with which photographic data can be assembled and the broad economic advantage of the photographic method are additional features that favor this method.

22–5. Summary

Bottom samples and sea-bottom photographs were collected with a combined camera and grab sampler at forty stations on the continental slope between Georges Bank and Cape Hatteras. Some quantitative comparisons were made of the photographs and grab samples, and observations were made on the distribution, size, body position, and related ecological aspects of the more common species in the photographs.

Hyalinoecia tubicola, a large tube-dwelling polychaete, was the most common animal in the photographs. A total of 177 specimens appeared in photographs from eight stations. The average density at the eight stations was 46/m²; maximum density was an unusually high 272/m². This polychaete occurred along the upper part of the continental slope at 346 to 564 m, from southeast of Nantucket to the vicinity of Cape Hatteras. Elongate grooves in the sediment at fifteen stations were interpreted as tracks of *H. tubicola*. Judging by the few worms protruding from their tube, their activity was at a low level. The stacking of one tube across another was common.

Ophiomusium lymani, a large, handsome, brittle star, was common in photographs taken near the base of the continental slope (1,480 to 2,150 m) between Nantucket and the vicinity of Cape Hatteras. Their density averaged 3.9/m², considering only the stations where they were detected (0.8/m² at all forty stations), which is high for large animals at these great depths; maximum density was 11/m². This species crawls along the bottom by making lateral sweeping movements with four arms; the fifth arm trails behind. The distal portion of the arm does not appear to be used in locomotion. Spacing between individuals appears rarely to be less than a few centimeters between arm tips. Small brittle stars were common in the grab samples, but only a few were detected in the photographs.

Substantially larger numbers of *Ophiomusium* and *Hyalinoecia* were revealed by the photographs than were captured in the grab. This relationship may demonstrate the superiority of the photographic method over the grab method for determining the occurrence and density of large epibenthic animals. For quantitative studies of infauna and small to medium-sized epibenthic forms (such as small brittle stars) the grab method is better, as discussed in chap. 20.

Benthic animals were most numerous (302/m²) and formed the largest biomass (14 gm/m²) in grab samples from the upper part of the continental slope. They diminished sharply with increased water depth to a density of 46/m² and a weight of 3.7 gm/m² at depths below 2,000 m. The number of individuals in the total benthos, but not the biomass, decreased from north to south at all depths.

A quantitative comparison between grab-sample contents and the fauna estimated from photographs for four depth zones in three geographic areas disclosed similar trends in the estimates of biomass and number of individuals. The comparatively higher biomass estimates from photographic interpretations is attributed chiefly to inefficiency of the grab sampler in collecting the larger animals. Low densities of individuals were relatively lower in the photographs mainly because small epibenthic and infaunal animals could not be detected. The results demonstrate the usefulness of sea-bottom photographs in macrobenthic research, especially in studies dealing with the biomass and ecology of large surface-dwelling species.

References

Edgerton, H. E., 1963: Underwater photography. In: *The sea; ideas and observations on progress in the study of the seas.* Vol. 3, *The earth beneath the sea,* M. N. Hill (ed.), Interscience Publ., New York, 473–79.

Emery, K. O., A. S. Merrill, & J. V. A. Trumbull, 1965: Geology and biology of the sea floor as deduced from simultaneous photographs and samples. *Limnology and Oceanography,* **10,** no. 1, 1–21.

McIntyre, A. D., 1956: The use of trawling, grab, and camera in estimating marine benthos. *J. Mar. Biol. Ass. U.K.,* **25,** 419–29.

Menzies, R. J., L. Smith, & K. O. Emery, 1963: A combined underwater camera and bottom grab: a new tool for investigation of deep-sea benthos. *Int. Revue ges. Hydrobiol.,* **48,** no. 4, 529–45.

Parker, R. H., 1963: Zoogeography and ecology of some macroinvertebrates, particularly mollusks, in the Gulf of California and the continental slope off Mexico. *Vidensk. Medd. fra Dansk naturh. Forch.,* **126,** 1–178.

Pettibone, M. H., 1963: Marine polychaete worms of the New England region. I. Aphroditidae through Trochochaetidae. *Bull. U.S. Nat. Mus.,* no. 227 (part 1), 1–356.

Shipek, C. J., 1960: Photographic study of some deep-sea floor environments in the eastern Pacific. *Bull. Geol. Soc. Am.,* **71,** 1067–74.

Verrill, A. E., 1885: Results of the explorations made by the steamer *Albatross,* off the northern coast of the United States in 1883. Rept. U.S. Comm. Fish and Fisheries (1883), 503–699.

23. Deep-sea photography in the study of fishes*

Norman B. Marshall   *British Museum (Natural History), London, England*

Donald W. Bourne   *Woods Hole Oceanographic Institution, Woods Hole, Massachusetts*

Abstract

Photography complements trawling and other methods of investigation of the benthic fish fauna; it is particularly useful in discovering the habits of these animals, and as a basis for population studies.

Fishes are among the largest animals living on or near the deep-sea floor, most of them ranging in length from 6 in (15 cm) to 3 ft (90 cm). It is unlikely, considering their size and shape, that some may be hidden under the sediments, as are so many invertebrates. Even so, our introduction to photographs of bottom-dwelling fishes was not encouraging. Only after close scrutiny of many series of photographs (and given a good knowledge of fish systematics) did it become clear that certain lines of research are perfectly feasible and valuable.

We soon realized that bottom photographs rarely show enough detail for fishes to be identified specifically, but that most specimens are referable to family groups. Stereo pairs help greatly in such determinations; they generally give a clearer picture of form and fin pattern than do single shots, and often they reveal features which are otherwise quite invisible.

Photographs of fishes on the deep-sea floor have an immediate use in resolving an old uncertainty. Though many species of fish which appear to be bottom-dwelling in shallow seas have been taken repeatedly in bottom trawls, the question has persisted: might they have been caught by the net as it was hauled to the surface? Such suspect fishes can now be seen to live at the bottom. Rat-tails (Macrouridae), deep-sea cods (Moridae), halosaurs, brotulids, notacanths, and eels, swim close above the bottom (figs. 23–1, 23–2, 23–4, 23–5 and 23–7). Others, such as tripod fish (Bathypteroidae), lizard fish (Synodontidae), scorpion fish (Scorpaenidae), and skates (Rajidae), rest upon the sea floor (fig. 23–3). Before seeing this direct

evidence, members of the first group, all of which have swimbladders, seemed to have been able to swim with ease in aquatic space. But the species in the second group, which have no swimbladder, usually appeared closer to the bottom. It is reassuring, though not too surprising, to find that these conjectures are essentially right.

Study of the free-swimming species is useful in other ways. When evidence of water movements is present, as in the bending of sea-lilies (crinoids), or in the pattern of ripple marks in the sediments, individual fishes are generally seen to be headed into the current. When the camera's orientation has been recorded, the direction of bottom currents can be quite readily found from these biological indicators.

Further, certain fishes swim in characteristic ways. Halosaurs, and most of the rat-tails, tend to swim with a slight head-down inclination to the bottom (figs. 23–1, 23–2 and 23–5), a position related to their fin pattern. These fishes have a long and broad anal fin which, when moved from side to side or undulated, lifts the tail as well as driving the fish forward. In this posture, the tipped-down snout may be used for rooting in the sediments, and the photographs occasionally show rat-tails doing this (fig 23–5a); the jaws, set below the snout in halosaurs and most rat-tails, can readily seize burrowing organisms (such as polychaete worms) turned up in this way as well as food to be found on the surface of the sea floor.

Photographs sometimes remind us that much of the deep-sea fauna remains to be discovered. Some thirty pictures taken at depths between 1,115 and 2,195 in the Red Sea show a small galeoid shark which does not fit into any known family (fig. 23–6) (Marshall & Bourne, 1964). These, and other pictures, show for the first time that there are bottom-dwelling fishes in the depths of the Red Sea, despite temperatures there which are exceptionally high — only a few degrees below those at the surface.

It is also a photograph that extends the known range of another group of fishes, the Holocephali, into the Gulf of Aden. This picture, taken at 1,240 m, shows the long-nosed chimaera, *Harriotta;* previously, only an egg-case of *Harriotta* had been found in the Gulf.

* WHOI Contribution Number 1762.

This work was supported by National Science Foundation grant GB-20702, and by the Office of Naval Research, under contract Nonr-4029 NR 260–101.

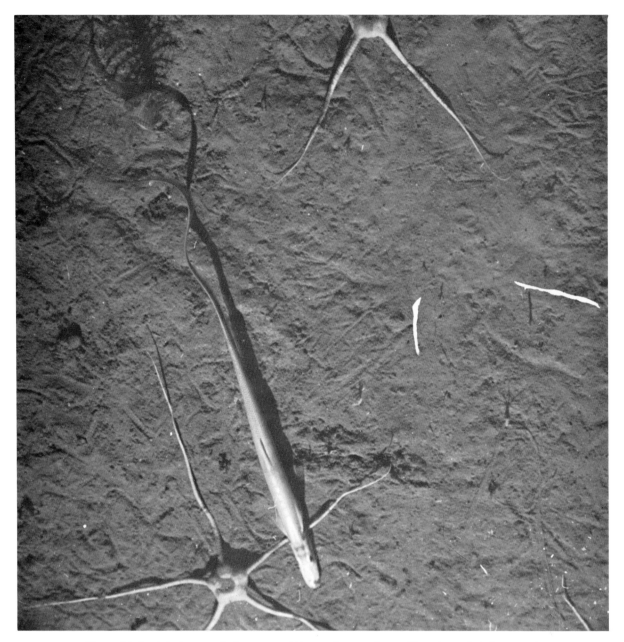

Figure 23–1. A halosaur of the genus *Aldrovandia*, together with a large brittle star. As shown in this picture, halosaurs generally swim by undulating the rear half of the tail, though exceptionally the whole body may be thrown into eel-like waves. In halosaurs, as in their very distant relatives the rat-tails, the long and broad anal fin, projecting snout, and inferior mouth, are adaptations for rooting in the ooze, a bit of which is stuck on the snout of this fish.

Where photographic coverage has been thorough, it frequently suggests which are the key fishes in local deep-sea faunas. For instance, the small shark mentioned in the last paragraph was by far the most photographed bottom fish in the Red Sea, while at the western end of the Gulf of Aden the commonest fishes were a kind of deep-sea cod (Morid) (fig. 23–8) having very long and mobile pelvic fin rays. Over the continental rise, south of Nova Scotia, at about 2,600 m, the most abundant fishes are rat-tails (figs. 23–2, 23–5a, 23–9), but at more southerly stations, on the Blake Plateau, brotulids and eels (figs. 23–4 and 23–7) predominate.

That the camera presents a useful panorama of fishes in an area seems likely, since deep-sea fishes seldom appear to be disturbed by the apparatus. An individual fish may be seen over the same bit of bottom in several

Figure 23–2. A rat-tail, probably *Lionurus carapinus*, swimming vigorously over the sediments of the continental rise south of Nova Scotia. This picture was taken at about 2,600 m by the Thorndike camera (Thorndike, 1959), which is triggered by contact with the bottom. Fishes and other organisms are more often photographed in close, oblique views with this camera than with the Edgerton stereo camera. (Photograph by courtesy of J. L. Worzel, Director of the Lamont Geological Observatory.)

shots in sequence, and it is rare to see the puffs of disturbed sediment raised by a startled fish. Films made during some deep dives of the bathyscaph *Trieste*, showing some of the species considered in this article, support these observations.

Though cameras are somewhat selective in the fishes they capture, they may be less so than trawls, and probably selective in a different way. Photographic surveys thus are valuable not only for recording the kinds of fishes present in a locality but also as an alternative to trawling in estimating population numbers. A good survey of this type requires many pictures and a reliable way of calculating the area of bottom which they represent. We have found continuous stereo mosaics (see fig. 23–9) especially good. In such strips, gradual changes in the character of the bottom may be seen reflected in different animal communities (the kinds of urchins present, for

example), and our preliminary studies suggest that the fishes live not at random on the bottom, but in clusters. This kind of information is difficult, if not impossible, to extract from scattered pictures.

The particular importance of stereo in population studies is that it furnishes an easy and accurate means of scaling the pictures (Boyce, 1964). Provided that similar cameras were used to make the pictures, that their optical axes were parallel and roughly normal to the bottom, and that the distance between these axes is known, this camera spacing can be shown to be the same as the width of that part of the field of one camera which is not overlapped by the field of the other. This being so, the dimensions of any particular picture (and often the sizes of the animals in it) can be quickly found with the aid of dividers.

In one well-surveyed area where we have applied these techniques, population densities figured for each of several tracks made over the same ground have agreed very closely, and the proportions among different kinds

Figure 23–3. The scorpion fishes (Scorpaenidae), such as this one confronting a shrimp, have no swimbladder and thus must spend most of their lives resting on the bottom. The broad pectoral fins aid these fishes in lunging after their prey, but they would soon be exhausted by sustained swimming. On this fish's right, in front of the pectoral fin, the pelvic fin may be seen in the forward position characteristic of this group.

Figure 23-4. A brotulid, probably an undescribed species of *Neobythites*, with a small rat-tail (*Hymenocephalus* sp.) in close attendance. Brotulids and rat-tails are the most diverse of bottom-dwelling, deep-sea fishes. Most species live over the continental slopes, particularly in the warmer parts of the ocean.

Figure 23-5a. A rat-tail rooting in the oozes. Inspection of their stomachs shows that much of the rat-tails' diet consists of small, burrowing invertebrates, such as polychaete worms and various crustaceans. The form of these fishes and the arrangement of their sense organs suggest the manner of their feeding and such photographs confirm it.

Figure 23-5b. A rat-tail of the subfamily *Bathygadinae*, taken at 1,240 m in the western Gulf of Aden. Unlike most rat-tails, the *Bathygadinae* have an anal fin no broader than the second dorsal, and this symmetry keeps the body not tipped tail upwards but horizontal, as these fishes swim (*cf.* figs. 23-2 & 23-5a). The mouth, in keeping with this posture, is terminal. Large lateral-line canals are easily seen on the head and flanks, here, while the characteristic elongated rays of the paired fins are revealed in the shadow. Stereo helps in detecting the long dorsal fin ray. One of the curious, shiny "marbles" common in our Gulf of Aden pictures may be seen just behind the fish's tail. These unidentified animals leave tracks in the sediment.

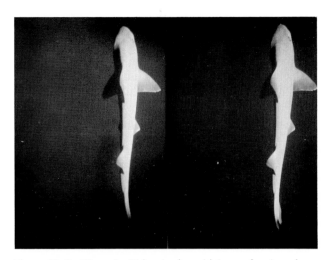

Figure 23-6. (Stereo). This shark, which so far has been captured only by the camera, is found in the depths of the Red Sea, where it was the most photographed fish. Its fin-pattern and the absence of spines show it to be a galeoid, but it does not fit into any of the known families of the order.

of fishes have also been consistent. Two successful trawls have since been made on this ground, and while these were too short for quantitative comparisons with the photographic stations they yielded many animals which appear in the pictures, and which now can be identified specifically.

Cameras and collecting gear such as trawls and grabs, used routinely together, seem a promising combination against the difficulties in exploring deep-benthic life.

References

Boyce, Robert Eugene, 1964: Simple scale determination on underwater stereo pairs. *Deep-Sea Res.*, 11, 89–91.

Marshall, N. B., & D. W. Bourne, 1964: A photographic survey of benthic fishes in the Red Sea and Gulf of Aden, with observations on their population density, diversity, and habits. *Bull. Mus. Comp. Zool.*, Harvard University, **132**, no. 2, 223–44.

Thorndike, E. M., 1959: Deep-sea cameras of the Lamont Observatory. *Deep-Sea Res.*, **5**, 234–37.

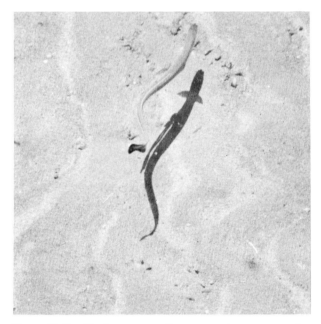

Figure 23-7a. Shadows, as well as stereo images, are often helpful in interpreting a photograph. Here, the pectoral fins are quite invisible except in the shadow, which also reveals what seems to be a parasite, attached to a swelling in the eel's side.

Figure 23-7b. The larger of the eels here is a notacanth; these deep-sea spiny eels share a number of characteristics with some other bottom fishes (rat-tails for example), which are equipped with a swimbladder. Among these are the fin pattern and protruding snout discussed in the text. (Photograph courtesy of Bruce C. Heezen, Lamont Geological Observatory.)

Figure 23–8. The fin-pattern typical at the tail end of morids is clearly shown by the shadow. Like their relatives the rat-tails, deep-sea cod have a well-developed swimbladder and cruise actively over the bottom. In this rather rare instance, the fish is swimming across the current (the direction of which is shown by the ripple marks in the sediments). Between the fish and its shadow is a small brittle star.

Figure 23–9. A halosaur, a deep-sea cod (black), and a rat-tail (white), living together 2,600 m deep, on the continental slope south of Nova Scotia. The rat-tail and the cod were swimming in the direction of travel of the camera and thus appear in successive shots. Long photographic sequences indicate that fishes and other animals are not randomly spread over the deep-sea floor, but occur in clusters which are sometimes seen to be associated with particular kinds of bottom. But continuous coverage shows large-scale patterns not only in the distribution of animals, but in physical features as well. For example, the rocks shown here are common in this locality, and are seen in the photo mosaic to be strewn in somewhat regular windrows; no hint of this order is to be found in isolated shots made on the same ground. Strips of bottom up to one mile long are covered in a single lowering, using the Edgerton stereo camera which made the negatives for this mosaic. With this equipment, the flash interval and the ship's drift determine the overlap between pictures, and a rudder on the framework keeps the camera's heading constant.

24. Lebensspuren photographed on the ocean floor[*]

Maurice Ewing *Lamont Geological Observatory, Columbia University, Palisades, New York*

Richard Arnold Davis *University of Iowa, Iowa City, Iowa*

Abstract

Photographs of the ocean bottom are an important source of information as to the identity of organisms responsible for the various lebensspuren. Photography also provides a tool for establishing the actual world-wide distribution of these lebensspuren. However, the establishment of the organic origin of features shown in sea-floor photographs and the study of such organic features are complicated by the inability of the camera to see into the third spatial dimension and to record the changes brought about through the passage of time.

Although the bases for their zoological classification are at present unclear, the photographed lebensspuren can be assigned to a number of purely morphologic groups, and these groups can be so arranged as to demonstrate at least geometric similarities.

24–1. Introduction

The term *Lebensspur* was coined by Abel over fifty years ago to designate any sedimentary structure produced by a living organism (Häntzschel, 1962); thus, tracks, trails, burrows, body imprints, etc. are all *Lebensspuren*. (Although the term is taken directly from the German, it is hereafter treated as though it were English, save for the fact that the German plural is retained.)

In the two-and-a-half decades since the first deep-sea photographs were obtained, numerous lebensspuren have been photographically recorded from the ocean floor. Ichnology (the study of tracks, trails, etc.) by ocean-bottom photographs suffers from many of the same drawbacks as paleoichnology, but, on the other hand, photographs offer some advantages not available to the paleontologist.

In some instances it is difficult to determine whether a feature shown in a photograph is a true lebensspur or is wholly inorganic, for rill marks, furrows produced by rolling pebbles or shells, etc., may all resemble lebensspuren. A more specific example is provided by the smooth lumps, approximately 2 cm in diameter, which are commonly pictured in ocean bottom photographs. Until recently the authors considered it probable that such lumps conceal buried organisms; however, the photographs obtained at one station (V–21–51, 30°84′N, 147°41′E, 6,139 m) and an associated dredge (V–21 dredge 4) indicate that at least some of these lumps may be buried volcanic bombs.

Features produced by a living organism but possibly not due to voluntary movement on the part of that organism (e. g., figs. 24–31; 24–32) are, perhaps, not lebensspuren in the strict sense of the word. Such features do, however, lie near the periphery of the concept of lebensspuren and, for that reason, are herein included.

One of the main drawbacks of underwater photography of lebensspuren is its inability to see all three spatial dimensions. Even stereo pairs cannot see beneath the sediment-water interface. For this reason assignment of photographed lebensspuren to taxa defined by reference to three-dimensional fossil material is difficult.

Until time-lapse photographic surveys can be carried out (see chap. 25) to measure the changes in the sea floor brought about through the passage of time, the age of a lebensspur shown in a photograph will be uncertain. How long this lebensspur will remain on the sea floor, likewise, can only be surmised. It will certainly be subject to one or more of the following, to at least some extent: accumulation of sediment on top of it, the action of bottom current (constant and/or intermittent), the activities of organisms, and slumping and compaction of the sediment. Of these, only the rate of sediment accumulation in a given geographic area can even be estimated at this time. In all cases the "life" of the lebensspur will be directly related to the relief of the lebensspur and the cohesiveness of the sediment of which the lebensspur is formed. The effect of other organisms on the lebensspur may be particularly confusing, for the sea floor is like a

[*] Lamont Geological Observatory Contribution Number 1082.

The authors are grateful to the Office of Naval Research, the National Science Foundation, and the American Chemical Society, whose support made the work reported in this chapter possible. They are also indebted to Daniel R. Dombroski, Jr., for preparing the photographs for publication.

Figure 24–1. V–19–31–1, 14°06′S, 96°12′W, 3,246–3,262 m, × 1/9) same lebensspur as a groove with parallel ridges in one place and as a branched groove just a short distance away.

palimpsest in that one organism after another may leave its spoor in the same place with greater or lesser erasure between organisms. Thus one may not even be sure as to whether a feature is a lebensspur or many superimposed lebensspuren.

Only very rarely can the maker of a fossil lebensspur be identified; on the other hand, sometimes an organism may be photographed in the act of making a lebenspur. (Understandably, but unfortunately for the purposes of ocean-bottom photographic interpretation, most zoologists are usually reluctant to attempt even the most general classification of an organism without an accompanying

specimen of that organism.) The number of photographs of associated organisms and lebensspuren may be artificially low because, until recently, the cameras used by the Lamont Observatory vessels were shutterless, and, therefore, the first few frames at a station were overexposed, and by the time the third or fourth frame was exposed on the sea floor, many of the motile organisms probably had moved out of the area. (The present camera design counteracts this effect by taking a few pictures on the descent.)

Another real advantage of photographic neoichnology is the possibility of discovering the actual worldwide distribution, both geographic and ecologic, of the various lebensspuren at a given "moment" in the earth's history.

Even though thousands of lebensspuren taxa have been described in over 1,100 references, the concept of the ichnologic species, or even of the genus, is muddled. Photographic evidence demonstrates that a lebensspur may radically change character within the space of a few feet (e.g., figs. 24–1 and 24–2). Moreover, more than one taxon of organisms may produce closely similar lebens-spuren (e.g., figs. 24–40 through 24–44).

In the course of preparing a report of *Vema* cruise 19 (1963), the authors erected a number of purely morpho-logic groups for the photographed lebenspuren. In addi-tion to these groups, some additional categories were recognized on photographs obtained on other cruises of *Vema* as well as on cruises of R/V *Conrad*. This chapter is mainly a presentation of photographs of representatives of these groups.

Some 2,300 photographs taken on *Vema* cruise 19 have been studied; these were obtained at 200 camera stations, which range geographically from just over 28.5°N to about 38°S and from approximately 65°W to nearly 3°E along the following route: New York, Bermuda, Virgin Islands, Panama, Peru, Tahiti, Samoa, Manila, Singapore, Ceylon (via the Straits of Sunda), Mombasa (Kenya), Cape Town, and Abidjan (Ivory Coast). The range of depth of stations at which photo-graphs were obtained is from about 30 m to about 5,950

m; the average camera-station depth is about 3,700 m. Nearly all topographic settings are represented among these 200 camera stations. The other *Vema* photographs, as well as those obtained by *Conrad*, are not part of this same coherent series and therefore no data tabulation for them will be attempted at this time.

In the following, each morphologic group is defined by one or more representative photographs and an accompanying description. Where available, photographs of the producing organism are shown. Each photograph is accompanied by a designation such as V–19–6–10; the letter signifies the ship (V for *Vema*, C for *Conrad*); the 19 is the voyage of the ship; the 6 is the camera station; the 10 is the frame (photograph) of that station. In addition to this designation, each photograph is accompanied by the following data in parentheses: the latitude and longitude of the camera station, the corrected depth at that station and the approximate scale of the photograph. For each group the range of depth (cor-rected), latitude and longitude (on *Vema* 19) are given. The number of photographs (of the 2,300) in which the group appears is also given, as well as the total number of stations at which representatives of the group were photographed. Remarks considered pertinent by the authors are likewise appended.

Although the authors do not claim to have made any-thing resembling a thorough perusal of ichnologic litera-ture, the "generic" name as given by Häntzschel (1962)

Figure 24–2. (V–19–16–4, 4°56′N, 78°16′W, 3,817 m, × 1/9) same lebensspur as a groove with parallel ridges in one place and as a broad ridge a short distance away.

Figure 24–3. (V–18–172–11, 29°40′S, 176°43′W, 4,872–5,068 m, × 1/9) Group IA1: ridge in the pattern of a single spiral which may depart into meandering or may reverse direction of coiling; cf. *Spirophycus*, Häntzschel, 1962; Jones (personal communication, 1964) suggested that the organism might be some kind of an enteropneust; 12°24′N to 29°31′S, 81°31′W west to 4°41′E, 2,715–5,951 m, 109 frames, 35 stations.

is provided for any group which has apparently been previously described. In many instances, partly because of the lack of information as to the third dimension, the ethologic nature of the photographed lebensspuren is unclear; for this reason the terms utilized by Seilacher (1964) are not applied.

Each group is identified by an upper case Roman numeral, an upper case Roman letter, and an Arabic numeral. The groups are arranged in order by these characters, the arrangement serving to place similar lebensspuren together.

None of the geometric groups recognized herein have been demonstrated to be the exclusive product of any particular animal taxon. For this reason no taxonomic name, in the zoological sense, can reasonably be applied to any of the groups. The number and letter identification given for each group does not necessarily reflect zoological affinities, but, rather, is meant only to indicate geometric similarity of the included lebensspuren.

Although location and depth data are given, these are not to be construed as being statistically valid generalizations, for, even at best, the 2,300 photographs do not constitute a representative sample of the ocean floor as a whole, or even of the latitude and longitude range of the 200 stations. This chapter is not an attempt at quantitative neoichnology, but is, rather, simply a qualitative presentation of photographs of lebensspuren in an orderly manner.

Following the group descriptions and photographs, an outline is given to aid the reader in locating a particular group by its morphologic characteristics. The position of a group in the outline conforms with its identification number.

24-2. Descriptions of lebensspuren groups

The following is an outline of the lebensspuren groups, representative photographs of which appear on the preceding pages. The purpose of the outline is to provide a

Figure 24-4. (V–20–14–3, 9°52′N, 92°32′W, 3,678–3,684 m, × 1/9) Group IA1: organism somewhat similar to that shown in the previous figure in process of making its trail.

Figure 24–5. (V–20–14–1, 15°05′N, 58°20′W, 3,821–3,751 m, × 1/9) Group IA1: organism just beginning to make trail.

Figure 24–6. (V–19–40–5, 17°02′S, 112°12′W, 3,191–3,186 m, × 1/9) Group IA2: ridge in the pattern of a double spiral; cf. *Spirorhaphe*, Fuchs, 1895; 15°10′S to 17°02′S, 100°21′W west to 112°12′W, 3,186–3,834 m, 7 frames, 2 stations.

means of comparing actual lebensspuren and lebensspuren photographs not reproduced herein with the geometric groups recognized here.

In order to use the outline some definitions need to be observed:

groove: a distinct concavity at least eight times as long as wide;

depression: a distinct concavity less than eight times as long as wide;

ridge: a distinct convexity at least eight times as long as wide;

lump: a distinct convexity less than eight times as long as wide;

narrow (deep, high): depth (height) at least as great as width;

broad (shallow, low): depth (height) not as great as width;

sculptured: other than merely smooth or randomly rough;

sculptured strip: a distinct path which is neither a distinct ridge nor a groove (but may be slightly convex), has a length at least eight times its width, and is sculptured (however, bands of uniform depressions or lumps are arbitrarily excluded).

These definitions are obviously arbitrary and, hence,

display the usual shortcomings of definitions, but they appear useful in describing the photographs at hand and in defining the various categories in the outline which follows.

The six basic subdivisions in the outline are: I. ridges and sets of ridges, II. lumps and sets of lumps, III. grooves and sets of grooves, IV. depressions and sets of depressions, V. sets of depressions and one or more grooves together, and VI. sculptured strips.

For some lebensspuren the question arises: At what point does a line of closely-spaced depressions become a groove (or lumps, a ridge)? For a group in which resolution of this type of question affects its outline position, the group is outlined in one place and notes are inserted at other probable locations to guide the reader to the outlined position of the group. In instances such as those in which a ridge is immediately paralleled by a groove on each side a nomenclatural problem becomes evident: should the lebensspur be considered a ridge with paralleling grooves, a groove with a median ridge, or two grooves with an intervening ridge? In cases in which the ridge is broad, flat, and at the level of the surrounding bottom, the lebensspur is considered to be two grooves. If the ridge is rounded and wider than the two grooves together, the lebensspur is called a ridge with paralleling grooves. If the grooves are wider than the ridge, the lebensspur is considered as a groove with a median ridge. (Similarly for two ridges and a groove.)

In any position in which possible confusion is foreseen by the authors a note is inserted to guide the reader.

So that the outline may be expanded easily, it does not follow correct outlining procedure in some cases.

24–3. Outline of lebensspuren groups

I. Ridges and sets of ridges
  A. Individual ridges
    1. Unsculptured ridges (see also III, A, 2, a & VI, B)
      a. Spiraling ridge
        (1) Single spiral (may depart into meandering or reverse coiling direction) IA1
        (2) Double spiral IA2
      b. Radiating ridges joined at center IA3
      c. Ridge forming a circle IA4
      d. Random path, may or may not be branched (see also VI, B, 1)
        (1) Without paralleling grooves; blends into surrounding bottom IA5
        (2) Without paralleling grooves; short; distinct from bottom IA6
    2. Sculptured ridge (see also VI, B)
      a. Like twisted rope in a single circle IA7
      b. Low ridge with a single longitudinal groove IA8
      c. Ridge paralleled by a shallow groove on each side; grooves and sides of ridge covered with numerous transverse grooves IA9
      d. Broad ridge with a median longitudinal ridge IA10
II. Lumps and sets of lumps
  A. Solitary lumps
    1. Unsculptured
      a. With an apical, round hole IIA1
      b. With a collapsed apex IIA2
      c. Dumbbell-shaped IIA3
    2. Sculptured
      a. Elongate, with irregular transverse grooves IIA4
      b. Elongate, resembling a coil lying on its side; commonly with a narrow projection from one end IIA5
      c. More or less equidimensional, but with a narrow projection at one place; resembles a pile of rope (not coiled) IIA6
      d. Somewhat elongate, with a groove running from the apex through one end and for a short distance on the adjacent floor IIA7
  B. Sets of lumps
    1. Random arrangement
      a. Same value as background IIB1
      b. Darker than background IIB2
    2. Single row of adjacent lumps IIB3
    3. Double row of lumps
      a. Lumps in each row directly opposite and adjacent to corresponding lumps in the other IIB4
    4. Numerous small, elongate, straight or bowed lumps randomly positioned along a 5-cm wide path IIB5
III. Grooves and sets of grooves
  A. Individual grooves
    1. Groove forming a ring
      a. Narrow groove
        (1) Around a small lump which bears a small apical hole IIIA1
        (2) Central area flat or somewhat concave, with a small, central hole IIIA2
        (3) Central area flat or somewhat concave, with a small, central lump IIIA3

b. Surrounded by a ridge which, along with the wall of the depression, is irregularly cut by radiating grooves.. IVA2

c. Very shallow, very smooth depression, with somewhat elevated periphery...................... IVA3

d. Deep hole of no particular size or shape, with or without ridges, etc.......................... IVA4

2. Elongate

a. 2 cm $\times$ 5 cm oval impression.. IVA5

b. 20-cm long depression with a ridge on either side; broadly expands at center to a width of 6 cm...... IVA6

c. 15-cm wide, without paralleling ridges.. IVA7

B. Groups of depressions

1. Depressions arranged in a circle

a. Numerous holes in a single row on the circumference.................................... IVB1

b. Numerous holes in an irregular circle around a large lump.................................. IVB2

c. Petal-shaped depressions radiating from a central hole.................................... IVB3

2. Linear arrangement of depressions (see also III, B, 2)

a. In one line (see also VI & III, A, 3, b)

(1) Identically oriented triangles of three depressions in a line, spaced about 20 cm apart........... IVB4

(2) 8-cm wide path of small depressions which don't touch one another........................ IVB5

(3) Single, straight line of holes spaced about their own diameter apart; the line is bordered on each side by a ridge.. IVB6

(4) 20-cm path of widely separated elongate impressions paralleling the path.................... IVB7

b. In two lines

(1) Randomly arranged in each 3-cm strip.. IVB8

(2) Two rows 4 cm apart; in approximately single file; each line of depressions may become a groove in places.. IVB9

(3) Straight, single rows of depressions which diverge.................................... IVB10

(4) Each row composed principally of groups of depressions of the form of arcs on opposite sides of concentric circles.. IVB11

c. In four lines

(1) Outer pair each composed of randomly arranged small depressions in a narrow path; in inner pair depressions are single file.. IVB12

(2) All four composed of depressions in single file.. IVB13

V. Sets of depressions and one or more grooves together (see also IV, B, 2 & III, A, 2, a)

A. Groove with paralleling row of depressions on each side; depressions are elongate normal to groove and are closely spaced.. VA1

B. Very shallow groove paralleled by a strip of closely spaced small depressions........................ VB1

VI. Sculptured strips

A. Numerous curved grooves forming a pattern somewhat similar to that of a braided stream............. VIA1

B. Strip of 2-cm depressions immediately adjacent to one another; paralleled on each side by a row of deeper depressions; may be slightly convex

1. Without median grooves.. VIB1

2. With a narrow median groove.. VIB2

3. With a broad median groove which bears a discontinuous median ridge........................ VIB3

C. 15-cm wide, somewhat convex, strip with an uneven groove paralleling it on each side and with an uneven median groove; the surface of the strip is irregularly undulatory.............................. VIC1

D. Strip ornamented with very closely spaced, minute lumps in curved lines transverse to strip; lines of lumps may pass into ridges.. VID1

E. Intricately patterned strip commonly with a low paralleling ridge on either side; some have a low median ridge.. VIE1

References

Bruun, Anton F., Sv. Greve, Hakon Mielche, & Ragnar Spärck, 1956: (eds.), *The Galathea deep sea expedition 1950-1952.* Macmillan and Co., New York, 296 p.

Dudley, Patricia: Department of Zoology, Barnard College, New York City.

Häntzschel, W., 1962: Trace fossils and problematica. In *Treatise on invertebrate paleontology.* R. C. Moore (ed.), vol. W, Geological Society of America and University of Kansas Press. W 177-W 245.

Heezen, Bruce C., & Charles D. Hollister, 1964: Deep-sea current evidence from abyssal sediments. *Marine Geology,* **1**, no. 2, 141-74.

Jones, M. L.: Associate Curator, Division of Marine Invertebrates, Smithsonian Institution, United States National Museum, Washington, D.C.

Matthews, D. J., 1939: Tables of the velocity of sound in pure water and the sea water. London, Hydrographic Department, Admiralty. 2nd ed. 52 p.

Pawson, D. L.: Associate Curator, Division of Marine Invertebrates, Smithsonian Institution, United States National Museum, Washington, D.C.

Seilacher, A., 1964: Biogenic sedimentary structures. In *Approaches to paleoceology.* Imbrie, J., and N. Newell (eds.), John Wiley and Sons, New York, 296-316.

Figure 24-7. (V-19-46-3, 17°00′S, 114°32′W, 3,147 m, × 1/6) Group IA3: pattern formed by a number of radiating ridges; only one photographed.

Figure 24-8. (V-19-40-10, 17°02′S, 112°12′W, 3,191-3,186 m, × 1/9) Group IA4: smooth ridge in the form of a circle; unidentified organism appears to be making the ridge; 8°14′N to 25°15′S, 100°21′W west to 3°20′E, 2,981-5,234 m, 13 frames, 11 stations.

Figure 24-9. (V–19–8–18, 14°10′N, 68°40′W, 4,740–4,737 m, × 1/12) Group IA5: unsculptured, smooth, or somewhat rough-surfaced ridge which has an irregular path and may or may not branch; blends into surrounding bottom; 28°32′N to 38°10′S, 57°09′W west to 3°20′E, 1,046–5,698 m, 241 frames, 89 stations.

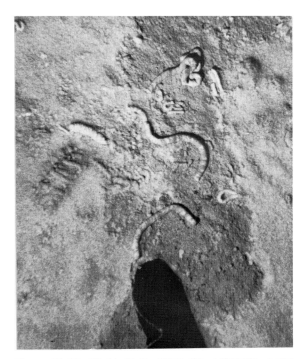

Figure 24-10. (V–14–28–8, 57°33.5′S, 17°22′W, 4,475 m, × 1/12) Group IA6: ridge which is distinct from the surrounding bottom; presumably faecal; (note representative of group IIA6 in the upper part of the frame); 28°32′N to 38°10′S, 57°09′W west to 3°20′E, 223–5,951 m, 943 frames, 128 stations.

Figure 24-11. (V–19–27–17, 12°16′S, 84°13′W, 4,361–4,367 m, × 1/9) Group IA7: ridge forming a small ring; ridge looks like a piece of twisted rope; 12°16′S to 34°20′S, 84°13′W west to 31°25′E, 4,111–4,367 m, 3 frames, 3 stations.

Figure 24-12. (V–19–25–5, 11°45′S, 79°12′W, 5,938–5,949 m, × 1/9) Group IA8: low ridge with a single longitudinal groove; unidentified organism appears to be making the trail; only one photographed.

Figure 24-13. (V–18–229–3, 16°48′S, 152°13′W, 3,937–3,935 m, × 1/12) Group IA9: ridge immediately paralleled on each side by a shallow groove; grooves and sides of ridge are striated normal to the ridge; 13°13′S, 92°53′W, 3,634–3,673 m, only one photographed.

Figure 24-14. (V–19–60–5, 15°39′S, 138°35′W, 3,939–3,941 m, × 1/12) Group IA10: broad ridge with a median, longitudinal ridge; cf. *Psammichnites*, Torell, 1870; 15°39′S to 17°50′S, 124°23′W west to 151°10′W, 3,713–4,017 m, 3 frames, 3 stations.

Figure 24-15. (V–18–108–2, 33°06′S, 22°48′W, 4,175–4,188 m, × 1/12) Group IIA1: smooth lump with an apical hole; 28°32′N to 34°25′S, 66°29′W west to 4°06′E, 139–5,951 m, 162 frames, 73 stations.

Figure 24-16. (V-19-1-20, 28°32'N, 67°57'W, 5,064–4,993 m, × 1/9) Group IIA2: lump which appears as though its apex has collapsed; 28°32'N to 34°20'S, 57°09'W west to 3°20'E, 185–5,949 m, 121 frames, 47 stations.

Figure 24-17. (V-19-30-6, 13°13'S, 92°53'W, 3,634–3,673 m, × 1/9) Group IIA3: dumbbell-shaped lump; cf. *Bifungites*, Desio, 1940; only one photographed.

Figure 24-18. (V-19-45-6, 17°00'S, 114°11'W, 3,177 m, × 1/6) Group IIA4: elongated lump with irregular transverse grooves; presumably faecal; 23°45'N to 17°00'S, 57°09'W west to 81°37'E, 3,177–5,104 m, 5 frames, 4 stations.

Figure 24-19. (V-19-158-18, 13°08'S, 44°09'E, 3,549 m, × 1/9) Group IIA5: elongated lump generally resembling a coil on its side; commonly has a narrow projection from one end; presumably faecal; 16°06'N to 36°11'S, 66°29'W west to 3°20'E, 1,900–5,949 m, 325 frames, 86 stations.

Figure 24–20. (V–19–48–14, 16°58'S, 115°12'W, 3,329–3,347 m, × 1/6) Group IIA6: more or less equidimensional lump resembling a pile of rope; has a narrow projection at one place; presumably faecal; 9°44'N to 29°37'S, 86°39'W west to 10°36'E, 2,578–5,422 m, 42 frames, 22 stations.

Figure 24–21. (V–19–56–9, 16°24'S, 127°38'W, 4,094 m, × 1/6) Group IIA7: small, elongated lump with a narrow groove extending from the apex, through one end, and out a short distance onto the surrounding bottom; 0°06'N to 25°46'S, 100°21'W west to 4°06'E, 3,100–4,552 m, 6 frames, 6 stations.

Figure 24–22. (V–19–46–10, 17°00'S, 114°32'W, 3,147 m, × 1/9) Group IIB1: equidimensional lumps of about the same value as the surrounding bottom; such lumps of one size or another are pictured in nearly every frame of the 2,300; some of these however, may be inorganic.

Figure 24–23. (V–19–145–19, 7°04'N, 60°55'E, 2,679–2,699 m, × 1/12) Group IIB2: equidimensional lumps which are darker than the surrounding bottom; 7°04'N to 33°22'S, 60°55'E west to 34°34'E, 1,999–4,136 m, 21 frames, 7 stations.

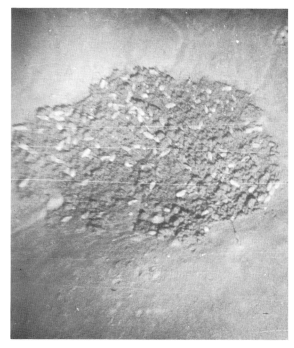

Figure 24–24. (V–19–164–10, 22°51′S, 42°10′E, 3,184 m, × 1/12) Group IIB2: dark lumps partly decomposed; white objects may be feeding organisms.

Figure 24–25. (V–19–52–11, 16°58′S, 116°48′W, 3,413 m, × 1/9) Group IIB3: single line of immediately adjacent lumps in single file; only one photographed.

Figure 24–26. (V–19–60–10, 15°39′S, 138°35′W, 3,939–3,941 m, × 1/9) Group IIB4: two immediately adjacent rows of lumps; lumps in each row are immediately adjacent to one another and are directly opposite those of the other row; only one photographed.

Figure 24–27. (V–19–8–10, 14°10′N, 68°40′W, 4,740–4,737 m, × 1/12) Group IIB5: numerous small, elongated, straight or bowed lumps randomly positioned along a path about 5 cm wide; cf. *Tomaculum*, Groom, 1902; two frames at this station only.

Figure 24-28. (V-19-2-6, 23°45′N, 57°09′W, 5,104–5,066 m, × 1/6) Group IIIA1: a narrow groove immediately surrounding a lump which bears a central hole; 28°32′N to 29° 37′S, 57°09′W west to 3°20′E, 3,592–5,302 m, 18 frames, 10 stations.

Figure 24-29. (V-19-5-12, 16°06′N, 66°29′W, 4,506–4,510 m, × 1/9) Group IIIA2: narrow groove immediately surrounding a flat to somewhat concave, circular area which bears a central hole; commonly the "wormlike" organism is present; 16°06′N to 29°27′S, 66°29′W west to 3°20′E, 2,520–5,363 m, 56 frames, 27 stations.

Figure 24-30. (V-19-193-5, 25°46′S, 4°06′E, 4,237–4,244 m, × 1/9) Group IIIA3: narrow groove immediately surrounding a flat to somewhat concave, circular area which bears a small central lump; 25°46′S to 27°15′S, 6°12′E west to 4°06′E, 4,237–4,865 m, 2 frames, 2 stations.

Figure 24-31. (V-19-156-6, 9°28′S, 43°19′E, 3,651–3,649 m, × 1/9) Group IIIA4: broad groove around a circular area 15 cm in diameter; Heezen & Hollister (1964) identify a similar organism as either a sea pen or a bryozoan colony; only one photographed.

Figure 24–32. (V–19–160–9, 16°56′S, 41°06′E, 2,496–2,476 m, × 1/12) Group IIIA5: broad groove curved on a radius of about 50 cm; unidentified organism appears to be making the groove; two photographed at this station only.

Figure 24–33. (V–19–31–1, 14°06′S, 96°12′W, 3,246–3,262 m, × 1/12) Group IIIA6: main groove with numerous equally spaced, similarly inclined branches on each side; the branches on one side are opposite those on the other side and at an angle of about 90° to them; the ridges paralleling the main branch may nearly unite; 13°13′S to 17°41′S, 92°53′W to 167°34′W, 3,177–5,422 m, 14 frames, 4 stations.

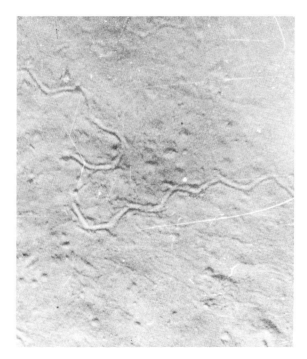

Figure 24–34. (V–19–57–2, 16°10′S, 129°52′W, 3,950–3,952 m, × 1/9) Group IIIA7: groove in the pattern of a "feather stitch"; 15°39′S to 16°24′S, 127°38′W west to 138°35′W, 3,939–4,094 m, 7 frames, 3 stations.

Figure 24–35. (V–19–97–9, 14°49′N, 119°37′E, 2,540 m, × 1/9) Group IIIA8: straight, narrow, very deep groove; only one photographed.

Figure 24-36. (V–19–3–22, 21°58′N, 67°11′W, 5,284–5,302 m, × 1/6) Group IIIA9: short groove, tapering in width from one end to the other; 28°32′N to 29°31′S, 67°11′W west to 3°20′E, 2,171–5,861 m, 29 frames, 17 stations.

Figure 24-37. (V–19–25–20, 11°45′S, 79°12′W, 5,938–5,949 m, × 1/9) Group IIIA10: very smooth, very shallow, branched or unbranched, groove; an unidentified organism appears to be making one of the grooves; 11°45′S to 35°50′S, 79°12′W west to 27°45′E, 3,309–5,949 m, 9 frames, 4 stations.

Figure 24-38. (V–19–136–5, 7°35′N, 74°13′E, 2,770–2,769 m, × 1/9) Group IIIA11: very shallow, irregularly meandering groove; five frames at this one station only.

Figure 24-39. (V–19–25–1, 11°45′S, 79°12′W, 5,938–5,949 m, × 1/9) Group IIIA12: shallow, very rough groove with or without a parallel ridge on each side; may be a variation of group IIIA13; 11°45′S to 14°53′S, 79°12′W west to 42°51′E, 3,250–5,949 m, 6 frames, 3 stations.

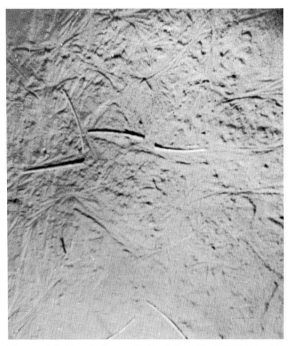

Figure 24-40. *Balanus* E-40, 39°34′N, 72°23′W, 450 m, × 1/14) Group IIIA13: narrow, relatively smooth unsculptured groove which may or may not branch and may or may not have parallel ridges; the animals seem to be quill worms; grooves referable to this group are pictured in most frames of V-19.

Figure 24-41. (V-19-10-8, 15°57′N, 71°05′W, 3,992–3,990 m, × 1/9) Group IIIA13: unidentified organism apparently is making the groove.

Figure 24-42. (V-18-122-6, 32°11′S, 13°25′E, 2,893–3,087 m, × 1/9) Group IIIA13: unidentified organism apparently is making the groove.

Figure 24-43. (V-19-25-6, 11°45′S, 79°12′W, 5,938–5,949 m, × 1/12) Group IIIA13: unidentified organism apparently is making the groove.

Figure 24-44. (V-19-11-1, 16°02'N, 72°51'W, 2,118-2,180 m, × 1/6) Group IIIA13: scaphopod-like organism is apparently making the groove.

Figure 24-45. (V-19-174-7, 35°50'S, 27°45'E, 4,656 m, × 1/9) Group IIIA14: broad groove with a parallel ridge on each side and with a median longitudinal ridge; 8°27'N to 36°11'S, 110°00'E west to 15°59'E, 1,046-4,656 m, 30 frames, 6 stations (See also fig. 24-93).

Figure 24-46. (V-19-9-10, 14°58'N, 68°54'W, 4,965-4,955 m, × 1/18) Group IIIA15: very shallow groove with a median ridge; 14°58'N to 16°58'S, 68°54'W west to 41°11'E, 2,578-4,965 m, 10 frames, 6 stations.

Figure 24-47. (V-19-71-1, 18°03'S, 159°43'W, 5,072-5,070 m, × 1/9) Group IIIA16: narrow groove with a low discontinuous median longitudinal ridge; 12°24'S to 32°36'S; 114°11'W west to 6°12'E; 1,465-5,072 m, 7 frames, 5 stations.

Figure 24-48. (V–19–191–5, 27°15′S, 6°12′E, 4,865 m, × 1/12) Group IIIA16: an unidentified organism is apparently making the groove.

Figure 24-49. (V–19–185–18, 32°35′S, 16°12′E, 1,567–1,465 m, × 1/9) Group IIIA17: narrow groove with numerous transverse partitions; only one photographed.

Figure 24-50. (V–19–118–7, 14°38′S, 101°20′E, 5,363 m, × 1/12) Group IIIB1: Asteroid-shaped impression; with maker; 13°42′N to 34°31′S, 78°16′W west to 3°20′E, 516–5,680 m, 48 frames, 30 stations.

Figure 24-51. (C–1–3–1, 31°27′N, 76°33′W, 2,734 m, × 1/9) Group IIIB1: asteroid-shaped impression; Pawson (personal communication, 1964) indicates that such impressions are probably made by *Astropecten* and that the sea star lies buried beneath the impression.

Figure 24-52. (V–18–289–7, 6°43′N, 86°30′W, 2,897 m, × 1/9) Group IIIB2: ophiuriod-shaped impression; 16°02′N to 36°11′S, 72°51′W west to 6°12′E, 2,118–5,422 m, 86 frames, 26 stations.

Figure 24-53. (V–19–32–8, 15°10′S, 100°21′W, 3,834–3,592 m, × 1/9) Group IIIB3: 15-cm long, narrow grooves, radiating from a small central hole; 23°45′N to 35°30′S, 57°09′W west to 4°06′E, 51–5,698 m, 156 frames, 59 stations.

Figure 24-54. (V–19–51–6, 16°57′S, 116°18′W, 3,413 m, × 1/9) Group IIIB3: partially completed radiating groove pattern.

Figure 24-55. (V–19–60–11, 15°39′S, 138°35′W, 3,939–3,941 m, × 1/9) Group IIIB4: 15-cm long, narrow grooves radiating from a 4-cm central lump; lump may actually be a puff of seadust; 14°09′N to 17°02′S, 111°53′W west to 118°57′E, 3,252–4,371 m, 8 frames, 6 stations.

Figure 24-56. (V–19–39–14, 17°02′S, 111°53′W, 3,255–3,252 m, × 1/6) Group IIIB5: 10-cm long, narrow grooves radiating from a center which bears neither a hole nor a lump; 28°32′N to 35°30′S, 66°29′W west to 3°20′E, 2,305–5,594 m, 32 frames, 26 stations.

Figure 24-57. (V–19–45–13, 17°00′S, 114°11′W, 3,177 m, × 1/9) Group IIIB6: 15-cm long narrow grooves radiating from a 15-cm diameter lump; the grooves appear concentrated into four (or sometimes two) distinct areas; 28°32′N to 17°00′S, 67°49′W west to 168°02′E, 3,177–5,088 m, 12 frames, 10 stations.

Figure 24-58. (V–19–47–9, 17°00′S, 114°53′W, 3,147–3,166 m, × 1/12) Group IIIB7: 40-cm long narrow grooves radiating from a small central hole; 15°57′N to 33°22′S, 71°05′W west to 34°34′E, 1,999–3,992 m, 4 frames, 4 stations.

Figure 24-59. (V–19–46–7, 17°00′S, 114°32′W, 3,147 m, × 1/9) Group IIIB8: 40-cm long narrow grooves, radiating from a small central lump; lump may be actually a puff of seadust; only one photographed.

Figure 24–60. (V–19–151–17, 2°40′S, 54°45′E, 4,184 m, × 1/12) Group IIIB9: long narrow grooves radiating from a 20-cm diameter depression; only one photographed.

Figure 24–61. (V–19–77–4, 16°40′S, 168°34′W, 5,086–5,088 m, × 1/12) Group IIIB10: long, narrow grooves radiating from a 25-cm lump; grooves are concentrated into two bunches; only one photographed.

Figure 24–62. (V–19–48–10, 16°58′S, 115°12′W, 3,329–3,347 m, × 1/9) Group IIIB11: broad grooves radiating from a small central hole; 14°49′N to 17°00′S, 114°53′W west to 101°41′E, 2,540–4,959 m, 4 frames, 4 stations.

Figure 24–63. (V–19–56–7, 16°24′S, 127°38′W, 4,094 m, × 1/9) Group IIIB12: broad grooves with a crude radiating pattern; there is no central structure; only one photographed.

Figure 24–64. (V–19–118–6, 14°38′S, 101°20′E, 5,363 m, × 1/9) Group IIIB13: two closely spaced narrow grooves; 11°12′N to 33°31′S, 84°13′W west to 4°06′ E, 659–5,680 m, 22 frames, 10 stations.

Figure 24–65. (V–20–55–7, 46°14′N, 136°30′W, 4,232 m, × 1/12) Group IIIB14: two widely spaced grooves, each with a single parallel ridge on its outer side; the strip between the grooves is rounded; 15°10′S, 100°-21′W, 3,834–3,592 m, 1 frame, 1 station.

Figure 24–66. (V–19–188–17, 29°37′S, 10°36′E, 4,488–4,484 m, × 1/9) Group IIIB15: two narrow grooves with a broad groove between them; only one photographed.

Figure 24–67. (V–19–191–7, 27°15′S, 6°12′E, 4,865 m, × 1/9) Group IVA1: apparently conical hole with a surrounding ridge; only one photographed.

Figure 24-68. (V–19–161–2, 17°54′S, 39°30′E, 2,307–2,305 m, × 1/18) Group IVA2· hole surrounded by a ridge which, along with the wall of the depression, is cut by irregular radiating grooves; only one photographed.

Figure 24-69. (V–19–31–1, 14°06′S, 96°12′W, 3,246–3,262 m, × 1/9) Group IVA3: a smooth, approximately equidimensional patch with slightly elevated periphery so as to be concave; 7°31′N to 20°56′S, 96°12′W west to 82°08′E, 185–5,363 m, 8 frames, 6 stations.

Figure 24-70. (V–19–20–2, 00°25′S, 82°04′W, 1,362–1,376 m, × 1/12) Group IVA4: deep, more or less equidimensional holes; such holes of one size or another are pictured in nearly every frame of the 2,300.

Figure 24-71. (V–19–30–4, 13°13′S, 92°53′W, 3,634–3,673 m, × 1/9) Group IVA4: hole with a pile of removed material.

Figure 24-72. (V-18-278-5, 14°31'N, 96°18'W, 3,720-3,674 m, × 1/9) Group IVA4: hole with its apparent maker; Dudley (personal communication, 1963) suggested that the organism is a primitive burrowing anemone.

Figure 24-73. (V-19-40-2, 17°02'S, 112°12'W, 3,191-3,186 m, × 1/9) Group IVA5: 2 × 5 cm depression; 14°06'S to 29°37'S, 96°12'W west to 10°36'E, 2,981-5,422 m, 19 frames, 12 stations.

Figure 24-74. (V-19-5-14, 16°06'N, 66°29'W, 4,506-4,510 m, × 1/9) Group IVA6: 20-cm long depression strongly expanded at its middle to a width of about 6 cm; the depression is paralleled by a ridge on each side; 16°06'N to 29°37'S, 66°29'W west to 4°06'E, 2,118-4,737 m, 13 frames, 11 stations.

Figure 24-75. (V-19-193-14, 25°46'S, 4°06'E, 4,237-4,244 m, × 1/9) Group IVA6: unidentified organism occupying a depression of this group.

Figure 24–76. (V–19–145–19, 7°04′N, 60°55′E, 2,679–2,699 m, × 1/12) Group IVA7: broad, elongate depression; dark material is presumably faecal; (Group IIB2); 7°04′N to 33°22′S, 60°55′E west to 34°34′E, 1,999–2,699 m, 3 frames, 2 stations.

Figure 24–77. (V–19–146–8, 6°52′N, 60°42′E, 3,345–3,356 m, × 1/12) Group IVB1: depressions in a single line around a flat, circular area; depressions are elongate normal to the line; 6°52′N to 36°11′S, 60°4′E2 west to 13°17′E, 3,054–4,574 m, 16 frames, 6 stations.

Figure 24–78. (C–8–48–6, 44°02′S, 93°53′E, 2,785–2,822 m, × 1/9) Group IVB2: depressions randomly arranged in a circle around a large lump; 16°58′S, 116°48′W, 3,413 m, 1 frame, 1 station.

Figure 24–79. (C–5–1–13, 32°26′N, 62°59′W, 4,497–4,501 m, × 1/9) Group IVB3: petal-shaped depressions radiating from a small central hole; 14°15′N to 34°20′S, 78°16′W west to 31°25′E, 3,100–5,698 m, 9 frames, 7 stations.

Figure 24–80. (V–19–16–3, 4°56′N, 78°16′W, 3,817 m, × 1/9) Group IVB3: petal-shaped depressions are more rounded; the depressions may be so closely spaced that the lebensspur appears to be a single, round depression with a central hole.

Figure 24–81. (V–19–113–4, 6°14′S, 104°50′E, 1,318–1,311 m, × 1/9) Group IVB4: a line of groups of three depressions; each group is in the pattern of a triangle; each triangle bears a similar orientation with respect to the trend of the line; only one photographed.

Figure 24–82. (V–16–48–14, 61°52′S, 91°12′W, 4,894 m, × 1/12) Group IVB5: path of numerous holes separated from one another; the organism is apparently the holothurian *Scotoplanes globose;* 11°59′S to 27°15′S, 81°31′W west to 6°12′E, 4,737–5,422 m, 6 frames, 3 stations.

Figure 24–83. (V–4–3–13, 37°20′N, 33°15′W, 1,723–1,774 m, × 1/9) Group IVB6: single, straight line of holes about their own diameter apart; the line is bordered on each side by a ridge; none photographed on V–19.

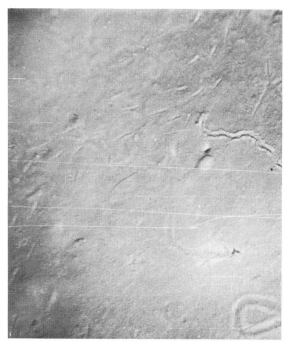

Figure 24–84. (V–19–164–11, 22°51′S, 42°10′E, 3,184 m, × 1/12) Group IVB7: a path of numerous, widely spaced, elongated depressions parallel the trend of the path; 11°29′S to 35°30′S, 109°12′E west to 6°12′E; 51–4,865 m, 4 frames, 4 stations.

Figure 24–85. (V–18–105–2, 35°38′S, 28°44′W, 4,354 m, × 1/12) Group IVB8: trail consisting of two parallel, linear areas of small holes; the holes are separated from one another; 23°45′N to 29°37′S, 57°09′W west to 10°36′E, 4,484–5,949 m, 6 frames, 6 stations.

Figure 24–86. (V–19–52–4, 16°58′S, 116°48′W, 3,413 m, × 1/12) Group IVB9: two parallel lines of depressions in approximately single file; in places the depressions may merge to form a discontinuous groove; 8°16′N to 16°58′S, 79°12′W west to 64°05′E, 1,311–5,949 m, 4 frames, 4 stations.

Figure 24–87. (V–19–25–18, 11°45′S, 79°12′W, 5,938–5,949 m, × 1/9) Group IVB9: the lines of depression are irregular grooves; an unidentified holothurian apparently is making the trail.

Figure 24–88. (V–19–191–3, 27°15'S, 6°12'E, 4,865 m, × 1/9) Group IVB10: two diverging lines of elongated depressions; depressions are in single file and are oriented approximately perpendicular to the trend of each line; only one photographed.

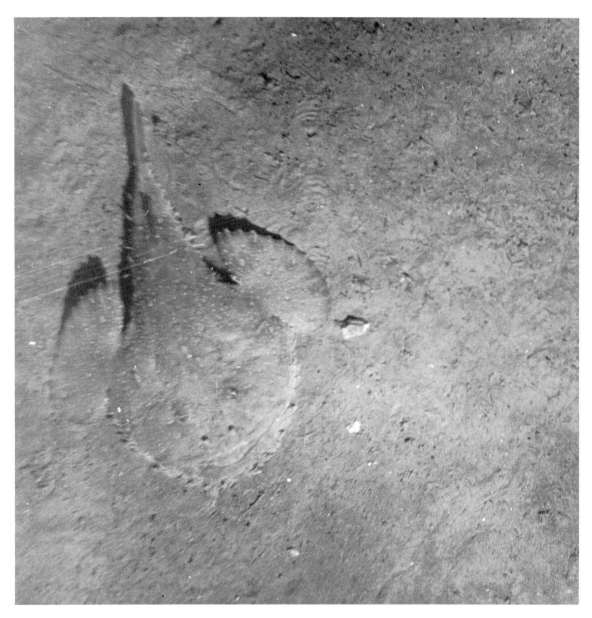

Figure 24–89. (V–19–176–12, 34°31'S, 17°44'E, 516–525 m × 1/6) Group IVB11: two widely separated lines of groups of depressions; each group consists of a number of concentric arcs of small diameter circles; the producing organism is similar to a goosefish (family: Lophiidae); only one photographed.

Figure 24–90. (V–18–229–8, 16°48′S, 152°13′W, 3,937–3,935 m, × 1/12) Group IVB12: four lines of depressions; the inner two consist of depressions in single file, so closely spaced that each line is almost a groove; outer lines are composed of depressions not in single file; none photographed on V–19.

Figure 24–91. (V–19–32–7, 15°10′S, 100°21′W, 3,834–3,592 m, × 1/9) Group IVB13: four lines each consisting of small depressions in single file; commonly the depressions in each line are so closely spaced as to form a groove; 15°10′S to 29°37′S, 100°21′W west to 10°36′E, 3,592–4,488 m, 5 frames, 4 stations.

Figure 24–92. (V–19–90–8, 11°12′N, 127°32′E, 5,594 m, × 1/12) Group VA1: broad groove with a single line of closely spaced depressions on each side; the depressions are elongated normal to the groove; 11°12′N to 29°31′S, 127°32′E west to 9°35′E, 4,870–5,594 m, 6 frames, 2 stations.

Figure 24–93. (V–19–80–2, 00°06′N, 170°46′E, 4,550–4,552 m, × 1/18) Group VA1: grooves like this appear associated with lebensspuren of this group and may be eroded representatives of the group; identical grooves occur associated with lebensspuren of Group IIIA14 and may be eroded representatives of that group.

Figure 24-94. (V–19–2–20, 23°45′N, 57°09′W, 5,104–5,066 m, × 1/12) Group VB1: very shallow groove bordered on each side by a strip of closely spaced, minute depressions; object in groove may be a holothurian or a lump of Group IIA4; 23°45′N to 29°37′S, 57°09′W west to 10° 36′ E, 4,484–5,104 m, 2 frames, 2 stations.

Figure 24-95. (V–18–296–14, 6°42′N, 80°42′W, 3,426–3,398 m, × 1/12) Group VIA1: A strip consisting of numerous grooves which form a pattern similar to that of a braided stream; none photographed on V–19.

Figure 24-96. (V–19–53–7, 16°59′S, 117°53′W, 3,442 m, × 1/9) Group VIB1; a flat or slightly convex strip consisting of immediately adjacent, shallow, 1.5-cm diameter depressions; the strip is bordered on either side by a single file of more prominent depressions; 13°30′N to 30°35′S, 84°13′W west to 10°36′E, 3,098–5,645 m, 38 frames, 18 stations.

Figure 24–97. (V–19–50–9, 16°58′S, 115°56′W, 3,329 m, × 1/12) Group VIB1: somewhat eroded.

Figure 24–98. (V–19–118–11, 14°38′S, 101°20′E, 5,363 m, × 1/9) Group VIB1: smooth, broad ridge with a parallel groove on each side; thus far only found at stations at which representatives of this group are pictured; presumably these ridges are partially eroded members of this group.

Figure 24–99. (V–18–299–10, 17°49′N, 83°14′W, 5,698–5,693 m, × 1/12) Group VIB2: closely similar to Group VIB1, but with a single median groove; may be only a variation of that group; 14°38′S to 33°22′S, 101°20′E west to 34°34′E, 1,999–5,363 m, 2 frames, 2 stations.

Figure 24–100. (V–19–81–4, 2°22′N, 168°02′E, 4,400-4,398 m, × 1/18) Group VIB3: closely similar to Group VIB1, but with a broad median groove, which bears a low, discontinuous ridge; only one photographed.

Figure 24–101. (V–18–54–7, 45°31′S, 51°55′W, 6,018 m, × 1/9) Group VIC1: somewhat convex strip with an uneven groove on each side and an uneven median, longitudinal groove; the surface of the strip is irregularly undulatory; unidentified organism apparently is making the trail; none photographed on V–19.

Figure 24–102. (V–19–56–6, 16°24′S, 127°38′W, 4,094 m, × 1/9) Group VID1: strip ornamented with very closely spaced, minute lumps in curved lines transverse to strip; the lines of lumps may pass into ridges in places; only one photographed.

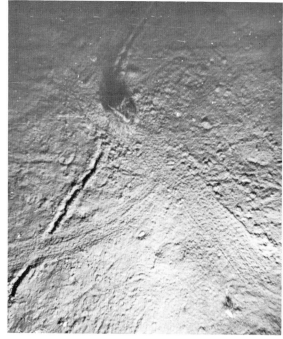

Figure 24–103. (V–18–296–2, 6°42′N, 80°42′W, 3,426–3,398 m, × 1/12) Group VIE1: a very intricate pattern of depressions, lumps, and grooves, may or may not have a parallel ridge on each side; may or may not have a low median, longitudinal ridge; none photographed on V–19.

Figure 24–104. (V–18–296–16, 6°42′N, 80°42′W, 3,426–3,398 m, × 1/18) Group VIE1: an unidentified organism may be making the strip.

25. Movements of benthonic organisms and bottom currents as measured from the bathyscaph *Trieste**

E. C. LaFond   *U.S. Navy Electronics Laboratory, San Diego, California*

Abstract

The ability of the bathyscaph *Trieste* to remain in one location for a protracted period of time afforded the unique opportunity to view and photograph the benthonic organisms in the San Diego Trough off Southern California. Both direct visual and photographic observations were made. Through time-lapse photography it has been possible to determine (a) water currents which varied in speed between 0.02 and 0.06 knots, and in direction from 335° to 040°; (b) that 20 per cent of a large brittle star population (*Ophiomusium lymani* and *Amphilepis sp.*) moved over a 44-min period and (c) that they moved in random directions, with speeds of 0.7 to 3.4 cm/min. Underwater photography revealed that the sea floor may be likened to a moving carpet, with a variety of organisms crawling on the surface.

25-1. Introduction

The bathyscaph *Trieste* was utilized, in its 84th dive on 25 October 1961, to (1) investigate the movements of benthonic organisms, and (2) measure ocean currents at maximum depth. The organisms that inhabit the sea floor have often been identified through photography, but normally a deep-sea camera does not remain long enough in one position to make repeated pictures revealing their movements. The *Trieste* (Piccard & Dietz, 1961) can remain stationary for long periods, making possible a study of sea-floor currents and the acquisition of oceanographic data on bottom water movement at a depth of 645 fm (1,180 m).

The subterranean region where the information was acquired was the San Diego Trough about 15 miles (24 km) west of San Diego, California (fig. 25-1). This deep-water location was chosen for the 84th dive of the *Trieste* because its proximity permitted the bathyscaph to

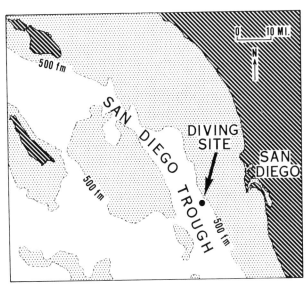

Figure 25-1. San Diego Trough off San Diego, California, showing location of *Trieste* Dive 84, where the study of brittle star movements and bottom current was made.

be towed out, dive, and return the same day to the U.S. Navy Electronics Laboratory dock in San Diego Bay (LaFond, 1962).

The San Diego Trough is a broad sea valley, varying in depth from 500 fm (900 m) in the north to 700 fm (1,300 m) in the south, and connecting with deeper basins at both ends (fig. 25-1). The sea floor of the trough is of fine silt and is relatively flat. The temperature and oxygen factors indicate that the valley flow comes from the south (Emery, 1960). The sediments from canyons, which cut into the east slope of the trough, flow through meandering channels and veer toward the south.

25-2. Equipment

Cameras and lights

Direct observations were conducted through the bathyscaph's main viewing port, which looks forward. Indirect observations were made possible by a closed circuit tele-

* This work was supported by USN *Buships* under Project SR 004 03 03, Task 0539. Assistance in preparing the report was furnished by Katherine G. LaFond and Eric G. Barham.

Figure 25–2. The camera and lighting arrangement of bathyscaph *Trieste* used for oceanographic research by USNEL. Encased cameras are mounted just forward of the forward shot tub.

vision camera, which is set sternward behind the gondola. Motion-picture cameras, both 16-mm and 35-mm, were located in front of the gondola. The 35-mm Edgerton deep-sea, single-frame movie camera was mounted obliquely and directed forward at a 45° angle toward the sea-floor area under maximum light from the bathyscaph floodlights (fig. 25–2). The Edgerton camera shutter was operated by the observers in the gondola of the *Trieste*. The camera's field of view was guided by looking through the gondola viewing port past the shot tub of the bathyscaph.

Twin floodlights, both forward and aft, projected their beams directly ahead of the main viewing port and into the center of the viewing area and camera field. The two lights most useful to the investigation were affixed to the undercarriage of the bathyscaph flotation chamber, forward of the gondola and ballast tub, and directed obliquely down and toward the area under the bow.

Inclinometer

For the measurement of bottom currents, an inclinometer, consisting of nylon-yarn streamers attached to a grid, was employed. The grid consisted of a group of 15-cm cubes joined to form one 45-cm cube, and was composed of 6-mm rods (fig. 25–3). This device, with its heavy streamers 5 mm in diameter and 12 cm long, was suspended between the forward ballast tub and the gondola and was in full view through the port. Immersed, the streamers are slightly heavier than water and hang downward from their point of attachment. The yarns follow the water movement in any horizontal direction. There is practically no vertical motion because the inclinometer is near the sea floor.

The primary asset of such an underwater inclinometer is its relative immunity to malfunction. Its other advantages are a quick response and a capacity to measure accurately the low-order-of-magnitude currents. The angle subtended by the yarn, for a speed interval of 0.01 to 0.08 kn, was determined in a towing tank by prior tests. The change in angle per 0.01 kn change in speed is about 7.5°. Thus the speed of the current was measured by the angle the yarn made with the vertical, and can be calculated to a reliability of ±0.01 kn. The direction of flow was the direction to which the yarn tended with reference to the gyro compass in the sphere.

In order to measure the motion of the yarn, and thus the water, toward or away from the observer in the bathyscaph, mirrors were attached to each side of the grid and their reflecting surfaces were held at an angle of 45° to the line of sight.

The motion of the inclinometer streamers was photographed through the viewing port by the 16-mm motion picture camera. Although the period of observation was short, the results were reliable, and, at the depth used, would be very difficult to obtain from a surface ship using suspended instruments.

25–3. Results

The descent to the sea floor

During the descent of the bathyscaph, pelagic organisms could be observed in the water column through the gondola viewing port, either by floodlight or by their own bioluminescence. Many species were seen in the column, but their concentration in the deep scattering layer, at 140 fm (260 m) to 180 fm (330 m), was estimated to be ten times larger than in most of the other depths. The photographing of these nearly transparent organisms — siphonophores, euphausiids, isopods — both living and dead, proved very difficult.

As the *Trieste's* gondola settled into the bottom at a depth of 3,870 feet, smooth layers of soft, brown and gray mud curled up to a height of about 12 cm around it. The subsurface mud was gray, but the undisturbed surface was covered with a fine, brown organic material that contrasted sharply in color. The sea floor, unusually flat, was marked by small, worm-tube projections measuring from 1 to 2 cm. Other mounds of biologically extruded substrata measured 2 to 3 cm in diameter. The remains of older, inactive borings were also visible. The floor was flat, devoid of noticeable ripples.

Measurements of current

After the bathyscaph had been on the bottom for 10 min, the water around the gondola gradually cleared, and the base of the inclinometer grid, suspended in front of the viewing port, was seen to be 15 cm from the sea floor.

After the *Trieste* had settled about 60 cm into the sea floor, it remained motionless during the period of observation and so formed a reliable reference for motion studies. The viewing port looked forward at 175°, or nearly south, and was especially suitable because of its parallel orientation with the San Diego Trough.

The water at the sea floor, though turbid on impact, became especially clear in time, although some particles could be seen moving with the slow current. Three short scenes of the motion, recorded on 16-mm film (47, 75, and 187 sec) were taken over a 25-min period. When these films were projected on a screen, the angles of the inclinometer streamers were determined. The east-west angle of the yarn varied from 20°E (left) to 0°W (right). The north-south component ranged from 10° to 40°N. By reading both angles every second, and computing

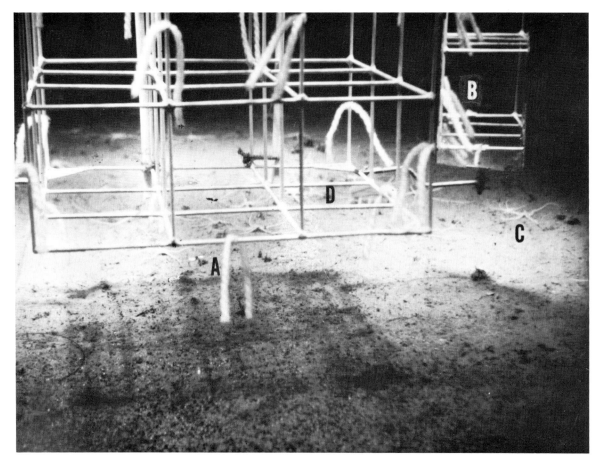

Figure 25–3. The three-dimensional grid, an 45-cm cube, attached to the bathyscaph and used for measuring the speed and direction of bottom currents, rests six inches from sea floor. (A) attached nylon yarn streamers used to determine water motion; (B) mirror to view streamers from side; (C) brittle stars; (D) sea cucumber.

the resultant angle with reference to the vertical and the compass direction toward which it was deflected, the speed and direction of currents a few inches from the bottom, at a depth of 645 fm (1,180 m), were accurately obtained.

Speed. The flow fluctuated from 0.02 to 0.06 kn and averaged 0.042 kn (fig. 25–4). The fluctuations in speed took place within a span of only a few seconds. The various time intervals, from maximum to minimum, or vice versa, of an oscillation in speed were 19, 32, 17, 4, 15, 9, 8, 21, 4, and 13 sec.

Direction. The direction varied from 355° to 040°. Stronger currents were observed in the more northerly direction (LaFond, 1962).

Benthonic life

A large, black sable fish (*Anoplopoma fimbria*), about 75 cm long, moved close to the gondola viewing port, then veered off, but remained within the fringe of the light beam for the two hours that the bathyscaph was on the bottom. In addition to this fish, the main benthonic organisms visible were brittle stars and sea cucumbers. Two sea cucumbers (*Scotophanes sp.*) and some fifty to sixty brittle stars could be seen through the port on the flat, muddy floor. The stern view, through closed-circuit television, revealed similar organisms.

Brittle stars. About fifteen brittle stars per square meter were visible. Dredging in this area (Barham, no date) revealed that the brittle stars represented at least two different species. These were identified in photographs and visually observed through the gondola port. One brittle star (*Ophiomusium lymani*) was larger (large specimens measured up to 20 cm) and rested on the sea floor, propped on the proximal part of its tentacles. Their disks tended to stand up from the surface, throwing shadows. They appeared to be extracting food particles from the water. The movement of these organisms was slow and independent, and in some cases they came to a stop while partially over one another. The other brittle-

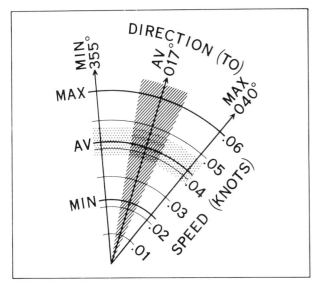

Figure 25–4. Measured current speed and direction on bottom of San Diego Trough.

star species (*Amphilepis sp.*) was smaller, darker, and often partly buried in the sediment silt. They were probably foraging for food. These are the most active travelers, often covering a distance of several centimeters at one time before stopping.

The close-up observations of benthonic life were recorded by the 16-mm camera. The sea floor farther ahead was covered by the 35-mm camera, which photographed the scene about every minute over a period of 44 min. This gave a total of forty-six photographs in 35-mm Ektachrome color.

By comparing consecutive photographs of the same bottom area, and noting the change in the position of the organisms, it was possible to determine their direction and speed of motion. Six of the repeated scenes are shown in fig. 25–5. The movements are grouped in 7- to 13-min intervals. In the first eleven photographs of the lighted area (about 5 m²), twenty brittle stars were seen to move. In this first 8-min period the average distance of movement was 19 cm. The general direction was toward the gondola; this motion was in the same direction in which the bottom current was flowing. Similar changes in posi-

tion were noted during each of the next successive few minutes (table 25–1). There was also a relative difference in movement between the large, white brittle stars, which moved more slowly and thus covered shorter distances, and the smaller, darker brittle stars. A composite of the total movement of individual organisms is shown in fig 25–6.

Sea Cucumbers. On dive 84, five sea cucumbers (*Scotoplanes sp.*) were visible: two in the television screen to the stern, three through the port forward. Two of these three in the forward area were photographed. This amounted to about one sea cucumber per 2.5 sq m. All were very sluggish in their movements. In general, each of the sea cucumbers moved toward the northeast, but their meanderings looked aimless, and both of those photographed were almost stationary part of the time, one at the beginning and one at the end of the period. They moved 31 and 34 cm, respectively, in 44 min, or 0.6 cm per min.

25–4. Discussion

The movement of benthonic organisms on the sea floor was quicker upon the immediate settling of the bathyscaph than later. Although the number of brittle stars that moved increased, the distance traveled decreased with time from an initial average of 19 cm per 8 min to about 15 cm per 8 min about 20 min later. This movement was probably influenced by the *Trieste*, which stirred up sediment and caused agitation in the water. The available food, stirred up from the sea floor or resulting from the convective circulation of the descending submersible, must also have had some effect in stimulating activity. In addition, the small pressure waves set up by the gondola as it reached a position in the soft bottom may have stimulated the movements of the brittle stars. The light, too, may have disturbed some organisms, but there appeared to have been as much activity in as out of the maximum light beam. As the bathyscaph settled and the water cleared, the number of brittle stars visible increased.

The directions of movement were in general fairly

TABLE 25–1. Summary of brittle star movements on sea floor.

Photo number	Time (min)	Number of brittle stars visible in photo	Number of brittle stars moved	Average distance moved	Average speed cm/min	General direction moved
1–11	0–8	83	20	19 cm	2.4	North (toward (bathyscaph)
11–21	8–15	94	26	24 cm	3.4	North (toward bathyscaph)
21–31	15–28	101	31	28 cm	2.2	Random directions
31–41	28–36	100	30	15 cm	1.9	Random directions
41–46	36–44	107	21	5 cm	0.7	Random directions

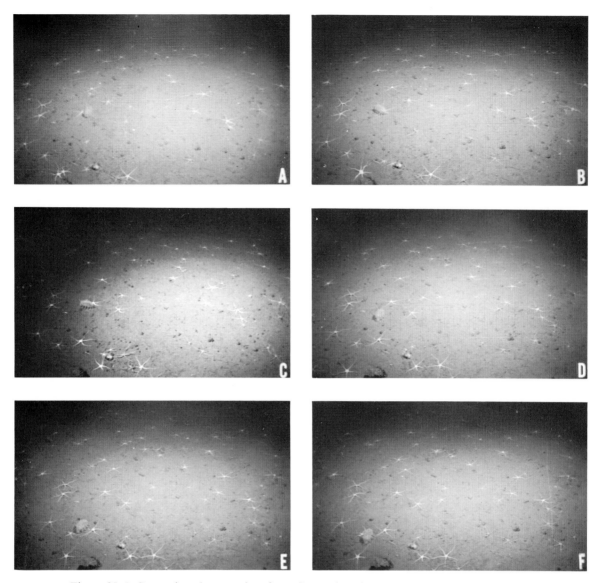

Figure 25–5. Successive photographs of sea floor taken from bathyscaph *Trieste* at 8–13-min intervals, showing the change in positions of benthonic organisms.

random. In some cases, however, there was a more or less continuous direction of movement from minute to minute. In the first 15 min, the general direction was toward the gondola, which was with the current. During the last 29 min, there was no dominant direction of movement.

Photographs taken on other dives of the bathyscaph have shown that large populations of brittle stars populate wide areas of the San Diego Trough. They are especially abundant along the deeper floors of canyons that drain into the trough. A photograph taken in the Coronado Fan Valley (fig. 25–7) shows about twice as many small brittle stars per square meter as were observed in the main part of the trough on Dive 84. Many in the Coronado Canyon were seen to overlap one another. Also

sea cucumbers in canyons are numerous. It is believed that the high benthonic populations in submarine canyons are due to the large quantity of trapped organic material in the canyons. High production is also aided by the vertical mixing of the water that occurs when oscillating internal waves and tidal currents create a funneling circulation in the canyon constriction.

25–5. Summary

The investigation described in this report demonstrated that benthonic organisms, sea-bottom currents, and pelagic life can be studied more efficiently from a deep-

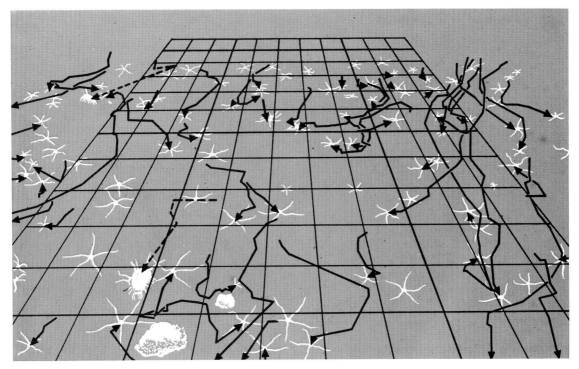

Figure 25–6. Trajectories of brittle stars over the sea floor in a period of 44 min. Grid in 20 cm squares.

diving submersible such as the bathyscaph *Trieste*, from which the technology of underwater time-lapse photographs and an *in situ* close-up environmental scrutiny can be more reliably used, than is possible from a ship.

From the study it can be seen that activity on the sea floor may be compared to a moving carpet. If the 20-cm diameter brittle stars all moved as measured, and covered a 20-cm swathe over the bottom without retracing other previous routes, it would require only 45 min to cover the entire floor in this area. Thus the bottom is discerned to have much benthonic activity — all parts are continuously traversed. Brittle stars move sporadically on the floor of the San Diego Trough, and about one-third were observed to move at some time during 7- to 13-min intervals. Those that did move averaged 0.7 to 3.4 cm

per min, and there was a tendency for them to move a little more with the current than against it.

References

Barham, E. G., no date: Observations of benthonic fauna of the San Diego Trough. Manuscript.

Emery, K. O., 1960: *The sea off Southern California; a modern habitat of petroleum*. John Wiley and Sons, Inc., New York, 366 p.

LaFond, E. C., 1962: Deep current measurements with the bathyscaph *Trieste*. *Deep-Sea Res.*, **9**, 115–116.

———, 1962: Dive eighty-four. *Sea Frontiers*, **8**, no. 2, 94–102.

Piccard, J., & R. S. Dietz, 1961: *Seven miles down*. Putnam, London, 1–249.

Figure 25–7. Population density of brittle stars at 580 fm in Coronado Fan Valley (photograph by R. F. Dill).

Appendix

The appendix is a table of principal facts about the deep-sea photographs reproduced throughout this book.

Key: — = data not furnished
c. = (*circa*) approximately

Chapter 1.

Fig.	Position Latitude	Position Longitude	Date	Ship	Cruise	Station	Depth m	Depth fm	Photographers	Equipment	Scale
1–4a	29°59.5′N	72°30′W	2 March 1940	Atlantis	94	C–13	c.5,300	c.2,900	M. Ewing, Vine	Camera #1	c. 3′ × 5′
1–4b	"	"			"		"	"			
1–4c	41°44′N	66°38′W	6 June 1940	"	98	C–28	58	32	M. Ewing	Camera #2A	c. 3′ × 4′
1–4d	41°55′N	66°55′W	"	"	"	C–29	"	"			
1–6a	41°34′N	66°24′W	23 June 1940	"	100	C–46–2	93	51	M. Ewing, R. H. Ewing	Camera #2B	"
1–6c	42°09′N	66°30′W	24 June 1940	"	"	C–48–1	99	54	"	"	"
1–6d				"	"	C–48–2	"	"			
1–7a	39°35′N	70°57′W	16 July 1940	"	101	C–65	832	455	M. Ewing, Vine, Worzel	"	c. 4′ × 5.5′
1–7b	39°52′N	71°07′W	"	"	"	C–66	594	325	"	"	
1–11b	41°30′N	70°40′W	16 Oct. 1942	Anton Dohrn	local	none	—	11	Worzel, Tirey	Hydrofoil camera	c. 3.6′ × 3.6′
1–11c	41°31′N	70°38.5′W	23 Oct. 1942	"	"	"	—	6			
1–12a	34°13′N	76°34′W	— May 1942 82W–7	"	24	U352	31.4	17.2	M. Ewing, Worzel, J. Ewing	Modification of cameras #10 and 11	c. 5.4′ × 7.6′
1–12b	"	"	7 June 1942 98W–19	"	"	unident.	—	—	"	"	"
1–12c	34°34′N	76°14′W	9 June 1942 102W–6	"	"	Atlas	35.4	19.3	"	"	"
1–12d	34°49′N	75°54′W	10 June 1942 105W–6	"	"	Proteus	25.0	13.7	"	"	"
1–13a	40°18′N	74°55′W	13 July 1942 111W–8	"	"	Mohawk	22.9	12.5	Worzel, J. Ewing, Tirey	Modification of camera #11	c. 2.9′ × 4.1′
1–13b	40°40′N	69°41′W	234–W–8	"	28(?)	Nantucket Lightship	—	38	M. Ewing, Worzel, Tirey	—	
1–13c	34°39′N	75°48′W	7 Aug. 1943 339 WR–17	Gentian	none	Manuela	44.2	24.2	M. Ewing, R. Tirey	Robot camera	c. 7.2′ × 7.2′
1–13d	39°48′N	72°49′W	12 June 1943 326 WL–11	"	"	Bidwind	58	32	"	"	c. 9.0′ × 9.0′
1–14a	34°49′N	75°54′W	21 July 1943 332WL11	Gentian	"	E. M. Clarke	25.0	13.7	"	"	"
1–14b	35°24′N	75°21′W	7–14 June 1944	Gentian	"	City of Atlanta	23.8	13.0	"	"	c. 7.2′ × 7.2′
1–14c	34°31′N	76°02′W	4 Aug. 1943	Gentian 338WR21	"	Tamaulipas	42.6	23.7	"	"	"
1–14d	"	"	"	Gentian 338WR7A	"	"	"	"	"	"	"
1–14e	40°22′N	72°20′W	13 June 1943	Gentian 327WL9	"	Coimbra	54.9	30.0	"	"	"
1–14f	36°50′N	75°23′W	8 June 1944	Gentian	"	Francis E. Powell	27.1	14.8	"	"	"
1–16a	34°02′N	30°15′W	23 Aug. 1947	Atlantis	150	—	730	400	M. Ewing, Edwards, Knight	Argoflex camera	c. 6.6′ × 6.6′
1–16b	"	"	27 Aug. 1947	"	"	—	"	"	"	"	"
1–16c	"	"	"	"	"	—	2,000	1,100	"	"	"
1–16d	42°13′N	70°15′W	29 Aug. 1947	Balanus	13	F40	21.4	11.7	Heezen, Northrup	"	"
1–18a	39°28′N	71°41′W	20 Aug. 1947	"	12	E–25	2,600	1,420	J. Ewing, Heezen, Northrup	"	c. 5.5′ × 5.5′
1–18b	39°39′N	71°37′W	"	"	1	E–22	2,120	1,160	"	"	"
1–18c	39°29′N	72°17.5′W	21 Aug. 1947	"	12	E–34	1,240	678	"	"	"
1–18d	39°56.5′N	68°58′W	11 Sept. 1947	"	15	H–27	1,958	1,071	Heezen, Northrup	"	"
1–19a	39°46′N	70°50′W	8 Dec. 1947	Atlantis	151	151–3	1,830	1,000	D. M. Owen	"	"
1–19b	39°06′N	25°53′E	1 March 1948	"	"	151–9	207	113	"	Argoflex camera	"
1–19c	37°50′N	26°25′E	9 Feb. 1948	"	"	151–14	1,103	603	"	"	"
1–19d	30°37′N	59°07′W	12 June 1948	"	"	151–73	5,534	3,206	"	"	"
1–20a	34°04′N	30°17′W	25 Aug. 1948	"	152	152–96	375	205	M. Ewing, Heezen	"	"
1–20b	34°05′N	30°11′W	26 Aug. 1948	"	"	152–98	412	225	"	"	"
1–20c	34°07′N	30°13′W	"	"	"	152–101	348	190	"	"	"
1–20d	34°14′N	30°12′W	"	"	"	152–110	750	410	"	"	"

Chapter 4.

Fig.	Position Latitude	Position Longitude	Date	Ship	Cruise	Station	Depth m	Depth fm	Photographers	Equipment	Scale
4–3a	12°05′N	60°25′W	27 Feb. 1955	*Cary N*	89	—	30	16	Johnson & Owen	35 mm Edgerton deep-sea camera, using acoustic focusing	length of fish c. 17.8 cm
4–4	near Pantelleria Island off Sicily		—	*Calypso*	—	—	—	—	Edgerton & Cousteau	35 mm Edgerton deep-sea camera	—
4–10b	32°57′N	18°10′W	26 July 1959	*Chain*	7	EC–14	2,917	1,595	Pratt	2 35-mm Edgerton deep-sea cameras (in stereo)	rock c. 7 cm diameter
4–11b	31°54′N	65°12′W	11 Oct. 1959	`"`	9	P	1,950	1,066	Raymond & Davis	`"`	width of picture c. 2.4 m at bottom, 2.9 m at top
4–13	Mid-Atlantic Ridge Rift		1959	Troika, towed by *Calypso*	—	—	—	—	Cousteau	`"`	—
4–14	Caribbean Sea west of Guadaloupe		1959	`"`	—	—	—	—	Cousteau & Edgerton	`"`	—

Chapter 5.

Fig.	Position Latitude	Position Longitude	Date	Ship	Cruise	Station	Depth m	Depth fm	Photographers	Equipment	Scale
5–2a	35°26.7′N	75°24.0′W	July, 1963	*Explorer*	OPR–438	11	18.3	10	Schuldt	Edgerton deep-sea camera	1 : 7
5–3a	38°15.0′N	71°21.0′W	Oct. 1961	`"`	10,000 –827	3	2,750	1,504	Hale	`"`	1 : 7

Chapter 6.

Fig.	Position Latitude	Position Longitude	Date	Ship	Cruise	Station	Depth m	Depth fm	Photographers	Equipment	Scale
6–4	41°44′N	64°57′W	11 July 1964	*Mizar*	—	—	2,500	1,350	Patterson	described in chap. 6	1″ = 3.3 meters
6–6	`"`	`"`	11 Sept. 1964	`"`	—	—	`"`	`"`	Brundage	`"`	1″ = 0.36 meter
6–7	`"`	`"`	23 Aug. 1964	*Gilliss*	—	—	`"`	`"`	Worthington	`"`	1″ = 0.635 meter
6–8	`"`	`"`	12 Sept. 1964	*Mizar*	—	—	`"`	`"`	Brundage	`"`	1″ = `"` `"`
6–9	`"`	`"`	14 July 1964	`"`	—	—	`"`	`"`	Patterson	`"`	1″ = `"` `"`
6–10	`"`	`"`	17 July 1964	`"`	—	—	`"`	`"`		`"`	1″ = `"` `"`
6–11	`"`	`"`	11 Sept. 1964	`"`	—	—	`"`	`"`	Brundage	`"`	1″ = 0.61 meter
6–12	`"`	`"`	3 June 1963	*Gilliss*	—	—	`"`	`"`	Walczak	`"`	1″ = 0.075 meter
6–13	`"`	`"`	20 Aug. 1964	*Mizar*	—	—	`"`	`"`	Brundage	`"`	1″ = 0.40 meter
6–14	`"`	`"`	23 July 1963	*Gillis*	—	—	`"`	`"`	Green	`"`	1″ = 0.13 `"`
6–15	`"`	`"`	22 July 1963	`"`	—	—	`"`	`"`	`"`	`"`	1″ = 0.20 `"`
6–16	`"`	`"`	19 July 1963	`"`	—	—	`"`	`"`	`"`	`"`	1″ = 0.10 `"`
6–17	`"`	`"`	14 July 1964	*Mizar*	—	—	`"`	`"`	Patterson	`"`	1″ = 0.10 `"`
6–18	`"`	`"`	18 July 1964	`"`	—	—	`"`	`"`	`"`	`"`	1″ = 0.36 `"`
6–19	`"`	`"`	17 July 1964	`"`	—	—	`"`	`"`		`"`	1″ = 0.40 `"`
6–20	`"`	`"`	17 Aug. 1964	`"`	—	—	`"`	`"`	Brundage	`"`	1″ = 0.80 `"`
6–21	`"`	`"`	22 July 1963	*Gilliss*	—	—	`"`	`"`	Green	`"`	1″ = 0.20 `"`
6–22	`"`	`"`	24 July 1963	`"`	—	—	`"`	`"`	`"`	`"`	1″ = 0.20 `"`
6–23	`"`	`"`	14 July 1964	*Mizar*	—	—	`"`	`"`	Patterson	`"`	1″ = 0.80 `"`
6–24	`"`	`"`	23 July 1963	*Gilliss*	—	—	2,500	1,350	Green	`"`	1″ = 0.20 `"`
6–25	`"`	`"`	`"`	`"`	—	—	`"`	`"`	`"`	`"`	1″ = 0.56 `"`
6–26	`"`	`"`	12 Sept. 1964	*Mizar*	—	—	`"`	`"`	Brundage	`"`	1″ = 0.43 `"`
6–27	`"`	`"`	17 Aug. 1964	`"`	—	—	`"`	`"`	`"`	`"`	1″ = 0.43 `"`
6–28	`"`	`"`	22 July 1963	*Gilliss*	—	—	`"`	`"`	Green	`"`	1″ = 0.20 `"`
6–29	`"`	`"`	10 Sept. 1964	*Mizar*	—	—	`"`	`"`	Brundage	`"`	1″ = 1.5 `"`

Chapter 7.

Fig.	Position Latitude	Position Longitude	Date	Ship	Cruise	Station	Depth m	Depth fm	Photographers	Equipment	Scale
7–5	33°38′N	165°24′E	1 Sept. 1964	USNS *Charles H. Davis* (AGOR–5)	—	17	5,930	3,240	Shipek	NEL deep-sea oceanographic system	width of field of view 0.9 m

Chapter 8.

Fig.	Position Latitude	Position Longitude	Date	Ship	Cruise	Station	Depth m	Depth fm	Photographers	Equipment	Scale
8–5	39°40.5′N	70°38′W	24 May 1961	*Atlantis*	264	G#1	2,086	1,141	Owen	Stereoscopic pair of Robot cameras	area of photograph 1/7 sq m
8–6	39°45′N	70°44′W	`"`	`"`	`"`	F#1	1,500	820	`"`	`"`	`"`
8–7	38°45′N	70°13.5′W	21 May 1961	`"`	`"`	HH#3	2,379	1,301	`"`	`"`	`"`
8–8	39°29′N	70°34′W	3 Oct. 1961	`"`	273	GH#4	`"`	`"`	`"`	`"`	`"`
8–9	`"`	`"`	`"`	`"`	`"`	`"`	`"`	`"`	`"`	`"`	`"`
8–10	39°40.5′N	70°38′W	24 May 1961	`"`	264	G#1	2,086	1,141	`"`	`"`	life size
8–11	24°00′N	77°14′W	20 Dec. 1960	*Bear*	258	30	1,335	730	`"`	`"`	field of view 58 cm (23 in) from foreground to background
8–12	23°52′N	77°06′W	21 Dec. 1960	`"`	`"`	32	1,322	723	`"`	`"`	field of view 76 cm (30 in) foreground to background

Chapter 9.

Fig.	Position Latitude	Position Longitude	Date	Ship	Cruise	Station	Depth m	Depth fm	Photographers	Equipment	Scale
9–2	30°59.5′N	78°10.5′W	15 June 1961	*Atlantis*	266	5	823	450	Pratt	Edgerton deep-sea camera with current meter attached	area of photograph 40 in × 40 in (102 cm × 102 cm)
9–3	25°11′N	76°05′W	20 Jan. 1959	*Chain*	3	2	1,200	660	Bruce	`"`	area of photograph 10 in × 10 in (25.4 cm × 25.4 cm)

Chapter 11.

Fig.	Position Latitude	Longitude	Date	Ship	Cruise	Station	Depth m	Depth fm	Photographers	Equipment	Scale
11–1	28°53′N	43°20′W	8 Dec. 1961	*Chain*	21	Dredge	—	1,950	Hays, Norwalk, & Edgerton	Edgerton deep-sea camera	area of photograph *c.* 1.4 sq m
11–3	19°59′N	66°30′W	1 March 1960	"	11	C–8a	6,949	3,800	Edgerton, Hays, & Davis	"	width of photograph *c.* 1.5 m "
11–4	"	"	"	"	"	"	"	"	"	"	"
11–5	"	"	"	"	"	"	"	"	"	"	"
11–6	20°00′N	66°28′W	28 June 1961	"	19	C–9	6,220	3,400	Hersey, Johnston & Edgerton	"	"
11–8	18°49′N	66°33′W	23 Feb. 1965	"	46	7	3,500	1,900	Johnston & Welby	"	—
11–9	"	"	"	"	"	"	"	"	"	"	—
11–10	"	"	"	"	"	"	3,900	2,100	"	"	—
11–11	"	67°20.5′W	22 Feb. 1965	"	"	1	3,500	1,900	"	"	width of photograph *c.* 4.6 m
11–13	42°00′N	06°00′E	6 July 1959	"	7	EC 12	2,440	1,335	Pratt	"	width of photograph *c.* 0.79 m
11–14	39°30′N	"	27 June 1959	"	"	EC 8	2,370	1,492	"	"	width of photograph *c.* 0.49 m
11–15	31°55′N	65°13′W	6 Oct. 1959	"	9	H	2,433	1,330	Raymond & Davis	"	width of photograph *c.* 0.95 m
11–16	43°24′N	08°44′E	9 July 1964	"	43	87	2,269	1,280	Johnston	"	
11–17	40°27′N	12°49′E	17 Nov. 1961	"	21	9	3,416	1,868	Pratt	"	width of individual pictures *c.* 5 m
11–19	31°56′N	65°10′W	3 Oct. 1959	"	9	D	155	85	Raymond & Davis	"	width of photograph *c.* 0.79 m
11–20	"	"	"	"	"	"	230	126	"	"	width of photograph *c.* 4.8 m
11–21	"	"	"	"	"	"	390	213	"	"	width of photograph *c.* 1.8 m
11–22	"	"	"	"	"	"	400	219	"	"	width of photograph *c.* 1.5 m
11–23	"	"	"	"	"	"	450	246	"	"	width of photograph *c.* 1.4 m
11–24	"	"	"	"	"	"	710	388	"	"	width of photograph *c.* 0.79 m
11–25	"	"	"	"	"	"	745	407	"	"	width of photograph *c.* 1.2 m
11–26	"	"	"	"	"	"	765	418	"	"	width of photograph *c.* 1.5 m
11–27	"	"	"	"	"	"	770	421	"	"	width of photograph *c.* 2.0 m
11–28	"	"	"	"	"	"	810	443	"	"	width of photograph *c.* 1.1 m
11–29	"	"	"	"	"	"	855	468	"	"	"
11–30	"	"	"	"	"	"	"	"	"	"	width of photograph *c.* 1.2 m at bottom, 1.8 m at top
11–31	"	"	"	"	"	"	863	472	"	"	width of photograph *c.* 1.3 m at bottom, 1.9 m at top
11–32	"	"	"	"	"	"	957	523	"	"	width of photograph *c.* 0.64 m at bottom, 0.84 m at top
11–33	31°55′N	65°13′W	6 Oct. 1959	"	"	H	858	469	"	"	width of photograph *c.* 2.1 m
11–34	"	"	"	"	"	"	1,210	662	"	"	width of photograph *c.* 3.2 m
11–35	"	"	"	"	"	"	1,430	782	"	"	width of photograph *c.* 6.8 m
11–36	"	"	"	"	"	"	1,480	809	"	"	width of photograph *c.* 5.2 m
11–37	31°54.5′N	65°12′W	5 Oct. 1959	"	"	G	950	519	"	"	width of photograph *c.* 1.2 m "
11–38	"	"	"	"	"	"	1,048	573	"	"	"
11–40	8°33′S	58°51′E	16 May 1964	"	43	46	2,013	1,101	Johnston	"	scale on photo
11–41	"	"	"	"	"	"	"	"	"	"	"

Chapter 12.

Fig.	Position Latitude	Longitude	Date	Ship	Cruise	Station	Depth m	Depth fm	Photographers	Equipment	Scale
12–1	41°31′N	68°00′W	27 Aug. 1963	*Gosnold*	24	1219	33	18	Trumbull & Emery	Robot Star camera — 30 mm lens	trip weight diameter, 9 cm
12–2	41°12′N	69°16′W	26 Aug. 1963	"	"	1216	88	48	"	"	"
12–3	43°20′N	65°47′W	31 Aug. 1963	"	"	1240*b*	38	21	"	"	"
12–4	40°12′N	73°45′W	11 Oct. 1963	"	29	1283	61	33	"	"	"
12–5	40°41′N	72°16′W	10 Oct. 1963	"	"	1277	53	29	"	"	"

Chapter 13.

Fig.	Position Latitude	Longitude	Date	Ship	Cruise	Station	Depth m	Depth fm	Photographers	Equipment	Scale
13–3	33°32′N	62°25′W	18 Oct. 1960	*Atlantis*	260	8	1,320	720	Pratt	Edgerton deep-sea camera	width of photograph *c.* 6 ft. (1.8 m)
13–5	31°59′N	65°11′E	28 Sept. 1959	*Chain*	9	B	<50 m	<27 fm	Raymond & Pratt	"	width of photograph *c.* 4 ft (1.2 m)
13–6	33°33′N	62°24′W	20 Oct. 1960	*Atlantis*	260	9	1,500 –1,600	820 –880	Pratt	"	width of photograph *c.* 3 ft (0.9 m)
13–7	"	"	"	"	"	"	"	"	"	"	width of photograph *c.* 2 ft (0.6 m)

Chapter 13 (*Continued*)

Fig.	Position		Date	Ship	Cruise	Station	Depth		Photographers	Equipment	Scale
	Latitude	Longitude					m	fm			
13–8	29°48′N	28°40′W	8 Sept. 1961	*Chain*	21	5	420–1,100	230–600	Pratt	Edgerton deep-sea camera	width of photograph 5 ft–6 ft (1.5 m–1.8 m)
13–9	"	"	"	"	"	"	"	"	"	"	width of photograph *c.* 5 ft (1.5 m)
13–10	"	"	"	"	"	"	"	"	"	"	width of photograph *c.* 6 ft (1.8 m)
13–11	37°37.5′N	59°59′W	19 June 1962	*Atlantis*	280	"	4,750–4,900	2,600–2,660	"	"	
13–13	38°55′N	60°25′W	17 June 1962	"	"	2	2,750–3,750	1,500–2,050	"	"	width of photograph 8 ft–10 ft (2.4 m–3.1 m)
13–14	37°58′N	62°09′W	20 June 1962	"	"	6	1,550–2,300	850–1,250	"	"	width of photograph 5 ft–6 ft (1.5 m–1.8 m)
13–15	39°21′N	67°05′W	29 June 1962	"	281	2	2,500–2,550	1,370–1,390	"	"	width of photograph *c.* 4 ft (1.2 m)

Chapter 14.

Fig.	Position		Date	Ship	Cruise	Station	Depth		Photographers	Equipment	Scale
	Latitude	Longitude					m	fm			
14–3	41°29.15′N	70°36.50′W	10 Aug. 1954–16 Aug. 1954	*Asterias*	—	—	8.5	4.6	Owen	Specially designed time-lapse camera	wavelength of ripple marks *c.* 8 cm
14–4	"	"	"	"	—	—	"	"	"	"	"

Chapter 15.

Fig.	Position		Date	Ship	Cruise	Station	Depth		Photographers	Equipment	Scale
	Latitude	Longitude					m	fm			
15–3	39°46′N	70°44′W	24 May 1961	*Atlantis*	264	F#1	1,523	835	Owen	Owen multi-shot stereoscopic camera (see chap. 8)	width at bottom 39 cm width at top 46 cm, bottom to top 43 cm
15–4	"	"	"	"	"	"	"	"	"	"	"
15–5	"	"	"	"	"	"	"	"	"	"	"

Chapter 16.

Fig.	Position		Date	Ship	Cruise	Station	Depth		Photographers	Equipment	Scale
	Latitude	Longitude					m	fm			
16–1	42°07′N	70°18′W	17 June 1960	*Bear*	247	57	60	33	Owen	Owen multi-shot stereoscopic camera described in chap. 8.	field of view 58 cm square
16–2	42°06′N	70°21′W	1 June 1960	"	241	43	"	"	"	"	"
16–3	24°33′N	77°23′W	20 Dec. 1960	"	258	29	1,513	832	"	"	"
16–4	24°18′N	77°33′W	29 Dec. 1960	"	"	48	1,304	717	"	"	"

Chapter 17.

Fig.	Position		Date	Ship	Cruise	Station	Depth		Photographers	Equipment	Scale
	Latitude	Longitude					m	fm			
17–2	64°03′S	89°49′W	(data not available)	*Eltanin*	(data not available)		4,747	2,374	United States Antarctic Research program	Deep-sea camera described by Thorndike (1959)	Area of photograph *c.* 5 sq m
17–3	70°07′S	102°56′W	"				3,840	2,100			"
17–4	"	"	"				"	"			"
17–5	61°58′S	90°01′W	"				4,840	2,647			"
17–6	66°20′S	90°24′W	"				4,525	2,474			"
17–7	65°36′S	121°11′W	"				4,873	2,665			"
17–8	58°03′S	160°12′W	"				4,170	2,280			"
17–9	61°57′S	78°58′W	"				4,798	2,624			"
17–10	63°14′S	71°34′W	"				3,766	2,059			"
17–11	70°25′S	99°36′W	"				3,680	2,012			"
17–12	63°58′S	71°13′W	"				3,548	1,940			"
17–13	61°16′S	89°49′W	"				4,898	2,678			"
17–14	62°00′S	115°14′W	"				5,139	2,810			"
17–15	57°28′S	64°51′W	"				4,531	2,478			"
17–16	"	"	"				"	"			"
17–17	55°59′S	61°43′W	"				4,220	2,308			"
17–18	56°03′S	60°48′W	"				4,190	2,291			"
17–19	"	"	"				"	"			"
17–20	58°11′S	79°11′W	"				4,716	2,579			"
17–21	57°59′S	70°44′W	"				3,924	2,146			"
17–22	61°01′S	99°59′W	"				4,966	2,715			"
17–23	60°07′S	128°54′W	"				4,336	2,371			"
17–24	58°54′S	95°08′W	"				3,981	2,177			"
17–25	"	"	"				"	"			"
17–26	"	"	"				"	"			"
17–27	57°59′S	120°03′W	"				4,825	2,638			"
17–28	"	"	"				"	"			"
17–29	59°01′S	99°54′W	"				5,072	2,773			"
17–30	59°07′S	105°03′W	"				3,904	2,135			"
17–31	60°01′S	104°52′W	"				4,802	2,626			"
17–32	56°07′S	144°59′W	"				2,629	1,438			"
17–33	57°35′S	138°51′W	"				2,921	1,597			"
17–34	55°59′S	134°27′W	"				3,157	1,726			"
17–35	"	"	"				"	"			"
17–36	"	"	"				"	"			"
17–37	"	"	"				"	"			"

Chapter 18.

Fig.	Latitude	Longitude	Date	Ship	Cruise	Station	Depth m	Depth fm	Photographers	Equipment	Scale
18–8a	05°35.5′N	61°48.1′E	30 Aug. 1963	*Discovery*	2	5105	3,090	1,690	A. S. Laughton & J. M. Jopling	NIO deep-sea camera	Area of photograph is 3m × 4m
18–8b	"	"	"	"	"	"	3,160	1,730	"	"	"
18–8c	05°35.7′N	61°51.2′E	31 Aug. 1963	"	"	5109	2,340	1,280	"	"	"
18–8d	"	"	"	"	"	"	2,380	1,300	"	"	"
18–9a	"	"	"	"	"	"	2,430	1,330	"	"	"
18–9b	05°28.3′N	61°47.1′E	3 Sept. 1963	"	"	5121	2,210	1,210	"	"	"
18–9c	05°26.2′N	61°49.1′E	1 Sept. 1963	"	"	5113	1,770	970	"	"	"
18–9d	05°28.3′N	61°47.1′E	3 Sept. 1963	"	"	5121	2,100	1,150	"	"	"
18–10a	05°35.7′N	61°51.2′E	31 Aug. 1963	"	"	5109	2,320	1,270	"	"	"
18–10b	05°26.2′N	61°49.1′E	1 Sept. 1963	"	"	5113	1,960	1,070	"	"	"
18–10c	05°35.7′N	61°51.2′E	31 Aug. 1963	"	"	5109	2,320	1,270	"	"	"
18–10d	05°28.3′N	61°47.1′E	3 Sept. 1963	"	"	5121	2,120	1,160	"	"	"
18–11a	02°47.5′N	60°20.7′E	4–5 Sept. 1963	"	"	5127	4,100	2,240	"	"	"
18–11b	"	"	"	"	"	"	"	"	"	"	"
18–11c	02°46.2′N	60°02.5′E	6 Sept. 1963	"	"	5132	3,820	2,090	"	"	"
18–11d	"	"	"	"	"	"	3,950	2,160	"	"	"
18–12a	02°47.5′N	60°20.7′E	4–5 Sept. 1963	"	"	5127	4,080	2,230	"	"	"
18–12b	02°46.2′N	60°02.5′E	6 Sept. 1963	"	"	5132	3,860	2,110	"	"	"
18–12c	02°47.8′N	60°21.4′E	5 Sept. 1963	"	"	5128	4,170	2,280	"	"	"
18–12d	"	"	"	"	"	"	4,190	2,290	"	"	"
18–13a	02°46.2′N	60°02.5′E	6 Sept. 1963	"	"	5132	3,800	2,080	"	"	"
18–13b	"	"	"	"	"	"	3,790	2,070	"	"	"
18–13c	"	"	"	"	"	"	3,770	2,060	"	"	"
18–13d	02°46.0′N	59°53.0′E	7 Sept. 1963	"	"	5137	3,880	2,120	"	"	"
18–14a	10°15.8′N	56°00.3′E	9-10 Nov. 1963	"	"	5222	402	220	"	"	1½ m × 2 m
18–14b	"	"	"	"	"	"	411	225	"	"	"
18–14c	"	"	"	"	"	"	450	246	"	"	"
18–14d	"	"	"	"	"	"	472	258	"	"	"

Chapter 19.

Fig.	Latitude	Longitude	Date	Ship	Cruise	Station	Depth m	Depth fm	Photographers	Equipment	Scale
19–1a	31°56′N	65°10′W	3 Oct. 1959	*Chain*	9		750	430	Hersey *et al.*	Edgerton deep-sea camera	*c.* frame size 1.5×.9m
19–1b	"	"	3 Oct. 1959	"	"		800	437	"	"	*c.* frame size .7×.4m
19–1c	31°56′N	65°10′W	3 Oct. 1959	"	"		750	410	"	"	*c.* frame size 1.06×.6m
19–1d	"	"	3 Oct. 1959	"	"		450	246	"	"	*c.* frame size 1.06×.6m
19–2	76°59′S	167°36′E	26 Jan. 1959	*Endeavour*		A–468	110	60	Bullivant	NZO1 camera	Foreground 0.75m wide
19–3	32°32′N	117°26′W	7 Sept. 1962	Bathyscaph *Trieste*	Dive 110	—	1,280	700	Barham	Photographed through port, w/searchlight	× 0.2m *c.*
19–4a	31°56′N	65°10′W	3 Oct. 1959	*Chain*	9		300	164	Hersey *et al.*	Edgerton deep-sea camera	× 0.1m *c.*
19–4b	San Diego Trench		29 July 1962	*Trieste*	Dive 104		600	328	Martin and Dill	Photographed through port, w/searchlight	× 0.1m *c.*
19–5	77°30′S	166°30′E	Jan. 1960	*Endeavour*	Dive 110	A–518	75	41	Bullivant	NZO1 camera	× 0.5m *c.*
19–6a	32°32′N	117°26′W	7 Sept. 1962	*Trieste*	Dive 110	A–531	1280	700	Barham	Photographed t/port	× 0.15m *c.*
19–6b	75°02′S	178°10′E	9 Feb. 1960	*Endeavour*		A–531	348		Bullivant	NZO1 camera	× 0.3m *c.*
19–7	77°30′S	160°30′E	Jan. 1960	"		A–518	75	41	Bullivant	"	× 0.5m *c.*
19–8	77°30′S	164°38′E	17 Feb. 1960	"		A–538	260	142	"	"	Lower edge 0.6m. wide
19–9	77°30′S	164°38′E	17 Feb. 1960	"		A–538	260	142	"	"	Lower edge 0.75m. wide
19–10				*Chain*			500	273	Hersey *et al.*	Edgerton deep-sea camera	a × 0.2m *c.* b × 0.1m *c.*
19–11	75°02′S	178°10′E	9 Feb. 1960	*Endeavour*		A–531	360	197	Bullivant	NZO1 camera	× 0.3m *c.*

Chapter 20.

Fig.	Latitude	Longitude	Date	Ship	Cruise	Station	Depth m	Depth fm	Photographers	Equipment	Scale
20–3	35°10′N	72°00′W	Aug. 1957	*Crawford*	15		100	55	Breslau, Clarke, & Edgerton	Luminescence camera	Siphonophore *c.* 6 cm diameter
20–4	24°00′N	82°00′W	Feb. 1958	*Yamacraw*	8		300	164	"	"	Euphausid *c.* 2 cm long
20–5	28°16′N	79°41′W	"	"	"		10	5	"	"	Crenophore *c.* 3 cm diameter
20–6	24°58′N	85°01′W	"	"	"		100	55	"	"	Medusa *c.* 1 cm diameter
20–7	23°52′N	82°05′W	"	"	"		50	27	"	"	Euphausid *c.* 1½ cm long
20–8	24°58′N	85°01′W	"	"	"		360	197	"	"	Medusa *c.* 1 cm diameter
20–9	43°31′N	07°23′E	10 June 1958	*Calypso*			10	5	"	"	Arms of cephalopod *c.* 2 cm long
20–10	41°46′N	69°47′W	Aug. 1958	*Crawford*	20		1,000	547	"	"	Medusa *c.* 7 cm diameter
20–12	42°24′N	"	"	"	"		200	109	"	Interruption camera	Siphonophore *c.* 25 cm long

Chapter 21.

Fig.	Latitude	Longitude	Date	Ship	Cruise	Station	Depth m	Depth fm	Photographers	Equipment	Scale
21–2	40°20.5′N	70°47′W	25 May 1961	*Atlantis*	264	C	97	53	Owen	Owen multi-shot camera described in chap. 8	1/7 sq m
21–3	39°58.4′N	70°40.3′W	28 Aug. 1962	"	283	S3	300	164	"	"	"
21–4	"	"	"	"	"	"	"	"	"	"	"
21–5	39°50.5′N	70°35′W	25 May 1961	"	264	E#3	823	450	"	"	"
21–6	39°47′N	70°45′W	24 May 1961	"	"	F#1	1,500	820	"	"	"
21–7	"	"	"	"	"	"	"	"	"	"	"
21–8	"	"	"	"	"	"	"	"	"	"	"
21–9	"	"	"	"	"	"	"	"	"	"	"

Chapter 21 (*Continued*)

Fig.	Position		Date	Ship	Cruise	Station	Depth		Photographers	Equipment	Scale
	Latitude	Longitude					m	fm			
21–10	39°42′N	70°39′W	24 May 1961	*Atlantis*	264	G#1	2,086	1,141	Owen	Owen multi-shot camera described in chap. 8	1/7 sq m
21–11	"	"	"	"	"	"	"	"	"	"	"
21–12	39°25.5′N	70°35′W	27 Sept. 1961	"	273	GH#1	2,500	1,367	"	"	"
21–13	39°29′N	70°34′W	3 Oct. 1961	"	"	GH#4	2,469	1,350	"	"	"
21–14	"	"	"	"	"	"	"	"	"	"	"
21–15	35°35′N	67°25′W	28 Sept. 1961	"	"	LL#1	4,977	2,721	"	"	"
21–16	33°56.5′N	65°50.7′W	26 May 1962	"	277	NN#1	4,950	2,707	"	"	"
21–17	"	"	"	"	"	"	"	"	"	"	"

Chapter 22.

Fig.	Position		Date	Ship	Cruise	Station	Depth		Photographers	Equipment	Scale
	Latitude	Longitude					m	fm			
22–3	38°42.4′N	73°00.1′W	19 Oct. 1963	*Gosnold*	29	1358	450	246	Wigley & Emery	35 mm Robot Star camera combined with grab samples	scale bar on photo is 10 cm long
22–4	36°21.7′N	74°45.3′W	7 Aug. 1964	"	49	2078	365	200	"	"	"
22–7	30°05.4′N	72°08.2′W	15 Aug. 1964	"	"	2124	1,780	973	"	"	"
22–8	39°56.0′N	68°50.8′W	20 Aug. 1964	"	"	2161	1,625	889	"	"	"

Chapter 23.

Fig.	Position		Date	Ship	Cruise	Station	Depth		Photographers	Equipment	Scale
	Latitude	Longitude					m	fm			
23–1	39°58′N	68°57′W	12 Oct. 1960	*Atlantis*	260	2	1,500	800	Pratt	Edgerton deep-sea camera	fish *c.* 29 cm long
23–2							2,600	1,420		Camera described by Thorndike (1959)	
23–3											
23–4	30°51′N	78°06′W	16 July 1961	*Atlantis*	266	19	840–887	460–485	Pratt & Stetson	Edgerton deep-sea camera	
23–5a	41°44′N	64°51′W	25 May 1963	*Atlantis II*	6	4	2,600	1,420	Johnston	"	fish 9 cm long *c.*
23–5b	12°51′N	45°57′E	6 June 1958	*Atlantis*	242	1	1,240	678	Graham	"	fish *c.* 34 cm long
23–6				"	"					"	
23–7a	30°50′N	78°51′W	13 July 1961	"	266	15	805	440	Pratt & Stetson	"	fish *c.* 24 cm long
23–7b											
23–8	38°12.5′N	60°29′W	18 June 1962	*Atlantis*	280	3	1,550	850	Pratt	Edgerton deep-sea camera	fish *c.* 130 cm long
23–9	41°44′N	64°61′W	25 May 1963	*Atlantis II*	6	4	2,600	1,420	Johnston	"	1 inch corresponds to *c.* 1.5 meters

Chapter 24.

Fig.	Position		Date	Ship	Cruise	Station	Depth		Photographers	Equipment	Scale
	Latitude	Longitude					m	fm			
24–1	14°06′S	96°12′W	19 April 1963	*Vema*	19	31	3,246–3,262	1,775–1,784	available from Lamont Geological Observatory		× 1/9
24–2	04°56′N	78°16′W	2 April 1963	"	"	16	3,817	2,087	"		"
24–3	29°40′S	176°43′W	12 Aug. 1962	"	18	172	4,872–5,068	2,664–2,771	"		"
24–4	09°52′N	92°32′W	22 March 1964	"	20	14	3,678–3,684	2,011–2,014	"		"
24–5	"	"	"	"	"	"	3,821–3,751	2,089–2,051	"		"
24–6	17°02′S	112°12′W	24 April 1963	"	19	40	3,191–3,186	1,745–1,742	"		"
24–7	17°00′S	114°32′W	25 April 1963	"	"	46	3,147	1,721	"		× 1/6
24–8	17°02′S	112°12′W	24 April 1963	"	"	40	3,191–3,186	1,745–1,742	"		× 1/9
24–9	14°10′N	68°40′W	18 March 1963	"	"	8	4,740–4,737	2,592–2,590	"		× 1/12
24–10	57°33.5′S	17°22′W	16 March 1958	"	14	28	4,475	2,447	"		"
24–11	12°16′S	84°13′W	16 April 1963	"	19	27	4,361–4,367	2,385–2,388	"		× 1/9
24–12	11°45′S	79°12′W	14 April 1963	"	"	25	5,938–5,949	3,247–3,253	"		"
24–13	16°48′S	152°13′W	11 Sept. 1962	"	18	229	3,937–3,935	2,153–2,152	"		× 1/12
24–14	15°39′S	138°35′W	2 May 1963	"	19	60	3,939–3,941	2,154–2,155	"		"
24–15	33°06′S	22°48′W	17 May 1962	"	18	108	4,175–4,188	2,283–2,290	"		"
24–16	28°32′N	67°57′W	7 March 1963	"	19	1	5,064–4,993	2,769–2,730	"		× 1/9
24–17	13°13′S	92°53′W	18 April 1963	"	"	30	3,634–3,673	1,987–2,008	"		"
24–18	17°00′S	114°11′W	25 April 1963	"	"	45	3,177	1,737	"		× 1/6
24–19	13°08′S	44°09′E	31 Aug. 1963	"	"	158	3,549	1,941	"		× 1/9
24–20	16°58′S	115°12′W	25 April 1963	"	"	48	3,329–3,347	1,820–1,830	"		× 1/6
24–21	16°24′S	127°38′W	29 April 1963	"	"	56	4,094	2,239	"		"
24–22	17°00′S	114°32′W	25 April 1963	"	"	46	3,147	1,721	"		× 1/9
24–23	07°04′N	60°55′E	15 Aug. 1963	"	"	145	2,679–2,699	1,465–1,476	"		× 1/12
24–24	22°51′S	42°10′E	5 Sept. 1963	"	"	164	3,184	1,741	"		"
24–25	16°58′S	116°48′W	26 April 1963	"	"	52	3,413	1,866	"		× 1/9
24–26	15°39′S	138°35′W	2 May 1963	"	"	60	3,939–3,941	2,154–2,155	"		"
24–27	14°10′N	68°40′W	18 March 1963	"	"	8	4,740–4,737	2,592–2,590	"		× 1/12

Chapter 24 (*Continued*)

Fig.	Position Latitude	Position Longitude	Date	Ship	Cruise	Station	Depth m	Depth fm	Photographers	Equipment	Scale
24–28	23°45′N	67°09′W	10 March 1963	*Vema*	19	2	5,104–5,066	2,791–2,770	available from Lamont Geological Observatory		× 1/6
24–29	16°06′N	66°29′W	16 March 1963	"	"	5	4,506–4,510	2,464–2,466	"		× 1/9
24–30	25°46′S	04°06′E	30 Sept. 1963	"	"	193	4,237–4,244	2,317–2,321	"		"
24–31	09°28′S	43°19′E	30 Aug. 1963	"	"	156	3,651–3,649	1,996–1,995	"		"
24–32	16°56′S	41°06′E	2 Sept. 1963	"	"	160	2,496–2,476	1,365–1,354	"		× 1/12
24–33	14°06′S	96°12′W	19 April 1963	"	"	31	3,246–3,262	1,775–1,784	"		"
24–34	16°10′S	129°52′W	30 April 1963	"	"	57	3,950–3,952	2,160–2,161	"		× 1/9
24–35	14°49′N	119°37′E	26 June 1963	"	"	97	2,540	1,389	"		"
24–36	21°58′N	67°11′W	11 March 1963	"	"	3	5,284–5,302	2,889–2,899	"		× 1/6
24–37	11°45′S	79°12′W	14 April 1963	"	"	25	5,938–5,949	3,247–3,253	"		× 1/9
24–38	07°35′N	74°13′E	10 Aug. 1963	"	"	136	2,770–2,769	1,515–1,514	"		"
24–39	11°45′S	79°12′W	14 April 1963	"	"	25	5,938–5,949	3,247–3,253	"		"
24–40	39°34′N	72°23′W	21 Aug. 1957	*Balanus*	12	E	450	246	"		× 1/14
24–41	15°57′N	71°05′W	19 March 1963	*Vema*	19	10	3,992–3,990	2,183–2,182	"		× 1/9
24–42	32°11′S	13°25′E	29 May 1962	"	18	122	2,893–3,087	1,582–1,688	"		"
24–43	11°45′S	79°12′W	14 April 1963	"	19	25	5,938–5,949	3,247–3,253	"		× 1/12
24–44	16°02′N	72°51′W	20 March 1963	"	"	11	2,118–2,180	1,158–1,192	"		× 1/6
24–45	35°55′S	27°45′E	13 Sept. 1963	"	"	174	4,656	2,546	"		× 1/9
24–46	14°58′N	68°54′W	18 March 1963	"	"	9	4,965–4,955	2,715–2,709	"		× 1/18
24–47	18°03′S	159°43′W	13 May 1963	"	"	71	5,072–5,070	2,773–2,772	"		× 1/9
24–48	27°15′S	06°12′E	29 Sept. 1963	"	"	191	4,865	2,660	"		× 1/12
24–49	32°36′S	16°12′E	26 Sept. 1963	"	"	185	1,567–1,465	857–801	"		× 1/9
24–50	14°38′S	101°20′E	20 July 1963	"	"	118	5,363	2,933	"		× 1/12
24–51	31°27′N	76°33′W	18 Dec. 1962	*Conrad*	1	3	2,734	1,495	"		× 1/9
24–52	06°43′N	86°30′W	22 Oct. 1962	*Vema*	18	289	2,897	1,584	"		"
24–53	15°10′S	100°21′W	20 April 1963	"	19	32	3,834–3,592	2,096–1,964	"		"
24–54	16°57′S	116°18′W	26 April 1963	"	"	51	3,413	1,866	"		"
24–55	15°39′S	138°35′W	2 May 1963	"	"	60	3,939–3,941	2,154–2,155	"		"
24–56	17°02′S	111°53′W	24 April 1963	"	"	39	3,255–3,252	1,780–1,778	"		× 1/6
24–57	17°00′S	114°11′W	25 April 1963	"	"	45	3,177	1,737	"		× 1/9
24–58	"	114°53′W	"	"	"	47	3,147–3,166	1,721–1,731	"		× 1/12
24–59	17°00′S	114°32′W	25 April 1963	*Vema*	"	46	3,147	1,721	"		× 1/9
24–60	02°40′S	54°45′E	21 Aug. 1963	"	"	151	4,184	2,288	"		× 1/12
24–61	16°40′S	168°34′W	18 May 1963	"	"	77	5,086–5,088	2,781–2,782	"		"
24–62	16°58′S	115°12′W	25 April 1963	"	"	48	3,329–3,347	1,820–1,830	"		× 1/9
24–63	16°24′S	127°38′W	29 April 1963	"	"	56	4,094	2,239	"		"
24–64	14°38′S	101°20′E	20 July 1963	"	"	118	5,363	2,933	"		"
24–65	46°14′N	136°30′W	10 May 1964	"	20	55	4,232	2,314	"		× 1/12
24–66	29°37′S	10°36′E	28 Sept. 1963	"	19	188	4,488–4,484	2,454–2,452	"		× 1/9
24–67	27°15′S	06°12′E	29 Sept. 1963	"	"	191	4,865	2,660	"		"
24–68	17°54′S	39°30′E	3 Sept. 1963	"	"	161	2,307–2,305	1,262–1,260	"		× 1/18
24–69	14°06′S	96°12′W	19 April 1963	"	"	31	3,246–3,262	1,775–1,784	"		× 1/9
24–70	00°25′S	82°04′W	5 April 1963	"	"	20	1,362–1,376	745–752	"		× 1/12
24–71	13°13′S	92°53′W	18 April 1963	"	"	30	3,634–3,673	1,987–2,008	"		× 1/9
24–72	14°31′N	96°18′W	13 Oct. 1962	"	18	278	3,720–3,674	2,034–2,009	"		"
24–73	17°02′S	112°12′W	24 April 1963	"	19	40	3,191–3,186	1,745–1,742	"		"
24–74	16°06′N	66°29′W	16 March 1963	"	"	5	4,506–4,510	2,464–2,466	"		"
24–75	25°46′S	04°06′E	30 Sept. 1963	"	"	193	4,237–4,244	2,317–2,321	"		"
24–76	07°04′N	60°55′E	15 Aug. 1963	"	"	145	2,679–2,699	1,465–1,476	"		× 1/12
24–77	06°52′N	60°42′E		"	"	146	3,345–3,356	1,829–1,835	"		"
24–78	44°02′S	93°53′E	26 Jan. 1964	*Conrad*	8	48	2,785–2,822	1,523–1,543	"		× 1/9
24–79	32°26′N	62°59′W	1 April 1963	"	5	1	4,497–4,501	2,459–2,461	"		"
24–80	04°56′N	78°16′W	2 April 1963	*Vema*	19	16	3,817	2,087	"		"
24–81	06°14′S	104°50′E	15 July 1963	"	"	113	1,318–1,311	721–717	"		"
24–82	61°52′S	91°12′W	3 May 1960	"	16	48	4,894	2,676	"		× 1/12
24–83	37°20′N	33°15′W	26 July 1954	"	4	3	1,723–1,774	942–970	"		× 1/9
24–84	22°51′S	42°10′E	5 Sept. 1963	"	19	164	3,184	1,741	"		× 1/12
24–85	35°38′S	28°44′W	15 May 1962	"	18	105	4,354	2,381	"		"

Chapter 24 (*Continued*)

Fig.	Position		Date	Ship	Cruise	Station	Depth		Photographers	Equipment	Scale
	Latitude	Longitude					m	fm			
24–86	16°58′S	116°48′W	26 April 1963	*Vema*	19	52	3,413	1,866	available from Lamont Geological		✕ 1/12
24–87	11°45′S	79°12′W	14 April 1963	"	"	25	5,938– 5,949	3,247– 3,253	Observatory .		✕ 1/9
24–88	27°15′S	06°12′E	29 Sept. 1963	"	"	191	4,865	2,660	"		"
24–89	34°31′S	17°44′E	20 Sept. 1963	"	"	176	516– 525	282– 287	"		✕ 1/6
24–90	16°48′S	152°13′W	11 Sept. 1962	"	18	229	3,937– 3,935	2,043– 2,152	"		✕ 1/12
24–91	15°10′S	100°21′W	20 April 1963	"	19	32	3,834– 3,592	2,096– 1,964	"		✕ 1/9
24–92	11°12′N	128°45′E	12 June 1963	"	"	90	5,594	3,059	"		✕ 1/12
24–93	00°06′N	170°46′E	29 May 1963	"	"	80	4,550– 4,552	2,488– 2,489	"		✕ 1/18
24–94	23°45′N	67°09′W	10 March 1963	"	"	2	5,104– 5,066	2,791– 2,770	"		✕ 1/12
24–95	06°42′N	80°42′W	24 Oct. 1962	"	18	296	3,426– 3,398	1,873– 1,858	"		"
24–96	16°59′S	117°53′W	26 April 1963	"	19	53	3,442	1,882	"		✕ 1/9
24–97	16°58′S	115°56′W	26 April 1963	"	"	50	3,329	1,820	"		✕ 1/12
24–98	14°38′S	101°20′E	20 July 1963	"	"	118	5,363	2,933	"		✕ 1/9
24–99	17°49′N	83°14′W	31 Oct. 1962	"	18	299	5,698– 5,693	3,116– 3,113	"		✕ 1/12
24–100	02°22′N	168°02′E	30 May 1963	"	19	81	4,400– 4,398	2,406– 2,405	"		✕ 1/18
24–101	45°31′S	51°55′W	16 April 1962	"	18	54	6,018	3,291	"		✕ 1/9
24–102	16°24′S	127°38′W	29 April 1963	"	19	56	4,094	2,239	"		"
24–103	06°42′N	80°42′W	24 Oct. 1962	"	18	296	3,426– 3,398	1,873– 1,858	"		✕ 1/12
24–104	"	"	"	"	"	"	"	"	"		✕ 1/18

Chapter 25.

25–5	32°37′N	117°30′W	25 Oct. 1961	Bathyscaph *Trieste I*	Dive 84	—	1,180	645	La Fond	Edgerton deep-sea camera	Diameter of brittle stars *c.* 12 cm
25–7	32°27.5′N	117°24.2′W	3 Aug. 1962	"	Dive 105	—	1,061	580	Dill	"	Diameter of brittle stars *c.* 20 cm